Fluid Dynamics of the Midlatitude Atmosphere

Advancing Weather and Climate Science Series

Series Editors:
Peter Inness, University of Reading, UK
John A. Knox, University of Georgia, USA

Other titles in the series:

Mesoscale Meteorology in Midlatitudes
Paul Markowski and Yvette Richardson, Pennsylvania State University, USA
Published: February 2010
ISBN: 978-0-470-74213-6

Thermal Physics of the Atmosphere
Maarten H.P. Ambaum, University of Reading, UK
Published: April 2010
ISBN: 978-0-470-74515-1

The Atmosphere and Ocean: A Physical Introduction, 3rd Edition
Neil C. Wells, Southampton University, UK
Published: November 2011
ISBN: 978-0-470-69469-5

Time-Series Analysis in Meteorology and Climatology: An Introduction
Claude Duchon, University of Oklahoma, USA and
Robert Hale, Colorado State University, USA
Published: January 2012
ISBN: 978-0-470-97199-4

Operational Weather Forecasting
Peter Inness, University of Reading, UK and
William Beasley, University of Oklahoma, USA
Published: January 2013
ISBN: 978-0-470-71159-0

Fluid Dynamics of the Midlatitude Atmosphere

Brian J. Hoskins
University of Reading
Imperial College London
UK

Ian N. James
Emeritus Professor
University of Reading, UK

WILEY Blackwell

Registered Office
John Wiley & Sons, Ltd, The Atrium, Southern Gate, Chichester, West Sussex, PO19 8SQ, UK

Editorial Offices
9600 Garsington Road, Oxford, OX4 2DQ, UK
The Atrium, Southern Gate, Chichester, West Sussex, PO19 8SQ, UK
111 River Street, Hoboken, NJ 07030-5774, USA

For details of our global editorial offices, for customer services and for information about how
to apply for permission to reuse the copyright material in this book please see our website at
www.wiley.com/wiley-blackwell.

Library of Congress Cataloging-in-Publication Data

Hoskins, Brian.
Fluid dynamics of the midlatitude atmosphere / B.J. Hoskins & I.N. James.
    pages   cm
  Includes bibliographical references and index.
   ISBN 978-0-470-83369-8 (cloth) – ISBN 978-0-470-79519-4 (pbk.)   1. Atmospheric circulation–Study
and teaching.   2. Dynamic meteorology–Study and teaching.   3. Middle atmosphere–Study and teaching.
I. James, Ian N.   II. Title.   III. Title: Fluid dynamics of the mid latitude atmosphere.
   QC880.4.A8H67 2014
   551.51'5–dc23

                                        2014012262

A catalogue record for this book is available from the British Library.

Cover image courtesy of John Methven.

Set in 10.5/12.5pt Times by SPi Publisher Services, Pondicherry, India
Printed and bound in Malaysia by Vivar Printing Sdn Bhd

1   2014

Contents

Series foreword

Advancing Weather and Climate Science

Meteorology is a rapidly moving science. New developments in weather forecasting, climate science and observing techniques are happening all the time, as shown by the wealth of papers published in the various meteorological journals. Often these developments take many years to make it into academic textbooks, by which time the science itself has moved on. At the same time, the underpinning principles of atmospheric science are well understood but could be brought up to date in the light of the ever increasing volume of new and exciting observations and the underlying patterns of climate change that may affect so many aspects of weather and the climate system.

In this series, the Royal Meteorological Society, in conjunction with Wiley-Blackwell, is aiming to bring together both the underpinning principles and new developments in the science into a unified set of books suitable for undergraduate and postgraduate study as well as being a useful resource for the professional meteorologist or Earth system scientist. New developments in weather and climate sciences will be described together with a comprehensive survey of the underpinning principles, thoroughly updated for the 21st century. The series will build into a comprehensive teaching resource for the growing number of courses in weather and climate science at undergraduate and postgraduate level.

Series Editors

Peter Inness
University of Reading, UK

John A. Knox
University of Georgia, USA

Preface

We have developed this book from lectures given by us and others in the Department of Meteorology at the University of Reading. Since 1965, this department has been an important centre for the study of meteorology and atmospheric science. Indeed, for many years, it was the only independent department of meteorology in the United Kingdom able to offer a full range of undergraduate and postgraduate teaching in meteorology. Many scientists and meteorologists have spent time at Reading either as students or researchers. So our book is a record of one facet of the teaching they met at Reading, and it aspires to encapsulate something of the spirit of the Reading department.

During the early part of the twentieth century, meteorology made the transition from a largely descriptive, qualitative science to a firmly quantitative science. At the heart of that transition was the recognition that the structure and development of weather systems were essentially problems in fluid dynamics. In the 1920s, the scientists of the Bergen School recognized this but lacked the mathematical tools to link their descriptive models of cyclone development in terms of air masses and fronts to the basic equations of fluid dynamics. In the 1940s, modern dynamical meteorology was born out of the recognition by Eady, Charney and others that cyclone development could be viewed as a problem in fluid dynamical instability.

Even so, great simplifications proved necessary to render the problem tractable. Amongst these simplifications was the linearization of the governing equations. The highly nonlinear equations governing fluid dynamics were reduced to relatively simple linear forms whose solutions can be written in terms of traditional analytic functions. Even so, these simplified equations could only be solved for very simple idealized circumstances. For example, the work of both Charney and Eady was confined to flows which varied linearly in the vertical but had no variation in other directions. There was still a big gap between theory and observation. The development of the digital computer in the 1950s opened up the possibility of bridging this gap. While their nonlinearity and complexity rendered the governing equations resistant to analytical solution, the digital computer could generate numerous particular solutions to discretized analogues to the governing equations. So two separate branches of dynamical meteorology developed. On one hand, the drive for weather forecasts and, increasingly, for climate modelling led to the development of elaborate numerical models of atmospheric flow. As computer power increased, these models became more realistic, with higher resolution and fewer approximations or simplifications to the governing equations, and with more elaborate representation of the processes driving atmospheric motion such as radiative transfer, cloud processes and friction. On the other hand, more sophisticated mathematical techniques drove the development of analytical theory either to better approximations to the governing equations or to explore more complex physical scenarios. As a result, there was a growing gulf between numerical modelling, theoretical meteorology and observation.

Our approach at Reading, both in terms of teaching and research, started with the intention of bridging this gulf. The word 'model' had become somewhat limited in its use in meteorology: it tended to refer to the large and elaborate numerical weather prediction and global circulation models that were primary tools in many applications of meteorology. But the word has far wider meaning than that. A model is any abstraction of the real world, any representation in which certain complexities are eradicated or idealized. All of meteorology, indeed, all of science, deals in models. They may be very basic, starting with conceptual verbal or picture models such as the Norwegian frontal model of cyclone development. They may be exceedingly complex and include a plethora of different processes. The coupled atmosphere–ocean global circulation models now used to study climate change exemplify this sort of model. But between these extremes lie a hierarchy of models of differing degrees of complexity. These range from highly idealized analytic models, models with only very few degrees of freedom, through models of intermediate complexity, right up to fully elaborated numerical models. Good science involves the interaction between these different levels of model and with observations, finding out which elements of the observed world are captured or lost by the different levels of model complexity.

A very complex model may give a faithful representation of the observed atmosphere, but of itself it can lead to rather limited understanding. A very simple model is transparent in its working but generally gives but a crude imitation of a complex reality. Intermediate models, grounded in constant reference to observations and to other models in the hierarchy, can illuminate the transition from transparent simplicity to elaborate complexity.

Our book focuses on the simpler and intermediate complexity models in this hierarchy. Although we shall refer to the results of calculations using elaborate numerical models, we have not set out to describe such models. That is a major topic, bringing together dynamical meteorology and numerical analysis, and deserves a textbook of its own. However, we shall make use of results from numerical models in a number of places.

Our textbook is based upon various lecture courses that we have given to students, both postgraduate and advanced undergraduate, in the Department of Meteorology, University of Reading, over many years. Many of our postgraduate students came to Reading with a first degree in other quantitative disciplines, so our teaching assumed no prior acquaintance with fluid dynamics and the mathematical techniques used in that discipline. Neither did we assume any prior knowledge of meteorology or atmospheric science. We did assume basic knowledge of vector calculus and differential equations. However, in order to make this book self-contained, we have included an appendix which gives a brief introduction to the essential elements of vector calculus assumed in the main body of the text. So our intended readership is primarily postgraduate and advanced undergraduate students of meteorology. We hope others, particularly quantitative scientists who wish to become better acquainted with dynamical meteorology, will also find our book interesting.

Our text begins with an opening chapter which gives a broad brush survey of the structure of the atmosphere and the character of atmospheric flow, particularly in the midlatitude troposphere. This opening chapter introduces in a qualitative way a number of concepts which will be elaborated in subsequent chapters.

Then follows our first major theme: a basic introduction to classic fluid dynamics as applied to the Earth's atmosphere. After deriving the fundamental equations in Chapter 2, we introduce the various modifications that are needed for this application. Foremost among these are the roles that the rotation of the Earth and its spherical geometry, and the stable stratification of the atmosphere play. Perhaps the most important chapter in our first theme is Chapter 5, which develops the technique of scale analysis and applies it systematically to flows in the atmosphere. Our focus of interest is upon the synoptic scale weather systems of the midlatitudes, but the discussion points to how other situations might be approached.

Our second theme recognizes that atmospheric flow on the larger scales is dominated by rotation. The Earth rotates on its axis and individual fluid elements spin as they move around. Such spin is a primary property of the atmosphere or ocean, and our insight into atmospheric behaviour is developed by rewriting the equations in terms of spin or 'vorticity'. Equations describing the processes which generate and modify vorticity result, and we spend some time exploring these equations in simple contexts. These simple examples help to develop a language and a set of conceptual principles to explore more elaborate and more realistic examples. A powerful unifying concept is a quantity called 'potential vorticity', which is introduced in Chapter 10.

Our third theme makes up the remainder of our book. That theme is the dynamical understanding of middle latitude weather systems exploiting the near balance between certain terms in the governing equations. Such a balance links together dynamical, pressure and temperature fields and constrains their evolution. Maintaining a near-balanced state determines the response of the atmosphere to thermal and other forcing. With these concepts, we are able to discuss the evolution of weather systems as problems in fluid dynamical instability, and we are able to extend our discussion to more elaborate, nonlinear regimes. Frontal formation is revealed as an integral part of cyclone development, and at the same time, developing weather systems play a central part in determining the larger scale flow in which they are embedded. Through the concept of balance, potential vorticity is revealed as a primary concept in modern dynamical meteorology.

This book is intended as a readable textbook rather than a research monograph. We hope that the material is largely self-contained. Consequently, we have made no attempt to provide a comprehensive and exhaustive bibliography. Rather we have included some suggestions for further reading which will give the interested reader a starting point from which to explore the literature. Modern electronic databases and citation indices make such exploration much easier than it has been in the past.

Both authors would like to acknowledge with thanks the influence of those supervising their early research. Francis Bretherton introduced one of us (BH) to geophysical fluid dynamics, and in particular the importance and role of potential vorticity. Raymond Hide..... Discussions with many colleagues outside Reading, such as Michael McIntyre, have also been very important to us.

Our book aims to sum up an important component of the teaching and thinking of the Reading Department of Meteorology in its first 50 years. We owe a debt of gratitude to its staff and students over many years. In some cases, they have made very specific contributions. The authors have worked closely with some of them on research that has now become part of this book. A number of colleagues have generously allowed us access to their own lecture notes and material. In other cases, the influence of students and colleagues has been more diffuse and pervasive: Reading has provided a stimulating and energizing place to study atmospheric dynamics throughout our careers. We remember with gratitude the many conversations with colleagues and the questions from students that have helped to mould our thinking. It is nigh on impossible to name all these individuals and any such list would inevitably have omissions. So we hope colleagues and students, past and present, will accept this general thanks for all they have given us. However, some specific thanks are in order. The comments provided by John Methven on a first draft of much of the book were extremely valuable to us. We have had the help of a number of members of the department in providing us with data and generating diagrams from them. In particular, we wish to thank Laura Baker, Paul Berrisford, Johannes de Leeuw and Andrew Lomas for supplying data and pictures. We are particularly grateful to Ben Harvey for his work in extracting data and developing many plots specially designed for this book, and to Robert Lee who has done a most professional job making some of our crude sketches into publication-quality diagrams.

But above all, thanks to our research colleagues and students over many years. Their interest and enthusiasm has been a continuing stimulus to us both.

Select bibliography

Chapters 2–8: Many text books on fluid mechanics and dynamical meteorology give complementary or more extended cover of the material in these chapters. The texts by Holton (1992), Vallis (2006) and Martin (2006) are particularly recommended. Ambaum (2010) is a text on atmospheric thermodynamics.

Chapter 9: For the calculations underlying Figure 9.5 to Figure 9.7, see James (1980).

Chapter 10: Rossby (1940) introduced the concept of potential vorticity. For a review of the literature for potential vorticity, see Hoskins *et al.* (1985). Thorpe (1985) gives more detail on invertibility.

Chapter 11: The sketches of Figure 11.1 are based on some of the beautiful photographs in Van Dyke (1982). Vallis (2006) gives a fuller account of two- and three-dimensional turbulences in a geophysical context.

Chapter 12: Charney (1948) laid the foundations of the quasi-geostrophic approximation.

Chapter 13: Hoskins *et al.* (1978) discussed different forms of the ω-equation and introduced the **Q** vector. See Machenauer (1977), Section 13.5, for more details on initialization.

Chapter 14: The papers of Charney (1947) and Eady (1949) are the much cited beginnings of this subject. Charney and Stern (1962) set out the general conditions for baroclinic instability, and the unifying boundary condition was introduced by Bretherton (1966). Simmons and Hoskins (1977) described unstable baroclinic normal modes growing on realistic midlatitude zonal jets. Farrell (1982) is a good source for the initial value problem. See also Badger and Hoskins (2001) for simple initial value problems and mechanisms for baroclinic growth. Simmons and Hoskins (1979) described the up- and downstream development of baroclinic waves.

Chapter 15: Hoskins (1982) reviews the mathematical theory of frontogenesis. See Eliassen (1962) for a derivation of the Sawyer–Eliassen equation for frontal circulations and Hoskins (1977) for more on frontal transformation discussed in Section 15.3. The semi-geostrophic Eady model was presented in Hoskins and Bretherton (1972).

Chapter 16: The formation of fronts in developing baroclinc waves was discussed by Hoskins (1976). Nonlinear baroclinic lifecycle calculations were presented by Simmons and Hoskins (1978). Thorncroft *et al.* (1993) discussed the LC1 and LC2 lifecycles. Edmon *et al.* (1980) derived Eliassen–Palm sections for baroclinic lifecycle calculation as well as for the observed midlatitude troposphere.

Chapter 17: A fuller review of the use of isentropic maps of potential vorticity is given by Hoskins *et al*. (1985). Hoskins (1997) introduced the alternative map of θ on potential vorticity surfaces. Masato *et al*. (2012) discussed blocking in terms of Rossby wave breaking. See Stoelinga (1996) for a case study of the role of heating and friction in the potential vorticity of a developing weather system.

Chapter 18: Charney and Drazin (1961) is the seminal reference for Rossby wave propagation from the troposphere into the stratosphere. Edmon *et al*. (1980) discuss the Eliassen–Palm flux and its relationship to group velocity. Karoly and Hoskins (1982) demonstrate the three-dimensional propagation of Rossby waves. Dritschel and McIntyre (2008) review recent work on potential vorticity staircases. Zhu and Nakamura discuss Rossby wave propagation for multiple potential vorticity steps, and hence the transition to the uniform potential vorticity gradient case. Rhines (1975) discussed the role of the Rhines number in geostrophic turbulence. Williams (1978) is the first in a series of papers in which he demonstrated the circumstances needed to generate multiple jets in a range of planetary atmospheres.

References

Adamson, D. S., Belcher, S.E., Hoskins, B.J. and Plant, R.S. (2006) Boundary-layer friction in midlatitude cyclones. *Quarterly Journal of Royal Meteorological Society*, 132, 101–124.

Ambaum, M.H.P. (2010) *Thermal physics of the atmosphere*. Wiley, Chichester. 239pp.

Badger, J. & B.J. Hoskins (2001) Simple initial value problems and mechanisms for baroclinic growth. *Journal of Atmospheric Sciences*, 38, 38–49.

Bretherton, F. P. (1966) Baroclinic instability and the short wave cut-off. *Quarterly Journal of Royal Meteorological Society*, 92, 335–345.

Charney, J.G. (1947) The dynamics of long waves in a baroclinic westerly current. *Journal of Meteorology* 4, 135–163.

Charney, J.G. (1948) On the scale of atmospheric motions. *Geofysiske Publikasjoner*, 17(2).

Charney, J.G. and Drazin, P.G. (1961) The propagation of planetary scale disturbances from the lower into the upper stratosphere. *Journal of Geophysical Research*, 66, 83–109.

Charney, J. G. and Stern, M. (1962) On the stability of internal jets in a rotating atmosphere. *Journal of the Atmospheric Sciences*, 19, 159–172.

Dritschel, D.G. and McIntyre, M.E. (2008) Multiple jets as PV staircases: the Phillips effect and the resilience of eddy-transport barriers. *Journal of Atmospheric Sciences*, 65, 855–874.

Eady, E.T. (1949) Long waves and cyclone waves. *Tellus*, 1, 33–52.

Edmon, H.J., Hoskins, B.J. and McIntyre, M.E. (1980) Eliassen-Palm cross sections for the troposphere. *Journal of Atmospheric Sciences*, 37, 2600–2615 (see also corrigendum *ibid*, **38**, 1115)

Eliassen, A. (1962) On the vertical circulation in frontal regions. *Geofysiske Publikasjoner*, 24(4), 147–160.

Farrell, B. F. (1982) The initial growth of disturbances in a baroclinic flow. *Journal of the Atmospheric Sciences*, 39, 1663–1686.

Heifetz, E., Bishop, C. H. and Alpert, P. (1999) Counter-propagating Rossby waves in the barotropic Rayleigh model of shear instability. *Quarterly Journal of Royal Meteorological Society*, 125, 2835–2853.

Holton, J.R. (1992) *An introduction to dynamic meteorology* (3rd edition). Academic Press, San Diego. 511pp.

Hoskins, B.J. (1972) Non-Boussinesq effects and further development in a model of upper tropospheric frontogenesis. *Quarterly Journal of Royal Meteorological Society*, 98, 532–541.

Hoskins, B.J. (1974) The role of potential vorticity in symmetric stability and instability. *Quarterly Journal of Royal Meteorological Society*, 100, 480–482.

Hoskins, B.J. (1976) Baroclinic waves and frontogenesis. Part 1: Introduction and Eady waves. *Quarterly Journal of Royal Meteorological Society*, 102, 103–122.

Hoskins, B.J. (1982) The mathematical theory of frontogenesis. *Annual Review of Fluid Mechanics*, 14, 131–151.

Hoskins, B.J. (1997) A potential vorticity view of synoptic development. *Meteorological Applications* 4, 325–334.

Hoskins, B.J. and Bretherton, F.P. (1972) Atmospheric frontogenesis models: Mathematical formulation and solution. *Journal of the Atmospheric Sciences*, 29, 11–37.

Hoskins, B.J. and Pedder, M.A. (1980) The diagnosis of middle latitude synoptic development. *Quarterly Journal of Royal Meteorological Society* 106, 707–719.

Hoskins, B.J. and Revell, M.J. (2001) The most unstable long wavelength baroclinic instability modes. *Journal of the Atmospheric Sciences*, 38, 1498–1503.

Hoskins, B.J., Draghici, I. and Davies, H.C. (1978) A new look at the ω equation. *Quarterly Journal of Royal Meteorological Society*, 104, 31–38.

Hoskins, B.J., McIntyre, M.E. and Robinson, A.W. (1985) On the use and significance of isentropic potential vorticity maps. *Quarterly Journal of Royal Meteorological Society*, 111, 877–946.

James, I.N. (1980) The forces due to geostrophic flow over shallow topography. *Geophysical and Astrophysical Fluid Dynamics*, 14, 225–250.

Karoly, D.J. and Hoskins, B.J. (1982) Three dimensional propagation of planetary waves. *Journal of Meteorological Society of Japan*, 60, 109–122.

Machenauer, B. (1977) On the dynamics of gravity oscillations in a shallow water equation model, with applications to normal model initialisation. *Contributions to Atmospheric Physics*, 50, 253–271.

Marcus, P.S. and Shetty, S. (2011) Jupiter's zonal winds: are they bands of homogenized potential vorticity organized as a monotonic staircase? *Philosophical Transactions of the Royal Society*, 369, 771–795.

Martin, J.E. (2006) *Mid-latitude atmospheric dynamics: a first course*. John Wiley & Sons, Chichester, 324pp.

Masato, G., Hoskins, B.J., and Woollings, T.J. (2012) Wave-breaking characteristics of midlatitude blocking. *Quarterly Journal of Royal Meteorological Society*, 138, 1285–1296.

Methven, J., Hoskins, B.J., Heifetz, E. and Bishop, C.H. (2005) The counter-propagating Rossby-wave perspective on baroclinic instability. Part III: Primitive-equation disturbances on the sphere. *Quarterly Journal of Royal Meteorological Society*, 131, 1393–1424.

Rhines, P.B. (1975) Waves and turbulence on a beta plane. *Journal of Fluid Mechanics*, 69, 417–443.

Rossby, C. G. (1940) Planetary flow patterns in the atmosphere. *Quarterly Journal of Royal Meteorological Society*, 66, Suppl., 68–77.

Sanders, F. (1955) An investigation of the structure and dynamics of an intense surface frontal zone. *Journal of Meteorology*, 12, 542–552.

Shapiro, M.A. (1980) Turbulent mixing within tropopause folds as a mechanism for the exchange of chemical constituents between the stratosphere and troposphere. *Journal of the Atmospheric Sciences*, 37, 994–1004.

Simmons, A.J. and Hoskins B.J. (1977) Baroclinic instability on sphere – solutions with a more realistic tropopause. *Journal of the Atmospheric Sciences*, 34, 581–588.

Simmons, A.J. and Hoskins, B.J. (1979) The downstream and upstream development of unstable baroclinic waves. *Journal of the Atmospheric Sciences*, 36, 1239–1254.

Stoelinga, M.T. (1996) A potential vorticity based study of the role of diabatic heating and friction in a numerically simulated baroclinic cyclone. *Monthly Weather Review* 124, 849–874.

Thorncroft, C.D., Hoskins', B. J. and McIntyre, M. E. (1993) Two paradigms of baroclinic-wave life-cycle behaviour. *Quarterly Journal of the Royal Meteorological Society* (1993), 119, 17–55.

Thorpe, A. J. (1985) Diagnosis of balanced vortex structure using potential vorticity. *Journal of the Atmospheric Sciences*, 42, 397–406.

Vallis, G.K. (2006) *Atmospheric and oceanic fluid dynamics*. Cambridge University Press , Cambridge. 745pp.

Van Dyke, M. (1982) *An album of fluid motion*. Parabolic Press, Stanford, 176pp.

Williams, G.P. (1978) Planetary circulations: 1: Barotropic representation of Jovian and terrestrial turbulence. *Journal of the Atmospheric Sciences*, 35, 1399–1426.

Woollings, T., Hoskins, B.J., Blackburn, M. and Berrisford, P. (2008) A new Rossby wave–breaking interpretation of the North Atlantic oscillation. *Journal of the Atmospheric Sciences.*, 65, 609–626.

Zhu, D. and Nakamura, N. (2010) On the representation of Rossby waves on the β-plane by a piecewise uniform potential vorticity distribution. *Journal of Fluid Mechanics*, 664, 397–406.

The authors

Having gained mathematics degrees from Cambridge and spent some post-doc years in the United States, **Brian Hoskins** has been at the University of Reading for more than 40 years, being made a professor in 1981, and also more recently has led a climate institute at Imperial College London. His international activities have included being President of IAMAS and Vice-Chair of the JSC for WCRP. He is a member of the science academies of the United Kingdom, Europe, United States and China, he has received the top awards of both the Royal and American Meteorological Societies, the Vilhelm Bjerknes Medal of the EGU and the Buys Ballot Medal, and he was knighted in 2007.

From a background in physics and astronomy, **Ian James** worked in the geophysical fluid dynamics laboratory of the Meteorological Office before joining the University of Reading in 1979. During his 31 years in the Reading meteorology department, he has taught courses in dynamical meteorology and global atmospheric circulation. In 1998, he was awarded the Buchan Prize of the Royal Meteorological Society for his work on low-frequency atmospheric variability. He has been President of the Dynamical Meteorology Commission of IAMAS, Vice President of the Royal Meteorological Society and currently edits the journal *Atmospheric Science Letters*. He now serves as an Anglican priest in Cumbria.

1

Observed flow in the Earth's midlatitudes

1.1 Vertical structure

The aim of this book is to explore how the Earth's atmosphere is set into motion and to elucidate the nature of those motions. Essentially, we shall apply Newton's laws of motion to a continuous fluid, taking account of all the different forces that can act upon the air making up the atmosphere. The principle is simple enough, but the application of that principle rapidly becomes very complicated. One of the challenges of the subject is that apart from a very few highly idealized examples, the equations governing real atmospheric flows have no known solutions in terms of simple functions: the equations are highly nonlinear, and many important atmospheric processes cannot be represented by simple mathematical functions. The challenge of this book is therefore to expose basic principles without compromising or oversimplifying the underlying science.

As a preliminary, the scene will be set in this chapter with a description of the observed flow. This flow continually changes. Features first develop and then decay until they can no longer be seen. This itself makes life very difficult. We are faced with a three-dimensional flow which continually changes – a four-dimensional object. We have the problem of finding suitable ways even to describe such a flow adequately, perhaps using suitable averaging to reduce the number of dimensions.

As we examine flows in greater detail, we shall often find that certain terms in the governing equations dominate and almost cancel out. Such flows are described as 'balanced', and their characteristic is that their evolution is in some sense 'slow'. Examples include hydrostatic balance, in which the pressure gradient force on the fluid virtually balances the gravitational force, and geostrophic balance, in which pressure gradients forces balance Coriolis accelerations.

We shall show in Section 5.5 that vertical accelerations are generally very small compared to the pressure gradient and gravitational accelerations. This means that the atmosphere is generally very close to a state of hydrostatic balance (Equation 5.5). This equation indicates that pressure drops monotonically with height. For the

Fluid Dynamics of the Midlatitude Atmosphere, First Edition. Brian J. Hoskins & Ian N. James.
© 2014 John Wiley & Sons, Ltd. Published 2014 by John Wiley & Sons, Ltd.

special case of an isothermal atmosphere (temperature T constant), Equation 5.5 can be integrated to give

$$p = p_0 e^{-z/H} \quad \text{where } H = \frac{RT}{g}$$

where p is the pressure, and p_0 its value at $z = 0$. However, the temperature is not usually constant with height. So this relationship is only rather a rough guide to the variation of pressure with height. Traditionally, the various layers of the Earth's atmosphere have been classified by the way in which their temperature varies with height.

The variation of temperature with height gives important information about the stratification of the atmosphere. Suppose an air parcel is displaced so rapidly in the vertical that it is unable to exchange significant heat with its surroundings. Its motion is then described as adiabatic, and for such an air parcel,

$$\frac{\partial T}{\partial z} = -\frac{g}{c_p} = -\Gamma_a \tag{1.1}$$

where Γ_a is the 'adiabatic lapse rate' and is around $10\,\text{K km}^{-1}$ for air.

Now suppose that the general rate of change of temperature with height is different from Γ_a. Again, suppose that an air parcel is subject to an adiabatic vertical displacement. Then, the displaced parcel will oscillate about its home level with a frequency N, the Brunt–Väisälä frequency. A formula for N is

$$N^2 = \frac{g}{T}(\Gamma_a - \Gamma) \tag{1.2}$$

where $\Gamma = -\partial T/\partial z$ is the observed lapse rate. If $\Gamma < \Gamma_a$, the atmosphere is said to be 'stably stratified'. A vertically displaced parcel of air will simply oscillate about its home level. If $\Gamma_a < \Gamma$, then N^2 is negative, N is imaginary, and instead of oscillating, a displaced parcel will continue moving away from its home level. Such an atmosphere is said to be 'unstably stratified'. In fact, unstable stratification is rather rarely observed save in a thin layer immediately adjacent to a strongly heated surface. Usually, rapid mixing across any unstable layer quickly leads to a state where $\Gamma \approx \Gamma_a$.

If an air parcel contains water vapour, then it will become saturated if it cools sufficiently, for example, through ascending. In this case, condensation of water vapour will generally occur, and latent heat will be released into the air parcel. Consequently, the parcel will not cool as rapidly as it rises. In particular, the lapse rate for a saturated atmosphere will be less than the dry adiabatic lapse rate. Table 1.1 gives examples of the saturated adiabatic lapse rate for typical polar and tropical surface temperatures and at two levels in the atmosphere. Since the amount of water required to saturate the warm tropical atmosphere is much greater than for the polar atmosphere, the warming by latent heat release and the consequent reduction in lapse rate below the dry adiabatic value of about $10\,\text{K km}^{-1}$ are much larger. At higher levels where the atmosphere is colder, the water content is smaller, and so the reduction in the lapse rate is less at $50\,\text{kPa}$.

Table 1.1 Saturated adiabatic lapse rate between 100 and 70 kPa and between 70 and 50 kPa for typical low- and high-latitude surface temperatures, together with the corresponding values of N^2

Surface Temperature (°C)	SALR (K km^{-1})	N^2 (10^{-4} s^{-2})	SALR (K km^{-1})	N^2 (10^{-4} s^{-2})
	100–70 kPa	100–70 kPa	70–50 kPa	70–50 kPa
0	7.36	0.90	8.39	0.56
25	4.15	1.89	4.51	1.85

The midlatitude atmosphere comprises a number of layers, each characterized by particular lapse rate or Brunt–Väisälä frequency. The lowest kilometre or so is called the atmospheric boundary layer. Its structure can change markedly through the day. During the daytime over land, strong heating at the surface and turbulent mixing through the boundary layer result in a lapse rate close to adiabatic. In that case, N is near 0, and the boundary layer is said to be neutrally stratified. At night, rapid surface cooling creates a stable layer near the ground. The lower part of the boundary layer can have a rather small lapse rate; indeed, in some conditions, the temperature actually increases with height, a situation called an inversion. In these conditions, N can become very large, and we talk of a 'stable boundary layer'. Perhaps the most extreme example occurs over Antarctica. During the long winter night, intense cooling at the surface of the ice cap results in a continent-wide inversion of as much as 35—40 K in the lowest few hundred metres of the atmosphere.

Above the boundary layer, diurnal variations are much smaller. Temperature falls off with height, but not as rapidly as the dry adiabatic lapse rate. A typical observed lapse rate is around 6 K km^{-1}, and this persists to heights of around 10 km at high latitudes and 18 km in the tropics. Over most of the tropics and indeed over considerable parts of the higher latitudes, the observed lapse rate is in fact intermediate between the dry adiabatic lapse rate and the saturated adiabatic lapse rate, a situation described as 'conditional instability'. This layer of conditionally unstable air is called the 'troposphere' and is the main focus of interest in this book. The actual stratification, measured by N, is a crucial parameter determining the character of motions in the troposphere. A typical value of N is around 10^{-2} s^{-1}, intermediate between the dry neutral value of zero and the values associated with the saturated adiabatic lapse rate given in Table 1.1. This corresponds to buoyancy oscillations with a period of about 10 min, a very short time compared to the rotation period of the Earth or to the typical advection-based timescales characterizing atmospheric motion.

Despite its central importance for the dynamics of atmospheric flow, the atmospheric stratification is generally taken simply as an imposed parameter. There is not much explicit discussion in the literature about why it has its observed value. Purely radiative considerations would recognize that the atmosphere is heated primarily at its base, where sunlight is absorbed by the ground, and cooled in the upper troposphere, where the atmosphere becomes optically thin to thermal infrared radiation. In the absence of any other processes, such a distribution of heating and cooling would lead to a strongly unstable lapse rate. Rapid vertical mixing would erode super-adiabatic lapse rates and lead to a 'radiative-convective equilibrium'. In the case of the Earth, such convective mixing is dominated by moist convection and so would lead to a lapse rate close to the saturated adiabatic lapse rate in the lower troposphere. In fact, the observed stratification is not far different from the saturated adiabatic lapse rate in the tropical lower troposphere, as shown in Table 1.1. It is rather more stable than the saturated adiabatic lapse rate in the middle and upper troposphere. A further complication is that atmospheric motions in the midlatitudes tend to transport heat upwards, cooling the lower layers and heating the upper layers, thereby increasing the stratification. The actual stratification will be a result of all these different processes. Hence we must conclude that the stratification is as much a product of the dynamics as a constraint upon the dynamics.

At heights above 16 km or more in the tropics or around 8 km at high latitudes, the lapse rate changes abruptly. Temperature is roughly constant with height above these levels and actually increases with height above 30 km or so. This region of very stably stratified air is called the stratosphere. Aircraft measurements show that the transition, called the tropopause, really is very abrupt, perhaps only a few metres deep. Across it, the Brunt–Väisälä frequency increases by a factor of two or more. The existence of the tropopause has important consequences for the dynamics of the troposphere. In many ways, the sudden large change of stratification is analogous to a rigid boundary at the tropopause level. Vertical velocities drop by a significant factor across the tropopause. At the same time, many smaller-scale tropospheric flow features are confined by the tropopause, giving stratospheric flow a very different character from that in the troposphere.

The basic reason for the change in lapse rate is the absorption of ultraviolet components of the solar beam at high levels in the atmosphere, particularly absorption by ozone. Although the amount of solar radiation at these wavelengths is a very small fraction of the total, it is absorbed at levels where the atmospheric density is very low, hence producing a substantial effect on the temperature.

Further changes of lapse rate characterize the mesosphere, above about 50 km, and the thermosphere, above 80 km. Rather little solar radiation is absorbed in the mesosphere, making it relatively unstably stratified. In the thermosphere, very energetic, short-wavelength photons in the far ultraviolet are absorbed, giving a strong temperature response. However, these levels have very little direct impact on the dynamics of the troposphere and will not be discussed further in this book.

1.2 Horizontal structure

In much of this book, rather than temperature, it will prove convenient to use instead a variable called the 'potential temperature'. This is the temperature a sample of air would have if it were compressed without adding or extracting heat, that is, adiabatically, to some arbitrary standard pressure p_s. The potential temperature is related to the temperature by

$$\theta = T \left(\frac{p}{p_s} \right)^{-\frac{R}{c_p}}$$

(1.3)

A great advantage of potential temperature is that it is conserved by air parcels which move without gaining or losing heat. The standard pressure is often taken to be 100 kPa. In terms of the potential temperature, Equation 1.1 can be written as:

$$N^2 = \frac{g}{\theta} \frac{\partial \theta}{\partial z}$$

and for a dry adiabatic neutral atmosphere, $\partial \theta / \partial z = 0$. Figure 1.1 shows a cross section of the time mean potential temperature, averaged around latitude circles, for the northern hemisphere winter, that is, for the months of December, January and February. The potential temperature increases with height everywhere, that is, the atmosphere is stably stratified, and it generally decreases from the equator towards the pole. The potential temperature difference between the equator and pole is around 50 K near the surface and somewhat less in the upper troposphere. The surfaces of constant potential temperature, called isentropes, are nearly horizontal in the tropics but slope downwards from the poles with a typical inclination of 10 kPa in 10° of latitude, that is, of about 1 in 1000, in the midlatitudes. The change in stratification at the tropopause is clear, with the vertical gradient of potential temperature increasing sharply above 30 kPa at high latitudes and 15 kPa in the tropics. Above the tropopause, the slopes of the isentropes reverse sign as well as becoming smaller. The 350 K isentrope is nearly flat. The region of strong poleward temperature gradient in the troposphere extends from around 30° of latitude to around 70° of latitude in both hemispheres, and so is around 4000 km wide.

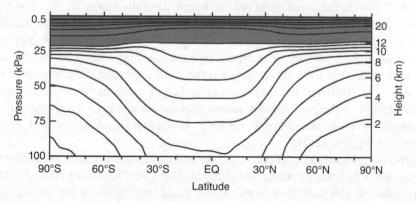

Figure 1.1 Pressure–latitude cross section of the time mean, zonal mean and zonal potential temperature for the December–January–February season. Contour interval 10 K up to 350 K and 50 K thereafter. Shading indicates values above 350 K

With potential temperature, and hence temperature, decreasing towards the pole within the troposphere, the separation between pressure surfaces will reduce with latitude, according to the hydrostatic relationship:

$$\Delta z = \frac{RT}{g} \cdot \Delta p$$

where Δz is the depth of a layer of air across which the pressure drop is Δp. Consequently, closer to the pole, where temperatures are lower throughout the troposphere than in the tropics, the upper-level pressure surfaces are lowered.

The westerly or zonal wind is related to the slope in height of a pressure surface:

$$u = -\frac{g}{f} \frac{\partial z}{\partial y}$$

where the gradient is calculated on a pressure surface and f is twice the local vertical component of the Earth's rotation. Therefore, the difference in height of a pressure surface between the polar and tropical regions means that there must be a westerly (eastward) wind in middle latitudes.

The change with height of the zonal wind is related to the potential temperature by

$$\frac{\partial u}{\partial z} = \frac{g}{f\theta} \frac{\partial \theta}{\partial y} \tag{1.4}$$

This equation is one form of the 'thermal wind relationship' and will be discussed in more detail in Chapter 12; a flow which satisfies Equation 1.4 is said to be in 'thermal wind balance'. Comparison of Figure 1.2 and Figure 1.1 suggests that the zonal wind is indeed close to thermal wind balance with the potential temperature field. In the midlatitudes, where θ decreases most rapidly towards the pole, the zonal wind is generally westerly and increases with height, reaching its largest values around the subtropical tropopause. This tropospheric jet reaches speeds of around $50\,\mathrm{m\,s^{-1}}$ in the zonal mean but can be substantially larger locally. The global average zonal wind is around $14\,\mathrm{m\,s^{-1}}$. In the tropics, the zonal mean winds are smaller and are easterly in the lower troposphere. At the surface, roughly equal areas of the Earth's surface are occupied by easterlies and westerlies: clearly, in the long-term mean, the total torque between the solid Earth and the atmosphere must integrate to 0. However, the long-term existence of mean easterlies in the tropics and westerlies in the midlatitudes means that the atmosphere itself must support horizontal stresses, transferring westerly momentum polewards and easterly momentum equatorwards.

Above the midlatitude tropopause, the vertical wind shear reverses, consistent with the reversal in poleward temperature gradients. The cross sections in this chapter all use pressure as a vertical coordinate, so that equal increments in the vertical correspond to equal increments of mass. Over 80% of the atmosphere lies beneath the tropopause and the stratosphere, and mesosphere is squashed into the topmost part of the plot. However, it can still be seen that the summer and tropical stratosphere are

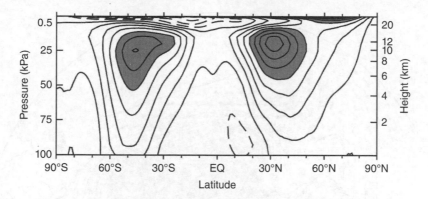

Figure 1.2 The time and zonal mean zonal wind for the December–January–February season. Contour interval 5 m s⁻¹, negative values dashed. Shading indicates winds in excess of 20 m s⁻¹

filled with easterlies. The winter hemisphere has a region of increasingly strong westerlies centred on around 60°N. This so-called 'polar night jet' runs close to the arctic circle, the latitude where the incoming solar radiation drops to zero at the winter solstice. The polar night jet runs around a core of sharply cooling polar air (the 'polar vortex') with the relatively warmer air of the midlatitudes to equatorward. This polar vortex is centred near to, but not exactly over, the North Pole.

Figure 1.1 and Figure 1.2 present time and zonal means. The time and spatial averaging implies that much of the dynamical detail has been smoothed away, whether those details be transient fluctuations in time or zonal variations around latitude circles. By way of contrast, Figure 1.3 shows the flow at 25 kPa for an arbitrary time. The general lowering of the 25 kPa pressure surface towards the colder pole is clear, as is the predominance of westerly winds throughout the midlatitudes. This large-scale structure, some 5000 km across, is sometimes called the 'polar vortex'. It is present throughout the year but is stronger during the winter season.

Around the periphery of the polar vortex, the pattern of the height field is very wavy, with a belt of strong winds, called the 'jet stream', meandering from low to high latitudes and back. Some five or six major troughs and ridges can be counted around a midlatitude latitude circle in this particular example. Such a spacing of the troughs and ridges corresponds to a typical wavelength of about 4000 km, which implies that the distance between a trough or ridge and a maximum of meridional wind is about a quarter of this, that is, about 1000 km. The diagram shows that although the time and zonal mean wind has a single maximum between equator and pole, the situation is more complicated locally. For example, in Figure 1.4, there are clearly two roughly parallel jet streams over the Middle East and Central Asia, but only one over the Pacific and North America.

Some of the features seen in Figure 1.3 are short lived or transient. Others have a much longer lifetime, and some persist throughout the season and recur from season to season. Figure 1.4 again shows the 25 kPa height field, but this time averaged over a 3-month period of December–February, corresponding to the Northern

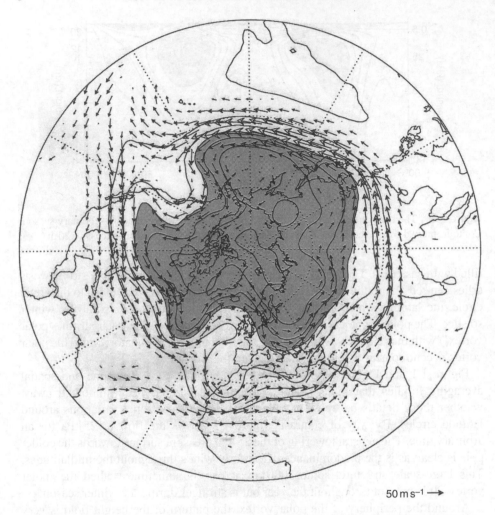

50 m s^{-1} →

Figure 1.3   Contours of the geopotential height of the 25 kPa surface on 10 February 2004, 00Z. Contour interval 250 m, values below 10 km shaded. The arrows represent the horizontal component of the wind

Hemisphere winter and then averaged over 40 such winter seasons. The field is much smoother than that shown in Figure 1.3, indicating that the short-term transients generally have a smaller length scale than the more permanent features of the flow. The most prominent features of Figure 1.4 are regions of tightly packed contours, that is, of high winds, over the western Pacific and Western Atlantic oceans, and the pair of troughs associated with them near the continental coasts. Downstream of these tight jets, there is a tendency to ridging of the contours and lighter westerly winds; these characterize the Eastern Pacific and Eastern Atlantic regions. The pattern is dominated by waves with two or three crests and troughs around a latitude circle, that is, with wavelengths of 9,000–13,000 km.

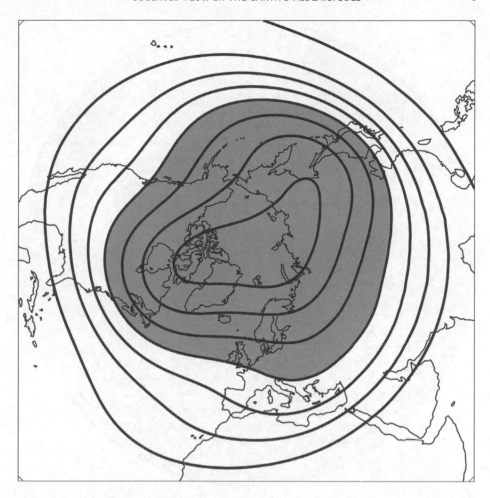

Figure 1.4 A 40 year mean of the DJF 20 kPa geopotential height field. Contour interval 200 m, and shading indicates values less than 10 km

A question which immediately arises concerns the origin of these semi-permanent features of the flow. Their very large scale, filling the entire hemisphere, suggests we might be looking at some sort of wave phenomenon. Their geographical location appears to be related to the distribution of oceans and continents. It turns out that the important influences are the forcing of the flow by major mountain ranges, particularly the Rockies and the Tibetan plateau and the remote impact of the heating of the atmosphere by deep convection in the tropics. A crucial ingredient in the large-scale response of the atmosphere is the way in which disturbances of some wavelengths can propagate away from the regions in which they are excited while others are confined and damped. These ideas will be developed in Chapters 9 and 18.

Apart from the main polar vortex itself, the flow shown in Figure 1.3 is characterized by a series of open troughs and ridges, that is, by a sinuous band of fluid snaking

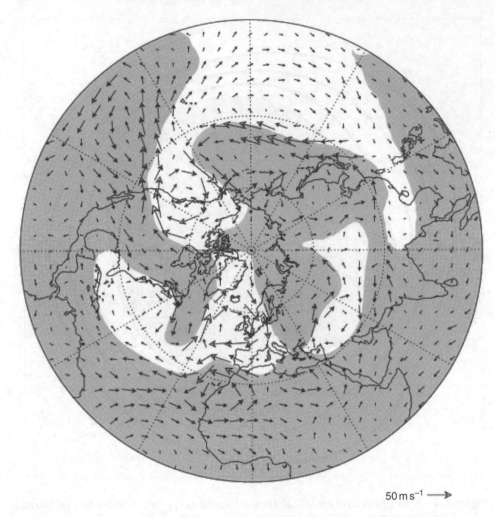

$50\,\mathrm{m\,s^{-1}}\longrightarrow$

Figure 1.5 Vectors of the horizontal component of wind, as in Figure 1.3, except that the zonal mean has been removed. Shading indicates regions where the 25 kPa geopotential height anomaly from its zonal mean is negative

from high to low latitudes and back. Note that this pattern is an extremely simple and effective way of transferring heat from the tropics to the pole. It also transports other tracers such as water vapour rather effectively in the meridional direction. However, this pattern does depend upon the frame of reference within which it is viewed. Figure 1.5 is the same as Figure 1.3 except that the zonal mean has been removed. For a small range of latitudes around any particular flow feature, removing the zonal mean in this way is roughly equivalent to viewing the flow in a frame of reference moving with the mean flow rather than one fixed relative to the solid Earth. With this change, the troughs and ridges have been transformed into closed cyclonic and anti-cyclonic eddies. The typical radius of the eddies is around 1000 km, and their typical circulation velocity is around $10\,\mathrm{m\,s^{-1}}$. This implies that fluid would circulate completely around the eddy every 0.7 days or so.

1.3 Transient activity

Compare Figure 1.3 and Figure 1.4. The flow at a random instant is very different from the time mean flow. Clearly, a good deal of the observed atmospheric flow is highly variable. Features develop, move and collapse in time. This transient part of the flow is extremely important. Indeed, the whole business of weather forecasting is concerned with predicting the evolution of the transient part of the flow. Fluctuations on a wide variety of timescales dominate the flow and make prediction of its future state extremely difficult. Much of this book will be devoted to exploring aspects of this transient behaviour, to understanding the mechanisms which mean that the flow is unstable and therefore unsteady. A first picture of the midlatitude transients is summarized by Figure 1.6a. This is a time series of the surface pressure, at 1 min intervals, observed at Reading, UK, a typical midlatitude location. Fluctuations on a wide range of timescales are present, but the time series is dominated by the large swings of pressure, as large as 1 kPa, as individual weather systems pass over the observing site. Smaller fluctuations are revealed with other periods. For example, during quieter periods of weather, a ripple with a period of 0.5 days and an amplitude of around 50 Pa is often observed. This is an atmospheric tide, raised in the atmosphere by gravitational and thermal effects. Generally, though, this semi-diurnal tide is swamped by the much larger fluctuations arising from the passage of major weather systems.

These different timescales are made clearer by a spectral analysis of the time series, shown in Figure 1.6b. The largest amplitudes are associated with low-frequency transients, with periods greater than a day. The spectrum at these periods is roughly white, with similar amplitude for all periods. As the period becomes shorter than about 1 day, the power drops off rapidly.

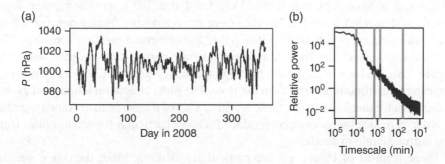

Figure 1.6   (a) A time series of the surface pressure averaged at 5 min intervals from a vibrating cylinder barometer at the University of Reading's Atmospheric Observatory, during 2008. (b) A spectrum of the time series shown in part (a) with vertical lines added to mark weekly, diurnal, semi-diurnal and hourly timescales. Courtesy of R.G. Harrison

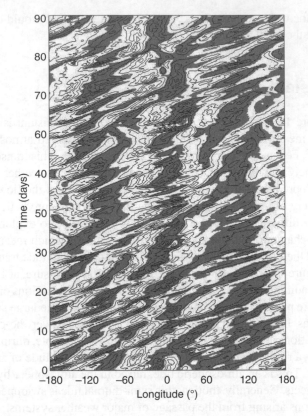

Figure 1.7 Longitude–time (Hovmöller) plot of poleward wind v averaged between 45°N and 55°N at 25 kPa for the December 1983 to February 1984 winter season. Shading indicates equatorward motion ($v<0$); contour interval 20 m s^{-1}

A clearer picture of the transients is given in Figure 1.7, which is a longitude-time (Hovmöller) plot of the flow at a particular level and latitude. Here, the poleward component of wind is plotted at the 25 kPa level, near the level of maximum mean winds, as shown in Figure 1.2. The data were extracted from routine meteorological analyses made every 6 hour as initial fields for numerical weather forecasts. The series therefore includes the white low-frequency part of the spectrum shown in Figure 1.6b, but little of the power law high frequency section. Changes of sign of v correspond to troughs and ridges in the flow. The plot is based on latitude 50°N, but in fact data for that latitude and those 5° either side of it have been averaged together in order to reduce the jerky effect produced when a particular feature migrates from one latitude circle to another.

Three features of Figure 1.7 are particularly striking. First, there is a general tendency of features to move from west to east. This is especially true of the smaller-scale features. The rate of progression is not constant but fluctuates. Nevertheless, typically features are moving so that they pass right around the latitude circle every 20 days or so, corresponding to a speed relative to the ground of around 15 m s^{-1}.

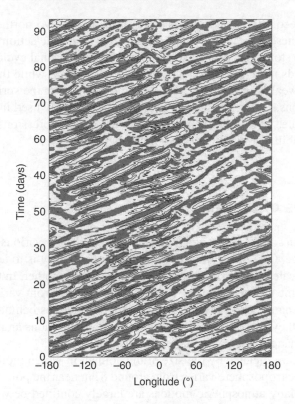

Figure 1.8 As Figure 1.7, except that the winds have been high-pass-filtered to include only those high-frequency transients with periods less than 6 days

Secondly, some features are relatively stationary and permanent. The transition from negative v to positive v, that is, from shaded to unshaded, represents a trough feature. One such feature, around longitude 120°E, is quasi-permanent and corresponds to the marked trough in the time mean flow near the eastern cost of Asia, shown in Figure 1.4. Thirdly, the most vigorous transient activity is not distributed uniformly around the latitude circle, but tends to be concentrated in certain preferred regions. One of these is for longitudes 80°W to 20°E. Another is from 120°W to 80°W. These regions are preferred locations for the cyclonic weather systems of the midlatitudes and are sometimes called storm tracks. They tend to occur at the western side of the Pacific and Atlantic Ocean basins.

A numerical filter can be applied to the data shown in Figure 1.7 to separate the low-frequency transients from the very-low-frequency transients. Figure 1.8 shows the effect of a 'high-pass' filter on the Hovmöller plot of Figure 1.7. At each time, the very low frequencies have been removed by subtracting a 3-day running mean of the data from the time series. This procedure acts as a 'poor man's filter', isolating the transients with periods of 6 days or less. More sophisticated filters can be devised, but the essential results are not much changed. The principal results are twofold. First, the quasi-stationary waves visible in Figure 1.7 have more or less disappeared. Secondly, great emphasis is placed on the two northern hemisphere storm tracks. These appear as distinct and separate under the action of such a filter. In fact, the two storm tracks have somewhat different seasonal cycles. The Atlantic storm track tends to be most pronounced in the midwinter, while the Pacific storm track is rather weaker in midwinter but more pronounced in the spring and autumn transition seasons. A 'high-pass' filter such as this is widely used in the analysis of atmospheric data. It tends to separate the midlatitude depressions, generated by baroclinic instability, from other types of circulation system.

1.4   Scales of motion

In the discussion so far, certain characteristic magnitudes for various quantities have begun to emerge. One of the objectives of dynamical theory is to account for such characteristic scales and to show how they relate to one another. In this section, we merely note some of the quantities which can be derived from various basic diagnostics. Later chapters will establish some of the principles behind these scalings and show how they can lead to insights into the approximations and simplifications which can be made to the governing equations.

Some scales are dictated by the size of the domain in which motions take place. The depth of the troposphere varies from around 8 km near the poles to 18 km in the deep tropics. Many atmospheric motions are largely confined between the surface and the tropopause, and so a maximum vertical scale of order 10 km emerges from such considerations. Similarly, in the horizontal, the Earth is a sphere of radius 6400 km. The distance from the pole to the equator is 10,000 km,[1] and this is effectively a maximum horizontal extent of atmospheric motion.

A comparison of these scales reveals the huge disparity between the vertical and horizontal scales. The atmosphere really is an extremely thin skin sitting upon the solid earth. Its thickness is in a similar proportion to the radius of the Earth as the skin of an apple is to its radius. In most depictions of the atmosphere, the vertical scale is orders of magnitude different from the horizontal scale.

The dimensions of many characteristic features of atmospheric flow are a good deal smaller than the maximum that the dimensions of the system would permit. For example, Figure 1.3 shows a band of air meandering between high and low latitudes. Something like six to eight such meanders can be counted around a latitude circle, corresponding to a wavelength of order 4000 km. A typical horizontal distance in a weather system might be one quarter of a wavelength, which is the distance between, for example, a 0 and a maximum or minimum of one of the horizontal components of the wind. Conveniently, this distance is about 1000 km in the troposphere.

Since they combine information about the time and space domains, the Hovmöller plots of Figure 1.7 and Figure 1.8 give some interesting information about timescales and speeds of flow features in the atmosphere. The diagrams shown apply to 50°N and 25 kPa. Reference to Figure 1.3 shows that the time and zonal mean wind at this level is around $20\,\mathrm{m\,s^{-1}}$, comparable to the speed at which disturbances move from west to east. However, if lines with a slope corresponding to an eastward speed of $20\,\mathrm{m\,s^{-1}}$ are drawn on the Hovmöller plots, it is clear that the speed at which features move around latitude circles is generally smaller than this. This is an important result. The synoptic features seen in the midlatitude flow are not simply swept around the globe by the mean winds blowing at the level. Air flows through the systems which should not therefore be thought of as material objects, but rather as organized features within the flow.

A number of timescales are associated with the observed transients. The simplest is the period of disturbances relative to the solid Earth, obtained from their speed and length scale. A length scale of $1000\,\mathrm{km}$ and a speed of $16\,\mathrm{m\,s^{-1}}$ imply a timescale of around 0.7 days, perhaps better rounded to 1 day. Other relevant timescales include the 'circulation timescale', the time taken for disturbances to pass right around the globe and return to their initial location, if they live that long. For $16\,\mathrm{m\,s^{-1}}$ at $50°\mathrm{N}$, this time-scale is around 20 days. Another relevant timescale is the lifetime of individual flow features, as observed for example, in a frame of reference moving with the disturbance. This is highly variable, and certain features, particularly in the high-pass-filtered data can be surprisingly long lived. But generally, most strong features can be traced unambiguously only for 5–10 days. Therefore, they do not usually circumnavigate the globe.

A final comment on the material of these last two sections is that it suggests two rather different ways of viewing atmospheric flows. In Section 1.2, we showed how the flow was made up of a large number of circulating eddies, with discrete masses of fluid orbiting around the eddy centre. There were small eddies with circulation times as short as a day or so, right up to the polar vortex itself, with its circulation time of 20 days or so. In this section, in contrast, we have emphasized the wave-like nature of atmospheric circulation systems. The Hovmöller plots revealed patterns in the flow with the actual fluid flow passing through them, or, equivalently, with the patterns propagating through the atmosphere. We shall refer to these two aspects of atmospheric flow as 'vortex-wave duality'. The two points of view are equally valid, and one may be preferred to another depending upon the circumstances. We shall return to this distinction throughout the book.

1.5 The Norwegian frontal model of cyclones

This chapter would not be complete without a brief description of a conceptual model of weather systems which originated in the 1920s but which is still widely, even universally, used in communicating information about weather analyses and forecasts. It is called the 'Norwegian frontal model', and most people will be familiar with it even if they do not know its name. The model was developed by Wilhelm and Jacob Bjerknes and colleagues at the University of Bergen in Norway after the First World War; it represents a summary of the large volume of meteorological analyses which had been accumulated at that time in Norway. Despites its

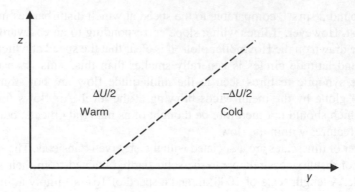

Figure 1.9   The polar front, according to the Norwegian frontal model

near-universal use today, it is important to recognize that the Norwegian model remains a conceptual model and no more; it proves useful for interpreting many observations, but equally, it can prove unhelpful in understanding others.

The basic hypothesis of the Bergen School was that the atmosphere consists of a number of distinctive air masses, each relatively uniform in the horizontal and each having properties which reflect their origins. Air masses are better distinguished in the lower and middle troposphere, above the atmospheric boundary layer where exchanges of heat, moisture and momentum with the underlying surface can modify the properties of the air mass rather readily. The model supposes that the interface between air masses is more or less sharp and discontinuous. Such an interface is termed a 'front'.

In particular, the Bergen school distinguished relatively cold, dry air of polar origin from warmer, moister air originating in low latitudes. They imagined an interface between the two air masses which girdles the entire Earth in the mid-latitudes and which they called the 'polar front'. If the polar front retreats, so that cold air is being replaced by warmer air at the surface, the front is called a 'warm front'. If on the other hand, the cold air is advancing, we use the term 'cold front'. The polar frontal surface slopes, so that the cold air forms a wedge obtruding beneath the warm air, as shown in Figure 1.9. The slope of the front depends upon the wind shear and temperature change ΔT (or density change $\Delta\rho$) across the front and is given by the 'Margules formula':

$$\left.\frac{\partial z}{\partial y}\right|_{\text{front}} = \frac{f\rho_0\Delta U}{g\Delta\rho} = -\frac{fT_0\Delta U}{g\Delta T} \tag{1.5}$$

where $f = 2\Omega\sin(\phi)$ is the Coriolis parameter, Ω being the rotation rate of the Earth and ϕ the latitude. The acceleration due to gravity is denoted by g and ρ_0, T_0 are a standard reference density and temperature respectively. This formula is in fact an application of the thermal wind equation, discussed in Section 11.2. The implication of the Margules formula is that, since the cold air must undercut the warm, that is,

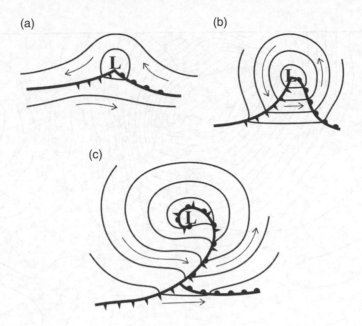

Figure 1.10 Schematic maps showing surface pressure and fronts during the development of a midlatitude depression, according to the Norwegian frontal model. (a) Developing phase, (b) mature phase and (c) occluding (decaying) phase

$\partial y/\partial z > 0$, then there must be a positive or cyclonic shear across the front. Putting in reasonable values for ΔU ($20\,\mathrm{m\,s^{-1}}$) and ΔT ($5\,\mathrm{K}$) gives a value for the slope of around 10^{-2}. So the polar front reaches from the surface to the tropopause over a horizontal distance of about $1000\,\mathrm{km}$.

Even when fronts can be analysed with confidence on a synoptic chart, a straight, zonally orientated front is a rarity. Such a front appears to be unstable. It develops kinks and ripples which amplify rapidly to give a series of cyclones and anticyclones. Figure 1.10 is a schematic illustration of the development observed. At the surface, the distorted polar front demarcates a triangular region of warm air, designated the 'warm sector', with a minimum of surface pressure at the apex of the triangle. To the east, the polar front becomes a warm front, as warm air advances. To the west, a cold front advances behind the system. Generally, the cold front moves to the east more rapidly than the warm front. So the warm sector is undercut by cold air and lifted away from the ground. This process is called 'occlusion', and it marks the end of the developing phase of the depression. Eventually, factors such as friction, latent heat release and so on cause the system to decay, returning the atmosphere to something like its initial state.

Like all models, the Norwegian frontal model is limited and provisional, despite its widespread use as a convention for reporting data. Figure 1.11 shows

(a)

(b)

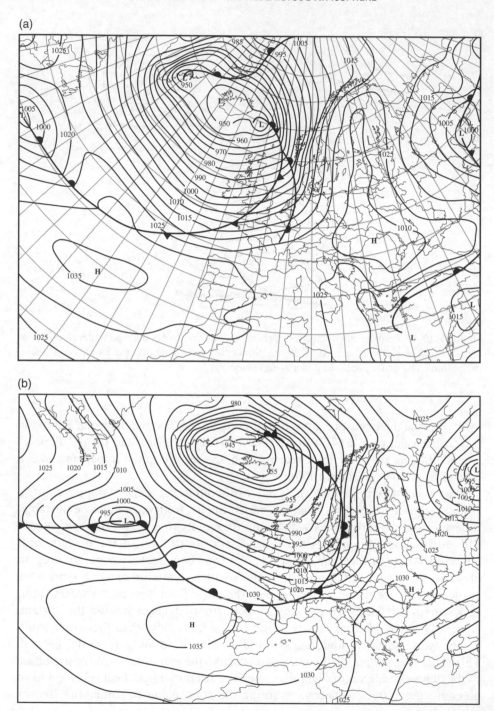

Figure 1.11 An example of an active developing cyclone, analysed according to the conventions of the Norwegian frontal model. The sequence shows maps of the surface pressure over the North-east Atlantic and Western Europe at 12 hourly intervals, beginning at 00Z on 18 February 1997. This case was part of the FASTEX experiment, intensive observing period 17

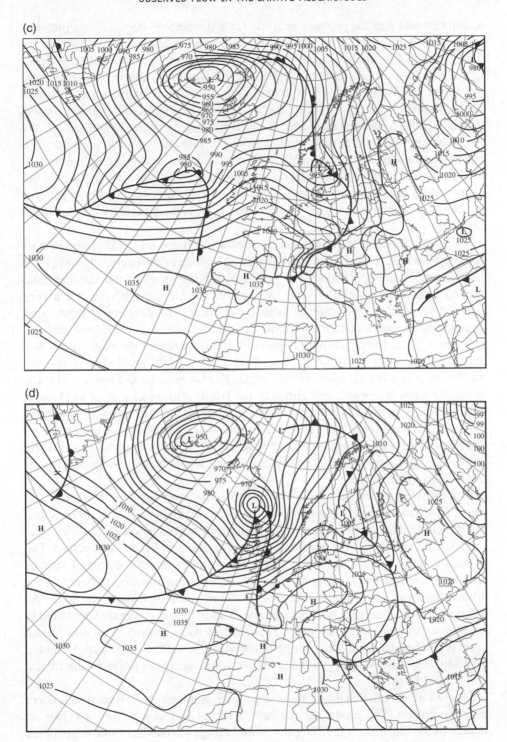

Figure 1.11 (*Continued*)

a conventional surface pressure analysis for a developing depression which has been studied extensively and which will provide a useful example of cyclone development. Fronts have been analysed, using the convention that they represent a near discontinuity of temperature and wind on this synoptic scale. However, the fronts are fragmentary. They do not form a complete unbroken line around the midlatitudes. In some places, no front could be identified. In other places, multiple fronts have been analysed. We shall see later in the book that, in fact, the development of a midlatitude cyclone does not depend upon the presence of a pre-existing polar front. It can occur in a region of uniform and gentle temperature gradients. In most circumstances, the formation of fronts is a product of the cyclogenesis rather than its cause. A deepening depression develops flow patterns which sharpen temperature gradients in some places and can actually lead to frontal formation.

The case of a developing depression shown in Figure 1.11 was observed during the FASTEX experiment and constitutes intensive observing period (IOP) 17. As well as the operational network of observing stations and satellite data, special additional observing systems operated during the IOPs, giving a very detailed view of the evolution of synoptic features. The developing depression shown here moved around the periphery of a large feature centred near Iceland. In Figure 1.11a, the developing depression is just crossing the edge of the plot and moving across the Atlantic. The depression is fairly shallow, and its cold and warm fronts form a wide, obtuse angle. Twelve hour later, Figure 1.11b, the depression has deepened, and although the fronts still make a wide angle with each other, a distinct warm sector has formed. The last analysis shown in Figure 1.11d, made 48 hour after the initial analysis, shows an intense, deep depression, which is beginning to occlude as the cold front overtakes the warm front. The tight isobars indicate that very strong winds were associated with the depression by this time.

The Norwegian frontal model was based largely on surface-based observations. Few routine upper-air observations were available. Indeed, even today, the typical separation of upper-air observing stations is several hundred kilometres. Such a spacing is far too large to resolve details of frontal structure. The picture which emerged from the Norwegian studies was of a frontal surface which extended from the ground through the depth of the troposphere, across which there was an abrupt change of temperature, humidity, wind and so on. As a full upper-air observing network became established, this picture was found to be an over-simplification. Figure 1.12 is taken from a classic study of a frontal zone published in 1955. The routine upper-air observations were supplemented by additional observations, so that a high-resolution cross section of the frontal zone could be made.

Figure 1.12 reveals that the tightest temperature gradients in the frontal zone occur near the ground. The frontal surface indeed slopes so that a wedge of colder air undercuts the warm air, as the Margules formula suggests, but the frontal temperature

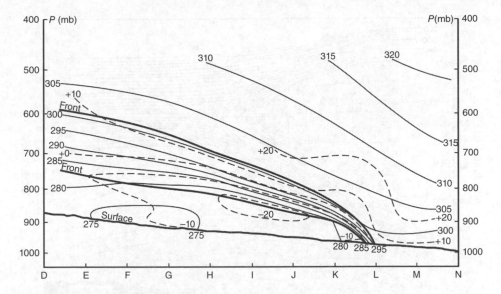

Figure 1.12   Cross section through an active frontal zone, showing contours of potential temperature (solid contours) and along the front component of the wind (dashed contours). From Sanders (1955), reproduced by permission of the American Meteorological Society

contrast becomes more diffuse with height. At heights of 300 m above the ground, the width of the frontal zone was no more than 25 km; at heights of 1200 m, it was 200–300 km. At the same time, the component of wind parallel to the front changes sharply across the front. Such a shear is implicit in the classic Norwegian school analyses, in which the isobars kink across the front. It is corroborated by a time series of the wind at an observing station as a front passes over; the wind strength and direction change abruptly and markedly as the front passes over the station.

Somewhat similar structures are observed in the upper troposphere, at the interface between the low static stability troposphere and the much more stable stratosphere. They have been called 'upper-level fronts' or 'troposphere folds'. Figure 1.13 shows a cross section through one such structure based on *in situ* aircraft observations. As in a surface front, the potential temperature contours are packed close together in the frontal region. An intrusion of stratospheric air descends along the frontal surface, suggesting that the front is associated with strong circulations in the plane perpendicular to the front. In this particular example, stratospheric air has descended four or more kilometres into the troposphere. Such upper-level fronts in the midlatitudes are thought to be an important mechanism for mixing stratospheric air into the troposphere and for the meridional circulation of the stratosphere.

We shall see in Chapter 15 that both surface and upper-level frontal structures can be reproduced using basic dynamical theory. The key to their formation is the circulation of the air required to maintain balanced flow, together with a basic

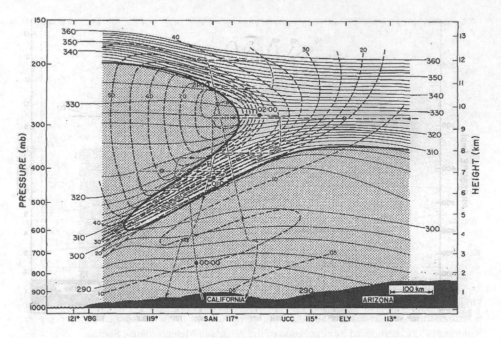

Figure 1.13 Cross section through an upper-level front, based on aircraft observations on 12 March 1978. From Shapiro (1980), reproduced by permission of the American Meteorological Society

mechanism called vortex stretching. Both these factors will be major themes throughout the succeeding chapters.

The Norwegian frontal model is limited in some respects. Nevertheless, the sort of analyses shown in Figure 1.11 are so extensively used that they do indeed provide a helpful starting point for the study of cyclonic weather systems in the midlatitudes. But the reader should always bear in mind that these, or any other analyses, are models, simplified idealizations of reality; they are at best a shorthand for more complex structures. They can in some circumstances be positively misleading.

This introductory chapter has surveyed some aspects of the observed temperatures and winds in the atmosphere and has introduced some indications of the variations of these quantities, both in space and time. We are now in a position to begin a systematic account of how basic physical principles determine these features. We shall start by formulating Newton's laws of motion and the laws of thermodynamics for a fluid such as air and then specifically recognize that the Earth's atmosphere is an extremely thin layer of fluid sitting on a spherical rotating planet. Motion in the atmosphere is strongly constrained by the rapid rotation of the Earth, and so much of the latter part of the book explores the consequences of these constraints.

Note

1. This round figure is no coincidence. The original definition of the metre was that it should be one ten-millionth of the distance from the pole to the equator along the Paris meridian. Later this was revised to be the length of a standard metal bar kept under standard conditions. Subsequently, it has been redefined as the distance travelled by light in a vacuum in 1/299,792,458th of a second.

Theme 1

Fluid dynamics of the midlatitude atmosphere

2
Fluid dynamics in an inertial frame of reference

2.1 Definition of fluid

Most of the observable universe consists of fluid, yet many of us have much greater familiarity with the mechanics of rigid bodies or point masses. The reasons are not hard to find: in developing the theory of fluid dynamics, we rapidly encounter very intractable nonlinear differential equations, we discover that some basic assumptions can turn out to be significantly limited, and from a physical perspective, we need to envisage systems in three dimensions, a task which our paper- and screen-based two-dimensional culture finds difficult. This chapter introduces some basic concepts in a rather general context, as a preliminary to the more specialized atmospheric applications discussed in later chapters. Here, the basic principles will be set out for an inertial or non-rotating frame of reference. Later chapters will address the problem of describing fluid flow when referred to rotating frames of reference, such as those embedded in the solid rotating Earth.

Our discussion begins by listing the forces acting upon a notional parcel of fluid, a 'parcel' being a coherent blob of fluid bounded by a surface, which may well be imaginary. These forces divide into three categories: forces which act tangentially to the surface of the parcel, forces which act in a direction normal to the surface of the parcel and forces which act throughout the volume of the parcel. The forces that act upon the surface of the parcel are measured by the force per unit area, designated as the 'stress'. Stress has units of $N\,m^{-2}$, a unit called the Pascal (Pa). We differentiate between 'tangential stresses' acting parallel to the surface, and 'normal stresses' which act perpendicularly to the surface. Forces which act throughout the volume of the fluid are called 'body forces' and are conveniently expressed in terms of force per unit mass, which of course has the units of acceleration, $m\,s^{-2}$.

Figure 2.1 shows a block of matter subject to tangential stresses at its opposite faces. A solid deforms when subject to such a stress; many common materials undergo a reversible deformation in which the strain, a dimensionless measure of the deformation, is proportional to the applied stress. This relationship is called

Fluid Dynamics of the Midlatitude Atmosphere, First Edition. Brian J. Hoskins & Ian N. James.
© 2014 John Wiley & Sons, Ltd. Published 2014 by John Wiley & Sons, Ltd.

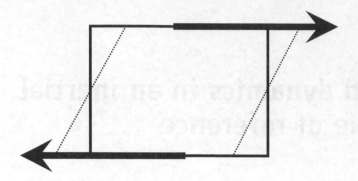

Figure 2.1   Matter subject to a tangential stress

Hooke's law, and it applies accurately to many common solids whose microscopic structure is crystalline. In such cases, the deformation is reversible: the sample springs back to its original shape when the stress is removed.

A fluid too deforms under the influence of tangential stresses, but it carries on deforming as long as the stress is applied. If the stress is removed, the fluid has no memory of its original shape and simply remains in its new configuration. Hooke's law is irrelevant, since the strain is indefinite. However, a corresponding law is 'Newton's law of viscosity', which states that the rate of strain is proportional to the stress. The constant of proportionality is called the viscosity of the fluid. Newton's law of viscosity may be written as:

$$F_T = \mu \frac{\partial u}{\partial z} \tag{2.1}$$

for the simple case illustrated in Figure 2.1. The constant of proportionality, μ, is called the 'dynamic coefficient of viscosity'. Fluids which obey Newton's law of viscosity are called 'Newtonian fluids'. Air and water are both examples of Newtonian fluids, provided the stress is not too large. Other fluids behave in more complex ways and are called 'non-Newtonian'. Examples include a water/cornflower mixture, whose viscosity increases sharply when the stress increases and which will fracture like a solid if the stresses are large enough. Other natural examples of non-Newtonian fluids include ice and the rocks making up the Earth's mantle.

Fluids come in two varieties, depending upon their reaction to normal stresses. Fluids such as air change their volume when subject to a normal stress, as shown in Figure 2.2, and are said to be compressible or gaseous. The relationship between the normal stress and the volume, which is linear for air, is called the 'equation of state' and will be discussed later on. Other fluids, such as water, scarcely change their volume at all when subject even to large normal stresses. Such fluids are called 'incompressible' or liquid. It turns out that at flow speeds which are small compared to the sound speed, compressible and incompressible fluids behave in virtually the same way, and the difference between liquids and gases can be unimportant.

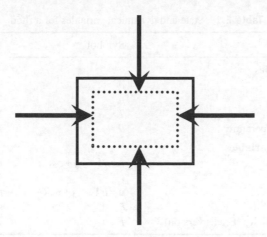

Figure 2.2 Fluid subject to normal stresses

The differences between solids, liquids and gases are related to their molecular structure. In a crystalline solid, the molecules are locked into fixed relative positions by intermolecular forces. In liquids, the molecules are so closely packed that they touch one another, so that normal stresses cannot force them any closer to each other, but they can easily slide past one another when subject to tangential shearing stresses. In the gaseous state, molecules are widely separated. Hence, the gas can both be compressed and sheared readily. It is the task of fluid physics to account quantitatively for the properties of fluids in terms of their molecular structures. In this book, we shall take such properties of fluids as their viscosity simply as given or as directly measured quantities.

2.2 Flow variables and the continuum hypothesis

The properties of fluids are described by a series of variables which in general vary with position and time. Such variables are called 'fields'. Two groups of fields are distinguished: those which describe the local properties of the fluid and those which describe its state of motion. The former are called 'state variables' and the latter 'dynamical variables'. Variables may either be scalar or vector quantities. Table 2.1 sets out some principal variables and establishes the notation that will be used in this book.

The crucial assumption made by fluid dynamics is that these fields can be differentiated, that is, the fluid can be subdivided indefinitely, and each sub-volume will retain well-defined values of the state and dynamical variables. This is the case as the size of the volume is reduced over many orders of magnitude. However, when the volume becomes so small that it holds only a relatively small number of molecules, it begins to break down. For example, temperature, which is a measure of the average

Table 2.1   State and dynamical variables for a fluid

Variable	Symbol	Units
State variables		
Density	ρ	$\mathrm{kg\,m^{-3}}$
Pressure	p	$\mathrm{Pa\,(N\,m^{-2})}$
Temperature	T	K
Potential temperature	θ	K
Dynamical variables		
Position vector	$\mathbf{r}=x\,\mathbf{i}+y\,\mathbf{j}+z\,\mathbf{k}$	m
Time	t	s
Velocity	$\mathbf{u}=u\,\mathbf{i}+v\,\mathbf{j}+w\,\mathbf{k}$	$\mathrm{m\,s^{-1}}$
Relative vorticity	$\boldsymbol{\xi}$	$\mathrm{s^{-1}}$
Potential vorticity (Ertel's version)	P	$\mathrm{K\,m^2\,kg^{-1}\,s^{-1}}$

Other variables will be defined throughout the text as needed.

kinetic, rotational and vibrational energy of individual molecules, begins to fluctuate randomly and eventually becomes meaningless as the volume shrinks to contain only a few molecules. However, under the 'continuum hypothesis', the assumption is that the large-scale behaviour is the same as if, as the volume of a fluid element tends to 0, the mean value of any fluid variable would tend towards a constant value.

2.3   Kinematics: characterizing fluid flow

'Kinematics' is concerned with the description of the flow field. How a flow comes to take its observed form and how it evolves is the topic of 'dynamics', the topic of most of this book. The two most basic properties of a flow are its direction and its speed. A compact way of illustrating its direction is by means of streamlines. A streamline is an imaginary line drawn through a flow at any instant which is everywhere tangential to the local velocity vector. Streamlines can be constructed for the flow at some instant t_0 from the differential equations:

$$\frac{\mathrm{d}x}{u(\mathbf{r},t_0)}=\frac{\mathrm{d}y}{v(\mathbf{r},t_0)}=\frac{\mathrm{d}z}{w(\mathbf{r},t_0)} \tag{2.2}$$

Streamlines cannot cross, for if they did, it would imply that the velocity field was multi-valued at the point of intersection. Streamlines must either extend to infinity or be closed loops. They cannot stop or start in the fluid interior, except in one special circumstance. That circumstance is a point where the velocity vector becomes 0, so the flow direction is indeterminate. Such a point is called a 'stagnation point'. Stagnation points can sometimes be identified on the grounds of symmetry and can be of some practical value. For example, consider a flow which approaches a bluff obstacle, whose surface is, at some point, perpendicular to the distant flow. Either side of this point, the

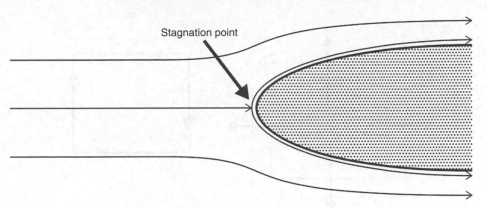

Figure 2.3 Streamlines of flow around a bluff body, illustrating the formation of a stagnation point

streamlines are deflected, and fluid passes to one side or other of the obstacle. But the streamline which intersects the point in the body which is perpendicular to the flow finds itself approaching a stagnation point. The flow becomes slower and slower, and the streamline ends in the stagnation point. At the same time, other streamlines leave the stagnation point in various directions. Figure 2.3 illustrates such a flow. The existence of a stagnation point on the upstream face of the body is exploited in a device such as the pitot tube, which can be used to infer flow speed from the pressure there.

Streamlines should not be confused with trajectories, which are the tracks followed by individual fluid elements. A trajectory may in principle be calculated by solving the following differential equations:

$$\frac{dx}{dt} = u(\mathbf{r},t); \quad \frac{dy}{dt} = v(\mathbf{r},t); \quad \frac{dz}{dt} = w(\mathbf{r},t) \tag{2.3}$$

At some initial time t_0, a trajectory lies on some specific streamline. Initially, the trajectory will develop parallel to that streamline. But generally, as time advances, the trajectory will diverge from the streamline since $\mathbf{u}(\mathbf{r}, t)$ becomes different from $\mathbf{u}(\mathbf{r}, t_0)$. Indeed, streamlines and trajectories may quickly become totally different. Only if the flow is steady, that is, $\mathbf{u}$ depends only upon $\mathbf{r}$ and not on t will trajectories and streamlines coincide. Many experimental techniques for visualizing laboratory flow, such as tracking neutrally buoyant beads or dye streaks, are in fact marking trajectories, and so the degree to which these differ from streamlines should be appreciated.

In the remainder of this section, we shall, for simplicity, mainly confine our attention to two-dimensional flows. The results can be generalized to three dimensions. The velocity vector in the x–y plane is denoted by $\mathbf{v}$:

$$\mathbf{v} = u\mathbf{i} + v\mathbf{j} \tag{2.4}$$

Consider an initially rectangular very small patch of fluid which is 'advected' by the flow, that is, it moves with the flow. Its evolution will be viewed in a frame of

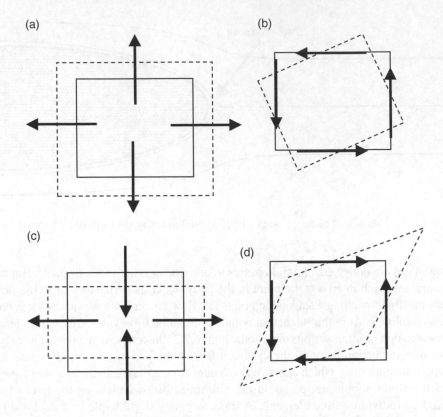

Figure 2.4 Two-dimensional flow kinematics: (a) divergent flow, (b) rotational flow, (c) first kind of deformation flow, and (d) second kind of deformation flow

reference which moves with the mean velocity of the rectangle. Variations of velocity across the patch will lead to changes in the orientation, size or shape of the rectangle. Figure 2.4 shows the possibilities.

The first case, Figure 2.4a, in which the area of the rectangle changes but its shape and orientation do not, is called pure divergent flow. In this case, the 'horizontal divergence' D is non-zero, where

$$D = \frac{\partial u}{\partial x} + \frac{\partial v}{\partial y} \qquad (2.5)$$

The orientation and shape of the fluid rectangle remain fixed, but its area A changes according to

$$D = \frac{1}{A}\frac{dA}{dt} \qquad (2.6)$$

If D is positive, the flow is said to be divergent, and the area of the patch increases with time. If D is negative, the flow is said to be convergent, and the area of the patch decreases. Pure divergent flow is unusual in the atmosphere. A pure divergent flow can be written in terms of a single scalar variable called the 'velocity potential' χ, defined by

$$u = \frac{\partial \chi}{\partial x}, \quad v = \frac{\partial \chi}{\partial y} \tag{2.7}$$

The flow is parallel to the gradient of χ, and the velocity potential and the divergence are related by

$$\nabla^2 \chi = D \tag{2.8}$$

which is an elliptic Poisson equation.

Case two, Figure 2.4b, is called 'pure rotational flow'. It is defined by ξ, a quantity called the vorticity, being non-zero as follows:

$$\xi = \frac{\partial v}{\partial x} - \frac{\partial u}{\partial y} \tag{2.9}$$

In such a case, the square simply rotates, with angular velocity $\xi/2$, with no change of size or shape. Note that a pure divergent flow, defined by Equation 2.7, has zero vorticity. As in the case of pure divergent flow, a rotational flow can be written in terms of a single scalar ψ called in this case the 'streamfunction':

$$u = -\frac{\partial \psi}{\partial y}, \quad v = \frac{\partial \psi}{\partial x} \tag{2.10}$$

so that the divergence is 0 and

$$\nabla^2 \psi = \xi \tag{2.11}$$

Substituting for the flow velocity components given by Equation 2.10 in the streamline equation, Equation 2.2, gives $\delta\psi=0$ along a streamline. Therefore, for pure rotational flow, lines of constant streamfunction ψ are indeed streamlines of the flow. The velocity vectors are parallel to the lines of constant ψ, and the speed of the flow is inversely proportional to the streamline spacing. Large-scale flows observed in the atmosphere or ocean are often close to pure rotational flow. They are characterized by circulating eddies or sinuous curving flows.

In fact, any arbitrary two-dimensional flow can be represented as the sum of a pure rotational flow and a pure divergent flow:

$$\mathbf{v} = \mathbf{v}_R + \mathbf{v}_D = \mathbf{k} \times \nabla \psi + \nabla \chi \tag{2.12}$$

so that the flow can be specified fully by the two scalars ψ and χ in place of the two velocity components u and v. The scalar fields ψ and χ can be obtained by determining D and ξ from Equations 2.5 and 2.9 and then solving the two Poisson Equations 2.8 and 2.11. This partitioning of the flow is called 'Helmholtz decomposition'. In the atmosphere, $|\mathbf{v}_D|$ is typically around 10% of $|\mathbf{v}_R|$.

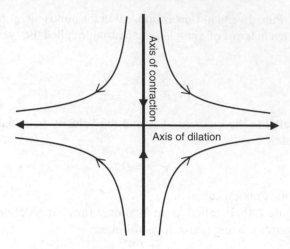

Figure 2.5   A deformation flow

One further generic type of flow can be identified. It can be visualized by considering the flow $u=Ax$, $v=-Ay$, corresponding to a streamfunction $\psi=-Axy$. It is illustrated in Figure 2.5. It has zero divergence and vorticity so that the area and orientation of the fluid parcel do not change, but its shape does. Such a flow leads to the sort of deformation shown in Figure 2.4c, and it is measured by

$$F_1 = \frac{\partial u}{\partial x} - \frac{\partial v}{\partial y} \qquad (2.13)$$

For the flow in Figure 2.5, $F_1=2A$. In this case, fluid elements are stretched parallel to the x-axis (the 'axis of dilation') and compressed parallel to the y-axis (the 'axis of contraction'). A second deformation flow, shown in Figure 2.4d, is like that shown in Figure 2.5, but with the axes of dilation and contraction inclined at an angle of 45° to the x- and y-axes. This is measured by

$$F_2 = \frac{\partial u}{\partial y} + \frac{\partial v}{\partial x}, \qquad (2.14)$$

If both F_1 and F_2 are non-zero, then the axis of deformation makes an arbitrary angle with the coordinate axes. In that case, the total deformation is

$$F = \sqrt{F_1^2 + F_2^2} \qquad (2.15)$$

and the axis of dilation makes an angle α with the x-axis, where

$$\alpha = \frac{1}{2}\tan^{-1}\left(\frac{F_2}{F_1}\right) \qquad (2.16)$$

Many geophysical flows exhibit strong and persistent deformation. As a result, initially compact blobs of fluid become drawn out into long streamers. An extreme development of this process, in which a secondary flow leads to infinite deformation in an infinitesimally small region in a finite time, is the process of frontogenesis. This is the subject of Chapter 15.

2.4   Governing physical principles

The study of atmospheric dynamics is based on a small number of fundamental physical principles which are straightforward to state but more complicated to write down in terms of precise mathematics. In the remainder of this chapter, we shall discuss these principles and their mathematical formulation.

The first fundamental principle which we will use is conservation of matter. We will assume that the fluid contains no sources or sinks of mass. Matter may be moved around, but it cannot be created or destroyed. In fact, there are circumstances when this is not exactly true. For example, if significant amounts of water vapour condense and fall out of the atmosphere as rain, the parcel of atmosphere involved loses some mass. The actual fractional mass change is rather small. For example, a wet day in the midlatitudes may deposit 10 mm of rain on the Earth's surface, which is $10 \, \text{kg} \, \text{m}^{-2} \, \text{day}^{-1}$. This compares with a column mass of tropospheric air of $8.2 \times 10^3 \, \text{kg} \, \text{m}^{-2}$. The timescale to change the atmospheric mass significantly as a result of heavy rain is therefore around 800 days. For most practical purposes, the mass loss (and the corresponding mass gain when water evaporates back into the atmosphere) can be ignored. In sophisticated models, it is accounted for explicitly.

The next fundamental principle is called the first law of thermodynamics, and it expresses a principle of energy conservation. Before discussing the first law of thermodynamics, we should note that the various state variables which describe the thermodynamic state of the atmosphere can be reduced in number by using the equation of state. Dry air behaves very closely to an ideal gas, and so its state variables are related by the ideal gas equation:

$$p = R\rho T \qquad (2.17)$$

Here R is the gas constant for dry air. It is related to the universal gas constant R^* by $R = R^*/m$, m being the mean molecular weight of dry air. Other state variables can be defined, and alternative equations of state derived. For example, the 'potential temperature' of air, defined as the temperature the air would have if it were compressed or expanded adiabatically to some specified reference pressure, is often useful as a thermodynamic state variable. It will be derived in Section 2.7, but for now, we anticipate that its definition is

$$\theta = T \left(\frac{p}{p_R} \right)^{-\kappa} \qquad (2.18)$$

where $\kappa = R/c_p$, c_p being the specific heat at constant pressure, a quantity to be discussed further in Section 2.7. In fact, this definition is an alternative equation of state; given pressure p and temperature T, it enables θ to be calculated. In general, for dry air, given any two state variables, all the rest can be calculated using the various forms of the ideal gas equation.

Having established the equation of state, the principle which defines the variation of the state variables is the first law of thermodynamics. This is essentially a statement of energy conservation. If heat is added to or removed from a parcel of air, the result must either be a change of internal energy or work being done by the parcel on its surroundings, or both. That is,

$$dq = dU + dW \tag{2.19}$$

dq denoting an infinitesimal increment of heat added to the air parcel. The internal energy U of the air is measured by its temperature. We write

$$dU = c_v dT \tag{2.20}$$

where c_v is the specific heat of air at constant volume. The work done by an air parcel on its surroundings involves the change of volume of the air parcel and the pressure acting upon it:

$$dW = pd\left(\frac{1}{\rho}\right) \tag{2.21}$$

for an ideal gas. The variable $1/\rho$, sometimes denoted α, is the specific volume, that is, the volume occupied by a unit mass of air. Thus in terms of more practical variables, the first law of thermodynamics can be written as follows:

$$dq = c_v dT + pd\left(\frac{1}{\rho}\right) \tag{2.22}$$

There are several useful alternative formulations of the first law of thermodynamics. These will be explored in Section 2.7.

The third principle is simply Newton's second law of motion, namely, that a parcel of fluid accelerates in response to the vector sum of the forces acting upon it. Newton's law may be written as:

$$\frac{d\mathbf{u}}{dt} = \sum_i \mathbf{F}_i \tag{2.23}$$

Here, the various forces acting are expressed as forces per unit mass. The forces per unit volume are simply $\rho\mathbf{F}_i$.

2.5 Lagrangian and Eulerian perspectives

Typically, a meteorological observing site monitors the values of various atmospheric properties by repeating measurements at more or less regular time intervals. The observing site is generally fixed relative to the solid Earth. When changes in some atmospheric property are observed, there are two possibilities:

1. The intrinsic properties of the air may have changed. For example, by adding heat to the air, its temperature may have increased.

2. Generally, the air is moving relative to the observing instrument. An observed change may simply reflect the replacement of the air originally located at the observing site by air with different properties from elsewhere. Such a process is called 'advection'.

In fact, for timescales of not more than a day or two, and for air away from the Earth's surface, advective changes generally tend to dominate over intrinsic changes.

This consideration suggests two complementary perspectives on atmospheric dynamics. From the point of view of most observers, the natural perspective is to view the atmosphere by means of observing sites at fixed points in space. Such a perspective is called the 'Eulerian' view. However, from the perspective of basic physics, it is more appropriate to focus on individual fluid elements and consider how various physical processes will modify the properties of the element. Such a perspective is called a 'Lagrangian' view. All the relevant physical laws, discussed in Section 2.4, are expressed in Lagrangian terms. Generally, the rate of change of any flow variable measured in an Eulerian frame of reference, that is, at a fixed point in space, will be different from that measured in a Lagrangian frame, that is, following the evolution of a fixed mass of fluid.

Consider any fluid quantity $Q(\mathbf{r},t)$. We denote the Eulerian rate of change using the standard notation of partial differential calculus:

$$\frac{\partial Q}{\partial t}$$

which indicates that Q is differentiated with respect to time while holding position $\mathbf{r}$ fixed. The Lagrangian rate of change of Q is denoted as:

$$\frac{DQ}{Dt}$$

This notation indicates that Q is differentiated with respect to time while allowing $\mathbf{r}$ to vary so as to follow the same fluid parcel. That is, the differentiation is carried out by comparing values of Q at different points along the parcel trajectory, as illustrated in Figure 2.6. The Lagrangian rate of change is therefore

$$\frac{DQ}{Dt} = \lim_{\delta t \to 0} \frac{Q(\mathbf{r} + \delta\mathbf{r}, t + \delta t) - Q(\mathbf{r},t)}{\delta t} \tag{2.24}$$

Expand $Q(\mathbf{r} + \delta\mathbf{r}, t + \delta t)$ by a Taylor series to first order:

$$\frac{DQ}{Dt} = \lim_{\delta t \to 0} \frac{Q(\mathbf{r},t) + \delta t\,(\partial Q/\partial t) + \delta\mathbf{r} \cdot \nabla Q + O\left(\delta t^2, |\delta\mathbf{r}|^2\right) - Q(\mathbf{r},t)}{\delta t}$$

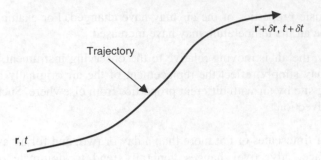

Figure 2.6   A parcel trajectory

Hence, taking the limit as δt tends to 0 gives

$$\frac{DQ}{Dt} = \frac{\partial Q}{\partial t} + \frac{D\mathbf{r}}{Dt} \cdot \nabla Q$$

But the rate of change of $\mathbf{r}$ along the trajectory with respect to time is simply the velocity of the fluid parcel (see Equation 2.3). Hence the final result is

$$\frac{DQ}{Dt} = \frac{\partial Q}{\partial t} + \mathbf{u} \cdot \nabla Q \qquad (2.25)$$

The second term on the right-hand side of this relationship is called the 'advection term'. It is the source of much of the complication in fluid dynamics, for it is a non-linear term, involving the product of the flow velocity with the gradient of other fluid properties. Except in certain special and idealized circumstances, the advection term is highly intractable, permitting no general solutions to the equations of fluid flow.

2.6   Mass conservation equation

This section leads to an Eulerian derivation for an equation expressing mass conservation. Figure 2.7 illustrates a fixed volume V in space, bounded by the surface A. The orientation of the bounding surface A is given by the outward-pointing normal unit vector $\mathbf{n}$. The total mass of fluid contained within the volume is

$$m = \int_V \rho \, dV$$

and the rate of change of mass within the volume is therefore

$$\frac{\partial m}{\partial t} = \int_V \left(\frac{\partial \rho}{\partial t} \right) dV$$

Note that the volume V is fixed in space, so it is appropriate to refer to the Eulerian time derivative of m. Conservation of matter means that any change of mass within

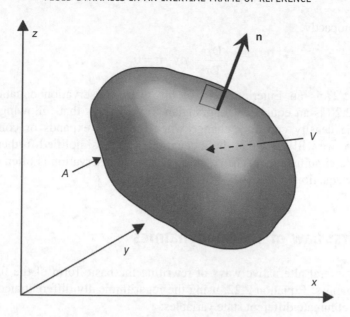

Figure 2.7   An imaginary surface, fixed in space

the volume V can only arise if there is a net exchange of matter between the volume and its surroundings. The total mass flux out of the volume is

$$F = \oint_A (\rho \mathbf{u}) \cdot \mathbf{n} dA$$

The principle of mass continuity requires

$$\frac{\partial m}{\partial t} = -F \qquad\qquad (2.26)$$

and so, using the Gauss theorem on the expression for F,

$$\int_V \left(\frac{\partial \rho}{\partial t} + \nabla \cdot (\rho \mathbf{u}) \right) dV = 0$$

Now since V is an arbitrary volume, and this relationship must hold for any possible such volume, the integrand in this equation must itself be zero. Consequently, the mass conservation equation becomes

$$\frac{\partial \rho}{\partial t} + \nabla \cdot (\rho \mathbf{u}) = 0 \qquad\qquad (2.27)$$

An alternative form is obtained by noting $\nabla \cdot (\rho \mathbf{u}) = \rho \nabla \cdot \mathbf{u} + \mathbf{u} \cdot \nabla \rho$ so that

$$\frac{\partial \rho}{\partial t} + \mathbf{u} \cdot \nabla \rho + \rho \nabla \cdot (\mathbf{u}) = 0$$

or more compactly

$$\frac{D\rho}{Dt} + \rho \nabla \cdot \mathbf{u} = 0 \qquad (2.28)$$

Equation 2.27 is an Eulerian form of the mass conservation equation, while Equation 2.28 is an equivalent Lagrangian form. It says that following the fluid motion, the density can only change when the fluid expands or contracts. In Section 4.5, we will show that these equations can be simplified further in many atmospheric circumstances. The equation of mass conservation is often called the 'continuity equation'.

2.7   First Law of Thermodynamics

There are several alternative ways of rewriting the basic form of the first law of thermodynamics, Equation 2.22, using the logarithmically differentiated equation of state to eliminate different state variables:

$$\frac{dp}{p} = \frac{dT}{T} + \frac{d\rho}{\rho} \qquad (2.29)$$

For example, a useful form involves the change of pressure to which a parcel of air is subject:

$$dq = c_p dT - \left(\frac{1}{\rho}\right) dp \qquad (2.30)$$

where $c_p = c_v + R$ is the specific heat of the air at constant pressure. If the process is adiabatic, that is, if no heat enters or leaves the air parcel, then from Equation 2.30, it follows that

$$\frac{dT}{dp} = \frac{R}{c_p} \frac{T}{p} \qquad (2.31)$$

for an ideal gas. This equation can be integrated to give

$$T = \theta \left(\frac{p}{p_0}\right)^\kappa \quad \text{where } \kappa = \frac{R}{c_p} \qquad (2.32)$$

Here, p_0 is some arbitrary reference pressure, and θ is called the 'potential temperature'. The reference pressure is usually taken to be 100 kPa. This is the result, Equation 1.3, quoted in Section 1.2, and also given in Equation 2.18.

An even more compact form of the first law of thermodynamics results if the definition of potential temperature is used to eliminate pressure:

$$dq = T ds \quad \text{where } s = c_p \ln(\theta) \qquad (2.33)$$

Indeed, s is yet another state variable and is called the specific entropy. Written in terms of potential temperature, Equation 2.33 is

$$dq = c_p \left(\frac{T}{\theta} \right) d\theta \qquad (2.34)$$

All these forms are Lagrangian; they are based on considerations of heat entering or leaving an isolated parcel of fluid. If the heat δq is added over a time δt, then taking the limit as δt tends to 0 gives a Lagrangian differential equation for the rate of change of potential temperature:

$$\frac{D\theta}{Dt} = \frac{1}{c_p} \left(\frac{\theta}{T} \right) \frac{Dq}{Dt} = \frac{1}{c_p} \left(\frac{\theta}{T} \right) \dot{q} = S \qquad (2.35)$$

The heating, $\dot{q}$, represents the net contribution of many different physical processes. These include convergence of radiative fluxes, turbulent transfer of heat from a bounding surface and latent heating when water substance changes phase. The 'source function' S is a convenient summary of these processes with the units of degrees of potential temperature per unit time. Typical values for S in clear air, when longwave cooling to space is generally the dominant process, are around $1\,\text{K}\,\text{day}^{-1}$. In many situations, the advective term is considerably larger than this, and so for these situations, the first law of thermodynamics amounts to a conservation law for potential temperature, at least for timescales of 2–3 days or less:

$$\frac{\partial \theta}{\partial t} + \mathbf{u} \cdot \nabla \theta = 0 \qquad (2.36)$$

2.8 Newton's Second Law of Motion

Two categories of force act upon fluid elements in the atmosphere. First are long-range 'body forces'. These act throughout the volume of the fluid on every element of mass present. The most important body force, and the only one we shall consider in this book, is the gravitational force. The force per unit mass is simply the gravitational acceleration:

$$\frac{D\mathbf{u}}{Dt} = \frac{\partial \mathbf{u}}{\partial t} + \mathbf{u} \cdot \nabla \mathbf{u} = \mathbf{g} \qquad (2.37)$$

Strictly, $\mathbf{g}$ varies with position. It falls off with distance from the centre of the Earth and also varies somewhat from place to place on the Earth's surface because of irregularities in the way in which the mass of the Earth is distributed. However, for our purposes, these variations are generally rather small. The gravitational force can be written in terms of the gradient of a scalar called the 'gravitational potential':

$$\mathbf{g} = -\nabla \Phi \qquad (2.38)$$

Φ has units of energy per unit mass. It can be thought of as the work done lifting a unit mass to a given level in the atmosphere. The actual potential always contains an arbitrary constant, depending upon where the zero level of potential is taken to be located. This arbitrariness does not matter, since the gravitational potential only operates in the equations through its gradient.

The second category of forces acting are short-range forces, generally acting over molecular scales between the molecules making up the fluid. Assuming the continuum hypothesis, they act upon the imaginary surfaces bounding a fluid element, and they constitute the tangential and normal stresses introduced in Section 2.1.

In circumstances where the tangential stresses are either small or confined to very limited volumes of the fluid, the normal stresses are isotropic, that is, they act equally in all directions. In this case, the pressure is defined to be minus the normal stress. To calculate the effect of normal stresses on the operation of Newton's second law of motion, consider a small cuboid of fluid, with sides of length δx, δy and δz. Its mass will be $\rho\delta x\delta y\delta z$. To begin, simply consider the pressure force acting on faces perpendicular to the x-axis, that is, to the component acting in the x-direction. The net force will be

$$\mathbf{F} = \mathbf{i}\left(p\delta y\delta z - (p+\delta p)\delta y\delta z\right)$$

Divide by the mass of the parcel to obtain the force per unit mass, or acceleration, due to pressure:

$$\frac{\mathbf{F}}{\rho\delta x\delta y\delta z} = -\mathbf{i}\frac{\delta p}{\rho\delta x}$$

In the limit of $\delta x \rightarrow 0$, the force in the x-direction is

$$F_x = -\frac{1}{\rho}\frac{\partial p}{\partial x}$$

Similar arguments hold for the effect of pressure on the faces perpendicular to the y- and z-axes. Finally, in vector form, the effect of normal stresses on the acceleration of fluid parcels can be written as:

$$\frac{D\mathbf{u}}{Dt} = -\frac{1}{\rho}\nabla p \qquad (2.39)$$

The actual pressure is irrelevant, since it is the gradient of pressure which accelerates the flow. This term is called the pressure gradient force. Note that it also is a nonlinear term, for both $(1/\rho)$ and p are flow variables. However, it is not so strong a nonlinearity as that of the advection term. For incompressible fluids, when ρ is a constant, the term becomes linear.

Putting the body force and the normal stresses together yields a form of the Newton's second law, also called the momentum equation, which is very widely used:

$$\frac{\partial\mathbf{u}}{\partial t} + \mathbf{u}\cdot\nabla\mathbf{u} = -\frac{1}{\rho}\nabla p - \nabla\Phi \qquad (2.40)$$

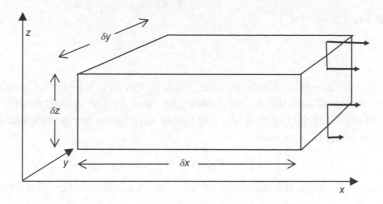

Figure 2.8 Tangential stresses acting on a fluid parcel. The arrows at the right-hand end of the cuboid represent the different velocity shears at its top and bottom surfaces

This equation is called Euler's equation. It presumes that tangential stresses are negligible. In reality, the situation is rather complex. Tangential stresses are indeed small over most of the volume of a geophysical fluid such as the atmosphere. However, there are tiny localized regions where they become large and can dominate the flow. For this reason, we will add the effect of tangential stresses to Equation 2.40, even if we can ignore them in many circumstances.

Consider a cuboid of fluid, as shown in Figure 2.8, and for the moment, simply consider the x-component of velocity varying in the z-direction. As in the case of the normal stress, what is important is the change of stress across the parcel rather than its actual value. By Newton's law of viscosity, the force acting in the x-direction on the top surface of the cuboid is

$$F_{\text{top}} = \mu \frac{\partial u}{\partial z}\bigg|_{\text{top}} \delta x \delta y$$

while the force acting on the bottom surface is

$$F_{\text{bottom}} = -\mu \frac{\partial u}{\partial z}\bigg|_{\text{bottom}} \delta x \delta y$$

The minus sign must be introduced because here we are considering the stress exerted by fluid below the cuboid on the surface of the cuboid. By Newton's third law, it must be equal in magnitude but opposite in direction from the stress exerted by the fluid in the cuboid on that below. The net force acting on the cuboid is the difference between these, and since the mass of the cuboid is $\rho \delta x \delta y \delta z$, the net force per unit mass has the following magnitude:

$$F_{\text{total}} = \frac{\mu}{\rho} \frac{\frac{\partial u}{\partial z}\big|_{\text{top}} - \frac{\partial u}{\partial z}\big|_{\text{bottom}}}{\delta z}$$

or, in the limit as $\delta z \to 0$,

$$F = v \frac{\partial^2 u}{\partial z^2}$$

Here, $\nu = \mu/\rho$ is called the 'kinematic coefficient of viscosity'. Similar formulae hold for the variation of u in the y- and z-directions, and for the variations of v and w. Putting all these results together, the full vector expression for the tangential stress gradient force per unit mass is

$$\mathbf{F}_T = v\nabla^2\mathbf{u} \tag{2.41}$$

Adding this force to the equation of motion, we obtain the 'Navier–Stokes' equation:

$$\frac{\partial \mathbf{u}}{\partial t} + \mathbf{u} \cdot \nabla\mathbf{u} = -\frac{1}{\rho}\nabla p - \nabla\Phi + v\nabla^2\mathbf{u} \tag{2.42}$$

For a compressible fluid, this is not quite the final result. An additional term representing the frictional resistance to expansion or compression needs to be included:

$$\frac{\partial \mathbf{u}}{\partial t} + \mathbf{u} \cdot \nabla\mathbf{u} = -\frac{1}{\rho}\nabla p - \nabla\Phi + v\left(\nabla^2\mathbf{u} + \frac{1}{3}\nabla(\nabla \cdot \mathbf{u})\right) \tag{2.43}$$

This additional compression term is rarely large in an atmospheric context and is not usually of any great importance for our purposes.

To summarize, through this chapter, we have assembled a set of differential equations:

$$\frac{D\theta}{Dt} = S \tag{2.35}$$

$$\frac{D\rho}{Dt} + \rho\nabla \cdot \mathbf{u} = 0 \tag{2.28}$$

$$\frac{D\mathbf{u}}{Dt} = -\frac{1}{\rho}\nabla p - \nabla\Phi + v\nabla^2\mathbf{u} \tag{2.43}$$

together with the equation of state which, by eliminating T between Equations 2.17 and 2.18, can be written in the form

$$p^{1-\kappa} = p_R^{-\kappa} R\rho\theta \tag{2.44}$$

These form a complete set of six scalar equations for the six scalar variables u, v, w, p, θ, and ρ. Equations 2.35, 2.28 and 2.43 describe the change of variables with time and so are called 'prognostic' equations. In principle, we have a mathematically complete description of the system, and we should be able to use these equations to start from any initial values of the variables to predict their subsequent values.

2.9 Bernoulli's Theorem

Bernoulli's theorem is one simple but powerful result which follows from the equations derived in the last section. Start from Euler's equation, Equation 2.40, and assume steady, inviscid and incompressible flow. Euler's equation then reduces to

$$\mathbf{u} \cdot \nabla \mathbf{u} = -\nabla \left(\frac{p}{\rho} \right) - \nabla \Phi$$

Use the vector identity

$$\nabla (\mathbf{A} \cdot \mathbf{B}) = (\mathbf{B} \cdot \nabla) \mathbf{A} + (\mathbf{A} \cdot \nabla) \mathbf{B} + \mathbf{B} \times (\nabla \times \mathbf{A}) + \mathbf{A} \times (\nabla \times \mathbf{B})$$

with $\mathbf{A} = \mathbf{B} = \mathbf{u}$ to rewrite the nonlinear advection term as follows:

$$\mathbf{u} \cdot \nabla \mathbf{u} = \nabla \left(\frac{\mathbf{u} \cdot \mathbf{u}}{2} \right) - \mathbf{u} \times (\nabla \times \mathbf{u}) \tag{2.45}$$

Use this identity and rearrange the Euler equation into the following form:

$$\mathbf{u} \times (\nabla \times \mathbf{u}) = \nabla \left(\frac{\mathbf{u} \cdot \mathbf{u}}{2} + \Phi + \frac{p}{\rho} \right) \tag{2.46}$$

Finally, take scalar product with $\mathbf{u}$ of this equation. Notice that the vector $\mathbf{u} \times (\nabla \times \mathbf{u})$ is perpendicular to $\mathbf{u}$, so its scalar product with $\mathbf{u}$ is 0. Therefore,

$$\mathbf{u} \cdot \nabla B = 0 \quad \text{where } B = \frac{\mathbf{u} \cdot \mathbf{u}}{2} + \Phi + \frac{p}{\rho} \tag{2.47}$$

The 'Bernoulli potential' B is conserved along a streamline. Conservation of B is essentially an energy relationship, relating the kinetic energy, potential energy and pressure energy along a streamline. The pressure term arises because we are dealing with a fluid parcel rather than a simple body. The theorem can be used to relate p, $\mathbf{u} \cdot \mathbf{u}$ and Φ along a streamline. For example, consider flow along a pipe of variable cross section. If the area of the pipe normal to the flow is $A(s)$, s being the distance along the streamline, and if the flow is steady, conservation of matter requires

$$A_0 u_0 = A(s) u(s)$$

Further, assume that the pipe is horizontal so Φ does not change and that the values of p and u at entry are p_0 and u_0, then

$$u_0^2 + \frac{p_0}{\rho} = \left(\frac{A_0}{A(s)} \right)^2 u_0^2 + \frac{p(s)}{\rho} \tag{2.48}$$

Thus, where the flow speed is larger, the pressure must be smaller and vice versa. Notice that Bernoulli's theorem can only be used to compare points on the same streamline. It cannot be used to compare the properties of points on different stream-lines. Neither can it be used to compare the properties of turbulent flow at different points, for in that case, the flow is unsteady and streamlines are no longer trajectories.

If $A(s)$ becomes so small that

$$\left(\frac{A_0}{A(s)}\right)^2 u_0^2 > u_0^2 + \frac{p_0}{\rho}$$

then $p(s)$ becomes negative. This is clearly unphysical and indicates a breakdown of the assumptions in Bernoulli's theorem. In fact, this criterion marks the transition to 'cavitating flow', in which voids open up and collapse in the fluid so that it becomes highly unsteady. Of course, Bernoulli's theorem is not applicable beyond the transi-tion to cavitating flow. A foaming mountain stream is a good natural example of a cavitating flow.

For compressible flow, such as characterizes the atmosphere, Equation 2.47 is incomplete. The reason for this is that if the density changes so that the volume of the parcel changes, the parcel does work against the pressure force exerted by its neigh-bours. To include this work, note that the first law of thermodynamics states that

$$\dot{q} = c_v \mathbf{u} \cdot \nabla T + p\mathbf{u} \cdot \nabla\left(\frac{1}{\rho}\right) \tag{2.49}$$

(see Equation 2.22) where $\dot{q}$ is the rate at which heat is added to the parcel. This equation states that the heating can be manifested either as a change of the internal energy of the parcel, or by the parcel doing work against its surroundings. As well as mechanical energy, account must be taken of the heat energy added to a parcel. So in place of Equation 2.46, we now have

$$\dot{q} = c_v \mathbf{u} \cdot \nabla T + p\mathbf{u} \cdot \nabla\left(\frac{1}{\rho}\right) + \frac{1}{\rho}\mathbf{u} \cdot \nabla p + \mathbf{u} \cdot \nabla\left(\frac{\mathbf{u} \cdot \mathbf{u}}{2} + \Phi\right) \tag{2.50}$$

This is now a total energy equation for the parcel, including the rates of change of both mechanical and internal energy. Assume zero heating, so the flow is adiabatic, and rearrange to give

$$\mathbf{u} \cdot \nabla\left(\frac{\mathbf{u} \cdot \mathbf{u}}{2} + \Phi + c_v T + \frac{p}{\rho}\right) = 0 \tag{2.51}$$

Now from the ideal gas equation, $p/\rho = RT$ and $c_v + R = c_p$, the specific heat at constant pressure, so that Equation 2.51 can finally be rewritten as:

$$\mathbf{u} \cdot \nabla B = 0 \quad \text{where } B = \mathbf{u} \cdot \mathbf{u}/2 + \Phi + c_p T \tag{2.52}$$

In words, Bernoulli's theorem for steady compressible flow is that the sum of kinetic energy, potential energy and 'enthalpy' for a parcel is conserved following the streamlines. Thus, if a parcel moves along a streamline to a place where the flow is accelerated (e.g., above an aerofoil), the temperature must fall to conserve B. Associated with this adiabatic cooling, the pressure must drop according to

$$Tp^{-\kappa} = \theta p_0^{-\kappa}$$

One may say that an aircraft flies because the air above the wing is colder than that below the wing, as a result of adiabatic expansion and compression. If the temperature change along a streamline is sufficiently small, Equation 2.52 reduces to the incompressible form, Equation 2.47. The condition for this to be the case is that the Mach number, the ratio of the typical flow speed to the sound speed, be small, a condition to which we shall return in Section 4.5. Strong winds in the lee of mountains are an example of Bernoulli's theorem acting in an atmospheric context. If a streamline descends from the top of a 1 km high mountain to near the surface, then Equation 2.52 suggests that winds as strong as $140\,\mathrm{m\,s^{-1}}$ are theoretically possible.

2.10   Heating and water vapour

Air parcels do not evolve exactly adiabatically, and so the heating and cooling terms on the right-hand side of the thermodynamic equation, Equation 2.35, can be significant. Air parcels both gain and lose heat by a variety of processes. Often, the rates of heating or cooling are sufficiently slow that the parcel motion can be treated as approximately adiabatic for periods shorter than 1 or 2 days. For example, clear air in the troposphere loses heat, eventually to space, by the emission of infrared thermal radiation. The typical cooling rate is around $1\text{--}2\,\mathrm{K\,day^{-1}}$, which is an order of magnitude smaller than the Eulerian rates of temperature change associated with a typical weather system. Similarly, the Earth–atmosphere system is heated by sunlight, most of which reaches the ground in clear conditions. A variety of processes then mix this heat through the depth of the atmosphere. A typical excess of shortwave heating over longwave cooling is around $10^2\,\mathrm{W\,m^{-2}}$. If this heat were mixed throughout the depth of the troposphere, it would result in a rate of temperature change of around $1\,\mathrm{K\,day^{-1}}$, comparable to the temperature change from longwave cooling. So on the basis of these calculations, one might conclude that for synoptic timescales of not more than a few days, the effects of heating and cooling will be no more than a small correction to the dynamics of air parcels.

However, this conclusion is of course misleading, and the reason it is misleading is because the discussion so far has neglected the role of water in the atmosphere. It is indeed remarkable that so much can be said about atmospheric dynamics without including the effects of water. After all, in the popular mind, the major application of meteorology is to predict rain and cloud events. Yet many of the major features of midlatitude weather systems can be elucidated without detailed discussion of the effects of moisture. But even if moisture does not change the character of basic dynamical processes, changes of phase of water substance have important effects on the details of the evolution of weather systems and on the vigour of certain processes.

   Water vapour is a highly variable atmospheric constituent. It can make up as much as 3% of the mass of an air parcel over the warm tropical ocean, while in the cold stratosphere, water vapour concentrations are measured in parts per million. The water vapour content of an air parcel can be expressed by quoting the specific humidity, denoted by q and defined as the mass of water vapour in an air parcel divided by the total mass of the air parcel. A sample of moist air is treated as a mixture of dry air and water vapour, each of which satisfies the ideal gas equation. Because the properties of water vapour are different from those of dry air, the equation of state for moist air differs slightly from that for dry air:

$$p = R_d \left(1 + q \left(\frac{R_v}{R_d} - 1 \right) \right) \rho T \qquad (2.53)$$

Here, R_d denotes the gas constant for dry air and R_v the gas constant for pure water vapour. The factor R_v/R_d, the ratio of the gas constants for water vapour and dry air, has a value of 1.61. Since q is of order 10^{-2} or smaller, the difference between the equation of state for dry air and moist air is not very large. Another measure of the water vapour content of the air is the partial pressure exerted by the water vapour component of the air, usually called the vapour pressure and denoted e. The specific humidity is related to the vapour pressure by

$$q = \frac{e}{\frac{R_v}{R_d}(p-e)+e} \simeq \frac{R_d}{R_v} \frac{e}{p} \qquad (2.54)$$

The approximate expression is valid in the limit where e is very small compared to p. Crucially, the maximum vapour pressure of an air parcel, and therefore its specific humidity, is limited. Beyond a maximum vapour pressure, called the saturated vapour pressure, water vapour will condense to form liquid droplets. For the purposes of our discussion, such condensation takes place virtually instantaneously when the air becomes saturated. The saturation vapour pressure e_s is given by the Clausius–Clapeyron equation:

$$\frac{de_s}{dT} = \frac{L}{T(\alpha_v - \alpha_l)} \qquad (2.55)$$

Here, $\alpha_v = \rho_v^{-1}$ and $\alpha_l = \rho_l^{-1}$ are the 'specific volumes' of water vapour and liquid water respectively, and L is the 'latent heat of condensation', the energy released when the water vapour condenses. Equation 2.55 can be integrated to give the dependence of saturated vapour pressure on temperature, an integration made simpler if L is assumed constant and if the specific volume of water vapour is assumed very large compared to the specific volume of liquid water. Then, for some temperature $T=T_0+\Delta T$,

$$e_s \simeq e_{s0} \exp\left(\frac{L}{R_v T_0^2} \Delta T \right) \qquad (2.56)$$

That is, the saturated vapour pressure, and hence the saturated specific humidity, increases roughly exponentially with temperature, doubling every $10\,K$ or so. For T_0 of $273\,K$, the saturated vapour pressure is $611\,Pa$, corresponding to a specific humidity of 3.8×10^{-3} at a pressure of $100\,kPa$.

The significance of water vapour in atmospheric dynamics is due to the very large value of the latent heat of condensation of water, L. This has a value of around 2.5×10^6 J kg^{-1}, which is huge, an order of magnitude greater than that of other common substances. So when air becomes saturated, typically the result of rising through the atmosphere and expanding adiabatically, large amounts of latent heat are released. The result is that the rising parcel of air is much warmer than it would be if it were unsaturated. If the specific humidity were to change by an amount Δq, then the latent heat of condensation released would be

$$\Delta Q = -L\Delta q$$

The minus sign indicates that heat is released when the specific humidity is reduced, and vice versa. To appreciate how important latent heat release can be, a few simple calculations are helpful. Suppose a weather system generates a rainfall of $10\,mm\,day^{-1}$, a relatively modest amount, which corresponds to $10\,kg\,m^{-2}$ of water falling in 24 hour. If the rain falls from a column of air reaching from the surface to, say, $40\,kPa$, the average change of specific humidity of air is around 1.7×10^{-3}. The latent heat released by condensation of this rainfall is 2.5×10^7 J m^{-2} day^{-1}, that is, about $290\,W\,m^{-2}$. This is roughly the same as the global mean insolation and three times larger than the typical net heating or cooling. The corresponding source term in the thermodynamic equation is

$$\dot{\theta} = -\left(\frac{p}{p_R}\right)^{-\kappa} \frac{L}{c_p} \frac{Dq}{Dt}$$

The saturated humidity mixing ratio q_s is a function both of temperature and pressure and varies most rapidly in the vertical. When vertical motion advects saturated air parcels, then the source term for the thermodynamic equation becomes

$$\dot{\theta} = \left(\frac{p}{p_R}\right)^{-\kappa} \frac{L}{c_p} \frac{\partial q_s}{\partial z} w \tag{2.57}$$

Consider the application of Equation 2.57 to a significant region of the atmosphere, rather than to an individual air parcel. In a numerical weather prediction model, such a region might be a grid box. Denote the area average by an overbar, and departures from the area average by a prime. Then, averaging Equation 2.57 over this area gives

$$\bar{\dot{\theta}} = \left(\frac{p}{p_R}\right)^{-\kappa} \frac{L}{c_p}\left(\frac{\partial \bar{q}_s}{\partial z}\bar{w} + \left(\frac{\partial q'_s}{\partial z}w'\right)\right) \tag{2.58}$$

Two extreme cases arise. First, if fluctuations across the region are negligible, then the latent heating and rainfall depend simply on the area averages of humidity, temperature and so on. Mean upward motion leads to condensation throughout the region, which is often referred to as 'large-scale rain'. Fronts and depressions are examples where the precipitation may be dominated by such a large-scale process. In the other extreme, the mean vertical motion is negligible, but large local fluctuations of vertical velocity and humidity mean that the area mean heating can be large. An example of this situation arises if the region is filled with deep cumulus clouds, with strong updrafts in the clouds and descent in the surrounding clear air. Precipitation in this case is referred to as 'convective rain'. In a numerical model, with a resolution too coarse to represent individual cumulus clouds, such convective precipitation has to be inferred from the large-scale resolved fields of temperature, humidity and wind. A number of such 'convective parametrization' schemes are in use, all with empirical elements, and it is difficult to argue from first principles which should be preferred. Parametrization of convective processes is a significant uncertainty in weather forecasting and climate modelling. This is particularly the case in the tropics where latent heat release from cumulus convection is often the dominant form of heating. Convective heating can also be very significant in the midlatitudes.

There are two particular complications introduced when water vapour condenses in the weather system. Before latent heat release takes place, an initially moist but unsaturated air parcel has to be lifted a certain height. Moist processes therefore do not fit easily into the linear instability framework used extensively in later chapters to discuss the development of weather systems. They begin to operate only when a disturbance reaches a finite amplitude.

Secondly, descent is not necessarily the reverse of ascent when phase changes of water vapour are involved. Consider a large-scale atmospheric wave. As clear air descends and warms, its relative humidity is reduced. The descending air remains clear, and no latent heat is released. Its dynamics remain much as discussed in chapter 14. In contrast, air in the ascending part of the waves quickly becomes cloudy and substantial heating takes place. In this way, a major asymmetry between ascent and descent develops. From a more Lagrangian perspective, focus on an individual air parcel. As it ascends, water condenses and heats it, and it becomes cloudy. When it descends, the liquid water droplets re-evaporate, cooling the air. If all the condensed liquid water remained suspended in the air parcel, the process would be reversible: the descending parcel would progress through the same thermodynamic states as the ascending parcel. But if a substantial fraction of the water vapour were rained out during the ascent, the descending air parcel would follow a very different set of states. In effect, it would be warmer, corresponding to the latent heat released by the condensation of the water which rained out of the parcel. The fraction of condensed water which is retained by an air mass as 'cloud liquid water' can vary considerably. In many situations, it is qualitatively adequate to assume that all liquid water rains out of the air immediately. More accurately, suspended liquid water must be treated as a separate variable. The processes whereby liquid water rains out bring us into the area of cloud microphysics, a major subject in its own right, and beyond the scope of this book.

Both these factors mean that the simple linear instability approach that will be developed in later chapters does not apply to systems where some of the air becomes saturated.

3

Rotating frames of reference

3.1   Vectors in a rotating frame of reference

Chapter 2 outlined the classical development of fluid mechanics, using the principles of mass conservation, the first law of thermodynamics and Newton's second law of motion. The first two principles are expressed as prognostic equations for scalar variables, namely, the density and potential temperature, respectively, and their form is independent of the frame of reference used to describe the fields. Newton's law, however, is a vector equation, expressing the rate of change of velocity as a function of the various forces acting upon fluid elements. It holds only in non-accelerating or 'inertial' frames of reference. This is a real problem for geophysical fluid dynamics, as to a first approximation the atmosphere and ocean move around with the Earth: the movement of a point on the surface of the Earth due to its rotation is more than $300\,\mathrm{m\,s^{-1}}$ at latitude 45°. Compared with this, the relative movements of the atmosphere, the winds, and of the ocean, the currents, are small, perhaps $15\,\mathrm{m\,s^{-1}}$ and $5\,\mathrm{cm\,s^{-1}}$, respectively. Therefore, the natural coordinate system for geophysical problems is one which is fixed in the rotating Earth and which is consequently rotating, that is, accelerating. In this chapter, the basic equations of motion will be modified so they hold in a uniformly rotating frame of reference.

First some notation is introduced. A subscript A denotes a quantity in an inertial or 'absolute' frame of reference, while the subscript R denotes a quantity in a uniformly rotating frame of reference. The axes in an inertial frame of reference are defined by orthogonal unit vectors $\mathbf{i}_A$, $\mathbf{j}_A$, $\mathbf{k}_A$. These vectors are constant; neither their magnitudes nor their directions change in time. A rotating frame of reference is defined by orthogonal unit vectors $\mathbf{i}_R$, $\mathbf{j}_R$, $\mathbf{k}_R$. The unit vectors in a rotating frame are not constant. Although their magnitudes do not change, their directions continually change as the coordinate system rotates. Note that scalars are invariant with respect

Fluid Dynamics of the Midlatitude Atmosphere, First Edition. Brian J. Hoskins & Ian N. James.
© 2014 John Wiley & Sons, Ltd. Published 2014 by John Wiley & Sons, Ltd.

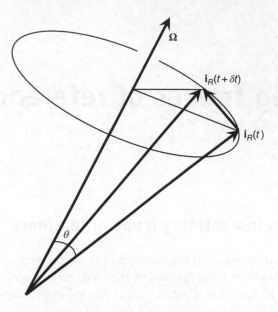

Figure 3.1   A rotating unit vector

to direction, and so their rates of change are the same in either a rotating or inertial frame of reference, or

$$\frac{d_A S}{dt} = \frac{d_R S}{dt}$$

(3.1)

S being any arbitrary scalar.

Figure 3.1 shows a rotating unit vector $\mathbf{i}_R$ which makes an angle θ with the rotation axis. The tip of the unit vector moves around the rotation axis on a circle of radius $s = \sin(\theta)$. In a short interval of time δt, the unit vector will be modified to

$$\mathbf{i}_R(t + \delta t) = \mathbf{i}_R(t) + \delta \mathbf{i}_R$$

(3.2)

where $\delta \mathbf{i}_R$ has magnitude $s\Omega\delta t$ and is at right angles to both $\mathbf{i}_R$ and $\mathbf{\Omega}$. In other words,

$$\delta \mathbf{i}_R = \mathbf{\Omega} \times \mathbf{i}_R \delta t$$

(3.3)

so that

$$\frac{d_A \mathbf{i}_R}{dt} = \lim_{\delta t \to 0} \frac{\delta \mathbf{i}_R}{\delta t} = \mathbf{\Omega} \times \mathbf{i}_R$$

(3.4)

This important result is the basis of what follows. Note that the dot product of the right-hand side of Equation 3.4 with $\mathbf{i}_R$ is 0, which is consistent with the length of $\mathbf{i}_R$ being constant (equal to 1). Identical results hold, of course, for $\mathbf{j}_R$ and $\mathbf{k}_R$.

From the absolute rate of change of a unit vector in a rotating system, it is straightforward to relate the rates of change of an arbitrary vector in inertial and rotating frames. Let

$$\mathbf{V} = V_1 \mathbf{i}_R + V_2 \mathbf{j}_R + V_3 \mathbf{k}_R \tag{3.5}$$

Now consider the absolute rate of change of $\mathbf{V}$:

$$\frac{d_A \mathbf{V}}{dt} = \frac{d_A V_1}{dt} \mathbf{i}_R + V_1 \frac{d_A \mathbf{i}_R}{dt} + \text{etc.} \tag{3.6}$$

Notice for the moment that it not specified whether Lagrangian or Eulerian rates of change are being considered. Since V_1, etc., are simply scalars, their rates of change are the same in either a rotating or an inertial frame and so

$$\frac{d_A \mathbf{V}}{dt} = \frac{d_R V_1}{dt} \mathbf{i}_R + V_1 (\Omega \times \mathbf{i}_R) + \text{etc.} \tag{3.7}$$

or gathering up terms

$$\frac{d_A \mathbf{V}}{dt} = \frac{d_R \mathbf{V}}{dt} + \Omega \times \mathbf{V} \tag{3.8}$$

If $\mathbf{V}$ is a constant in a rotating frame, then in an inertial frame, it must be in a direction that has rotated from the original direction, so that in this frame, $\mathbf{V}$ is actually changing. This is what is represented by the second term on the right-hand side of Equation 3.8. In what follows, this general result will be applied to different specific examples of the vector $\mathbf{V}$.

3.2 Velocity and Acceleration

The Lagrangian rate of change of position of a fluid parcel is simply its velocity $\mathbf{u}$. The velocity will depend upon whether the motion is referred to an inertial or rotating frame of reference. Letting the arbitrary vector $\mathbf{V}$ in Equation 3.8 be $\mathbf{r}$, the position vector, this result follows:

$$\mathbf{u}_A = \frac{D_A \mathbf{r}}{Dt} = \frac{D_R \mathbf{r}}{Dt} + \Omega \times \mathbf{r} = \mathbf{u}_R + \Omega \times \mathbf{r} \tag{3.9}$$

The adjustment to $\mathbf{u}_R$ to obtain $\mathbf{u}_A$ is generally large. The second term due to the rotation is of magnitude Ωs where s is the distance of the point in question from the axis of rotation and is directed around this axis. Using the numbers given before, for the atmosphere, the magnitude of $\mathbf{u}_R$ may be of order $15\,\mathrm{m\,s^{-1}}$ and that of $\Omega \times \mathbf{r}$ about $300\,\mathrm{m\,s^{-1}}$.

Equation 3.9 now leads to an expression for the acceleration of a fluid parcel in a rotating frame of reference. The acceleration in an absolute frame is

$$\frac{D_A \mathbf{u}_A}{Dt} = \frac{D_A}{Dt}(\mathbf{u}_R + \Omega \times \mathbf{r}) = \frac{D_A \mathbf{u}_R}{Dt} + \Omega \times \mathbf{u}_A \qquad (3.10)$$

Now write the entire right-hand side in terms of the rotating coordinates:

$$\frac{D_A \mathbf{u}_A}{Dt} = \frac{D_R \mathbf{u}_R}{Dt} + \Omega \times \mathbf{u}_R + \Omega \times (\mathbf{u}_R + \Omega \times \mathbf{r})$$
$$= \frac{D_R \mathbf{u}_R}{Dt} + 2\Omega \times \mathbf{u}_R + \Omega \times (\Omega \times \mathbf{r}) \qquad (3.11)$$

Thus, in order to transform the acceleration from a rotating frame of reference to an inertial frame of reference, two corrective accelerations must be added. The first is called the Coriolis acceleration and the second the centripetal acceleration. The centripetal term is quite familiar: it is of magnitude $\Omega^2 s$ and is directed inwards. The Coriolis term is less familiar. It can be seen that it comes equally from viewing the change of $\mathbf{u}_R$ in a rotating frame and from the relative rate of change of the velocity associated with the coordinate rotation, $\Omega \times \mathbf{r}$. It is always orthogonal to both Ω and $\mathbf{u}_R$ and to the left of this velocity if the rotation axis is considered to be pointing upwards.

3.3 The momentum equation in a rotating frame

The Navier–Stokes equation derived in Chapter 2 is straightforwardly modified to a rotating frame of reference. The expression just derived, Equation 3.11, replaces the acceleration in an inertial frame on the left-hand side of the Navier–Stokes equation. The terms on the right-hand side of the equation, involving the pressure gradient force, the gravitational force and the viscous stress, are all unchanged in a rotating frame of reference. It is conventional to move the corrective accelerations over to the right-hand side of the equation set and to refer to them as 'forces' or more correctly, 'pseudo-forces'. In this perspective, the centrifugal force acts 'outwards' and the Coriolis force acts to the 'right' of the relative velocity. Each has the dimension of a force per unit mass. The result is an equation which is very similar to the Navier–Stokes equation (and which reduces to it in the limit $\Omega \to 0$)

but which includes additional 'pseudo-forces' that account for the rotation of the frame of reference:

$$\frac{D_R \mathbf{u}_R}{Dt} = -\frac{1}{\rho}\nabla p - \nabla \Phi - 2\boldsymbol{\Omega} \times \mathbf{u}_R - \boldsymbol{\Omega} \times (\boldsymbol{\Omega} \times \mathbf{r}) + \nu \nabla^2 \mathbf{u}_R \qquad (3.12)$$

It is important to recognize that the Coriolis and centrifugal forces are not real physical forces, but simply corrections added to maintain the illusion that Newton's second law of motion holds in a rotating frame of reference. Even so, if the rotation is rapid, these new terms can become dominant.

This notation is rather cumbersome, and so in subsequent chapters, the A or R subscripts will generally be dropped. The context usually makes it clear whether a rotating or inertial frame of reference is implied.

When is it important to include the Coriolis and centrifugal forces? The answer to this question requires some simple scale analysis, the subject of Chapter 5. The arguments of that chapter will be anticipated here. Suppose the fluid motion has a characteristic velocity U which fluctuates over a characteristic length scale L. Then, the relative acceleration will have a typical magnitude U^2/L, the Coriolis force will have a typical magnitude ΩU and the centrifugal force a characteristic magnitude $\Omega^2 a$, a being the radius of the Earth. The ratio of the Coriolis force to the total observed acceleration is of order Ro^{-1} where the dimensionless combination of parameters $\mathrm{Ro} = U/\Omega L$ is called the 'Rossby number'. The ratio of the total acceleration to the centrifugal force can be written as $\mathrm{Ro}^{-2}(a/L)$. The ratio of typical weather system scales L to the radius of the Earth is always less than 1 and generally considerably less than 1. So if the Rossby number is of order 1 or smaller, then the effects of rotation on the momentum equation are substantial and important; the rotating form of the equation should be used in such circumstances.

3.4   The centrifugal pseudo-force

The centrifugal force depends only upon position and not on the state of motion of the fluid parcel. Consider a point on the Earth's surface, position vector $\mathbf{r}$, and at latitude ϕ. The vector $\boldsymbol{\Omega} \times \mathbf{r}$ has magnitude Ωs where $s = r\cos(\phi)$ is the perpendicular distance from the rotation axis (Figure 3.2). The vector $\boldsymbol{\Omega} \times \mathbf{r}$ is directed perpendicularly into the plane of the diagram. It follows that the centrifugal force itself is perpendicular both to $\boldsymbol{\Omega}$ and to $\boldsymbol{\Omega} \times \mathbf{r}$ and so must be directed radially outwards from the rotation axis. Its magnitude is $\Omega^2 s$. This is by no means a small force. At latitude 45°, for example, the horizontal component of the centrifugal force has magnitude $0.017\,\mathrm{N\,kg^{-1}}$. Compare this with the typical horizontal pressure gradient force, which is of the order of $0.0016\,\mathrm{N\,kg^{-1}}$, that is, an order of magnitude smaller.

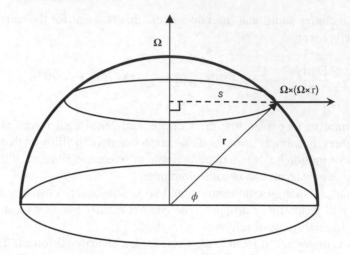

Figure 3.2 Illustrating the centrifugal pseudo-force

Although the centrifugal force may often be larger than other terms in the momentum equations, it is generally possible to avoid quoting it explicitly. The reason for this is that the centrifugal force is a so-called conservative force, meaning that its whole value depends solely upon position and that its curl is 0. Consequently, it can be expressed as the gradient of a scalar function of the following position:

$$-\Omega \times (\Omega \times \mathbf{r}) = \nabla \left(\frac{\Omega^2 s^2}{2} \right) \tag{3.13}$$

where $s = r\cos(\phi)$. But the gravitational force can also be expressed as the gradient of a scalar potential, Φ. The two scalars can be combined to form an 'effective' gravitational potential

$$\Phi_e = \Phi - \frac{\Omega^2 s^2}{2}, \tag{3.14}$$

and so the momentum equation can be written in a way in which the centrifugal force is not explicit:

$$\frac{D\mathbf{u}}{Dt} = -\frac{1}{\rho} \nabla p - \nabla \Phi_e - 2\Omega \times \mathbf{u} + \nu \nabla^2 \mathbf{u} \tag{3.15}$$

Although the centrifugal acceleration is often much larger than the horizontal accelerations, it is small compared with the gravitational acceleration for the Earth. Thus, the differences between $\nabla\Phi$ and $\nabla\Phi_e$ are small and can be ignored for many purposes.

The use of an effective gravitational potential means that the level surfaces, that is, the surfaces of constant potential, are not spheres but are slightly oblate, with the polar radius less than the equatorial radius. The surface of the Earth, whose mantle behaves as a fluid on geological timescales, is indeed approximately oblately spheroidal in shape, with a polar radius of 6357 km and an equatorial radius of 6378 km. The difference of 21 km is only about 1/300 of the radius. However, it is much bigger than any mountain on Earth! The magnitude of the effective gravitational acceleration, that is, $|\nabla\Phi_e|$, also varies from 9.83 m s^{-2} at the pole to 9.79 m s^{-2} at the equator. It is usually a very good approximation to ignore the slight oblateness of the Earth and consider it to be a sphere of radius 6371 km (having the same volume as the real oblate Earth) and the gravitational acceleration to be a uniform 9.81 m s^{-2}.

3.5   The Coriolis pseudo-force

The Coriolis force depends upon the velocity, but not on the location of a fluid parcel. Its magnitude is $2\Omega U$, and its direction is perpendicular to both Ω and $\mathbf{u}$. It is worth writing out the components of the Coriolis force relative to the local vertical direction on the Earth's surface. Assuming the Earth is a sphere, then at a latitude ϕ, the components of the Earth's angular velocity are

$$\Omega = \Omega\cos(\phi)\mathbf{j} + \Omega\sin(\phi)\mathbf{k}$$

Then, the Coriolis force is

$$-2\Omega \times \mathbf{u} = (2\Omega\sin(\phi)v - 2\Omega\cos(\phi)w)\mathbf{i} - 2\Omega\sin(\phi)u\mathbf{j} + 2\Omega\cos(\phi)u\mathbf{k}$$

Although $\mathbf{u}$ is usually nearly parallel to the ground, so that w is small, the Coriolis force generally has a substantial vertical component, save for the special case when $\mathbf{u}$ is directed towards or away from the pole. Figure 3.3 illustrates the Coriolis force for the case relevant to the atmosphere, when the velocity vector is nearly horizontal, with only relatively small vertical components. Such a horizontal wind is designated $\mathbf{v}$. The Coriolis force lies in a plane perpendicular to Ω and at right angles to $\mathbf{v}$. The horizontal components of the Coriolis force have a combined magnitude of $2\Omega U\sin(\phi)$, independent of wind direction. It has a maximum vertical component when $\mathbf{v}$ is directed to the east and a minimum (negative) vertical component when $\mathbf{v}$ is directed to the west. Because the Coriolis force acts at right angles to the velocity vector, it can do no work on a fluid parcel. That is, the speed of a fluid parcel subject only to the Coriolis force remains constant although the direction of motion will change uniformly in time. If no other forces act on a fluid parcel, it will move in a

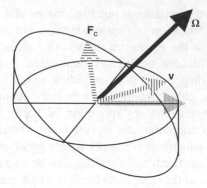

Figure 3.3   The relationship between the vectors Ω, v, and the Coriolis force, F_c

circular orbit, the plane of which is perpendicular to the rotation axis. The radius of the orbit is given by

$$\frac{U^2}{R} = 2\Omega U, \quad \text{i.e., } R = \frac{U}{2\Omega} \qquad (3.16)$$

and the period of the orbit is

$$\tau = \frac{2\pi R}{U} = \frac{\pi}{\Omega} \qquad (3.17)$$

In practice, because of the large gravitational acceleration on Earth, the only possible realization of such motion is in the local horizontal, and then it is the local vertical component of rotation, $\Omega\sin(\phi)$, that is relevant and replaces Ω in these expressions. Such motion is called 'inertial oscillation', and the orbit is called an 'inertial circle'. The length of the sidereal day is $2\pi/\Omega$, so the period of inertial oscillations is $12/\sin(\phi)$ hours.

At a latitude of 45°, the period is about 17 hour, and for a wind of $10\,\mathrm{m\,s^{-1}}$, the radius of the inertial circle would be 97 km. For an ocean current a 100 times slower, the radius would be about 1 km. Nearer the equator, the period and the radius become much larger. Inertial oscillations are not easily seen in the atmosphere, where the pressure gradient term is usually so large as to mask any inertial oscillations, but they can be observed in the ocean. At 30° latitude, the inertial period is 1 day, and it has sometimes been considered that there could be special behaviour because of this. However, no convincing evidence has been found. The Somali Jet is a very strong wind in the Northern Hemisphere summer taking a large amount of moisture from the Southern Indian Ocean into the Northern Indian Ocean and feeding it into the Indian Summer Monsoon. The tendency to move in inertial circles can sometimes mean that the air in the Somali Jet starts to move back towards the equator rather than moving into India. However, friction and heating usually act to stop this.

3.6 The Taylor–Proudman theorem

The impact of rapid rotation can be very dramatic. 'Rapid' rotation means a flow for which the Rossby number $Ro = U/fL$, introduced in Section 2.4, is small compared to 1. When the rotation is slow, so the Rossby number is large compared to 1, the effects of rotation are no more than a small correction and can often be ignored completely. So, in the context of flow in the Earth's midlatitude atmosphere, is the rotation of the Earth rapid or slow?

By way of example, in the Earth's midlatitude troposphere, a typical flow speed U is $10\,\mathrm{m\,s^{-1}}$ and a typical large length scale over which U varies is $10^6\,\mathrm{m}$. The rotation rate of the Earth is (to the nearest power of 10) around $10^{-4}\,\mathrm{s^{-1}}$. The corresponding Rossby number is around 0.1. So rotation dominates. For ocean flows, which are generally much slower, the Rossby number may be as small as 10^{-3}.

One might say that appreciable, or indeed, rapid rotation is the defining characteristic of geophysical fluid dynamics. Stratification also plays an important role, but rotation is crucial. In this section, some dramatic results in simple situations will be described. Later sections will show how these extreme examples relate to the flows observed in more realistic circumstances.

Consider a tank of fluid which has been standing on a rotating turntable for sufficient time to come to rest relative to the rotating frame of reference fixed in the turntable (Figure 3.4). It is then gently stirred to generate some weak motion relative to the turntable. We shall make three assumptions about the flow:

1. The relative motions are weak, in the sense that the Rossby number $U/\Omega L$ is small compared to 1.

2. The fluid is incompressible, with constant density, so that $(1/\rho)\nabla p \rightarrow \nabla(p/\rho)$.

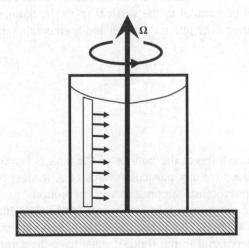

Figure 3.4 A tank of homogeneous fluid on a rotating turntable

3. The flow is inviscid in the sense that viscous stresses are much smaller than other forces acting.

The equations of motion for the tank of fluid are

$$\frac{D\mathbf{u}}{Dt} = -2\mathbf{\Omega} \times \mathbf{u} - \nabla\Phi_e - \frac{1}{\rho}\nabla p + \mathbf{F} \qquad (3.18)$$

Incorporating the three assumptions given earlier provides

$$0 = -2\mathbf{\Omega} \times \mathbf{u} - \nabla\Phi_e - \nabla\left(\frac{p}{\rho}\right) \qquad (3.19)$$

Take the curl of this equation; the gravitational and pressure gradient terms, being pure gradients, have zero curl, and so

$$\nabla \times (\mathbf{\Omega} \times \mathbf{u}) = 0 \qquad (3.20)$$

From the standard vector identity,

$$\nabla \times (\mathbf{A} \times \mathbf{B}) = (\mathbf{B} \cdot \nabla)\mathbf{A} - \mathbf{B}(\nabla \cdot \mathbf{A}) - (\mathbf{A} \cdot \nabla)\mathbf{B} + \mathbf{A}(\nabla \cdot \mathbf{B}) \qquad (3.21)$$

and noting that for incompressible fluid, the continuity equation is $\nabla \cdot \mathbf{u} = 0$, the curl of the equation of motion reduces to

$$(\mathbf{\Omega} \cdot \nabla)\mathbf{u} = 0 \qquad (3.22)$$

That is, the velocity vector $\mathbf{u}$ cannot vary in a direction parallel to the rotation axis. Let the rotation axis be parallel to the vertical (z-) axis, parallel to unit vector $\mathbf{k}$. Then, splitting Equation 3.22 into its vertical and horizontal components gives

$$\begin{aligned} \Omega\frac{\partial w}{\partial z} &= 0 \quad \text{(a)} \\[2mm] \Omega\frac{\partial \mathbf{v}}{\partial z} &= 0 \quad \text{(b)} \end{aligned} \qquad (3.23)$$

Now the boundary condition at the bottom of the tank is $w = 0$. Equation 3.23(a) implies that if w is zero for any particular value of z, it must be zero at all other values of z. Thus, rapid rotation suppresses vertical motion.

The constraints on the horizontal components of the flow implied by Equation 3.23(b) are not quite as strong. No particular value is implied for $\mathbf{v}$. However, whatever value $\mathbf{v}$ has at any point in the fluid, it must have the same value at all other points on a line parallel to the rotation axis which passes through that point. That

(a) (b)

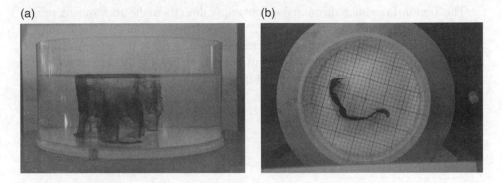

Figure 3.5 Taylor's inkwall experiment. (a) top view and (b) side view

means that the fluid must move as coherent columns orientated parallel to the rotation axis, with the same velocity at every level in the fluid.

The effect of rapid rotation on fluid flow expressed by the Taylor–Proudman theorem is both dramatic and counter-intuitive. It is as if the rotation imparts a degree of rigidity to the flow. Many beautiful experiments can be devised to illustrate the Taylor–Proudman theorem. One of the simplest is 'Taylor's inkwall experiment'. A tank of water on a turntable is allowed to spin up and is then lightly stirred. The flow is visualized by dropping a droplet of dense dye into the tank. As the dye falls, it leaves a vertical trail behind. After a short time, this trail is spread and distorted by the weak motions in the fluid. But the velocity field is the same at every level, according to the Taylor–Proudman theorem. So each column is pulled out into a curving thin sheet, with the same distortion at every level, a sheet which Taylor called an 'inkwall'. Seen from above, the round spot of dye left by the droplet is pulled and sheared into long curving streamers, but with a high degree of vertical coherence (Figure 3.5).

Another experiment which illustrates the Taylor–Proudman theorem consists of towing a shallow obstacle slowly across the base of a rotating tank. The tank is filled with water which has been allowed to spin up to rest in the rotating frame of reference of the tank. At a level below the summit of the shallow obstacle, the flow must part and move around the obstacle. But because the Taylor–Proudman theorem operates, the flow at every other level, even those well above the obstacle, must also pass around the obstacle edge. As a result, a column of fluid extends through the depth of the tank, from the obstacle to the surface, apparently attached to the obstacle. The rest of the flow passes around that column. It is as if the obstacle had been extended to fill the entire depth of the tank. Such a column is called a 'Taylor column'. There is some evidence for such structures in the ocean, where gentle currents pass over isolated seamounts on the oceans' abyssal plains. Taylor columns have been suggested as the origin of long-lived features in the outer fluid layers of the giant planets: Jupiter's 'Great Red Spot', in particular, has been interpreted in these terms.

The Taylor–Proudman theorem does not apply directly to the atmosphere, principally because the density is not constant. Also, the Rossby number, though small, is not infinitesimal. However, we can generalize the theorem to the atmospheric situation, and we can devise a continuum of behaviour linking the Taylor regime to more realistic situations. Chapter 12 and later chapters will address these issues.

4
The spherical Earth

4.1 Spherical polar coordinates

The Earth is very nearly a sphere, and so spherical polar coordinates are an appropriate frame of reference for atmospheric motions. In saying this, there are two matters which will need further discussion. First, the Earth is not exactly a sphere, but is slightly oblate. Section 3.4 discussed the slight distortion of geopotential surfaces by the centrifugal force and the associated small variations of the magnitude of the apparent gravitational acceleration. Secondly, the discussion benefits from recognizing that the vertical extent of the atmosphere is very much less than the horizontal extent. This leads to a shallow atmosphere approximation which will be discussed in Section 4.3.

The spherical polar coordinate system is illustrated in Figure 4.1. The coordinates are longitude λ, latitude ϕ and distance from the centre of the Earth, r. Each increases in the direction of the unit vectors $\hat{\mathbf{i}}$, $\hat{\mathbf{j}}$ and $\hat{\mathbf{k}}$ respectively. Some of the implications of using a coordinate system like this can be envisaged by thinking about what happens to the unit vectors as you move around a non-rotating Earth. First consider moving due north from the equator. The eastward direction stays the same, but the northward and upward vectors rotate so that by the time you nearly reach the pole, the upward direction is almost the same as the northward direction at the equator, and the northward direction is almost in the opposite direction to the local vertical at the equator. The rotation rate for these unit vectors is v/r where v is the northward speed. Secondly, consider moving around a latitude circle at latitude ϕ. As you move, the local direction away from the Earth's axis of rotation and the eastward direction rotate such that by the time you have moved a quarter way around the Earth, the local outward direction is parallel to the eastward direction at the starting point and the eastward direction is in the opposite direction to the vertical at the starting point. The radius of the latitudinal circle at latitude ϕ is $s = r\cos(\phi)$, and the rotation rate of the unit vectors is u/s where u is the eastward speed. In this case,

Fluid Dynamics of the Midlatitude Atmosphere, First Edition. Brian J. Hoskins & Ian N. James.
© 2014 John Wiley & Sons, Ltd. Published 2014 by John Wiley & Sons, Ltd.

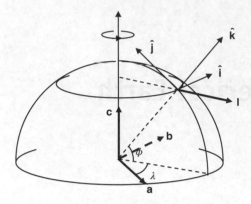

Figure 4.1　Spherical polar coordinates, showing the unit vectors $\hat{\mathbf{i}}, \hat{\mathbf{j}}$ and $\hat{\mathbf{k}}$ as well as **a**, **b**, **c** and **l**

the local northward direction also rotates. Therefore, using a spherical polar coordinate system will imply the introduction of additional coordinate rotation terms in the equations of motion.

The process of determining the equations of motion in spherical coordinates starts with a consideration of the variations of vector quantities. Consider a small displacement of position $\delta\mathbf{r}$ where (following the notation of Chapter 2) $\mathbf{r}$ is the position vector. The displacement is related to changes of λ, ϕ and r by

$$\delta\mathbf{r} = h_\lambda \delta\lambda \hat{\mathbf{i}} + h_\phi \delta\phi \hat{\mathbf{j}} + h_r \delta r \hat{\mathbf{k}} \tag{4.1}$$

Here,

$$h_\lambda = r\cos(\phi), \quad h_\phi = r, \quad h_r = 1 \tag{4.2}$$

are called the 'metrics' of the coordinate system.

An immediate application of this formalism is the velocity vector. The velocity in a spherical system is

$$\mathbf{u} = \frac{\mathrm{D}\mathbf{r}}{\mathrm{D}t} = h_\lambda \frac{\mathrm{D}\lambda}{\mathrm{D}t} \hat{\mathbf{i}} + h_\phi \frac{\mathrm{D}\phi}{\mathrm{D}t} \hat{\mathbf{j}} + h_r \frac{\mathrm{D}r}{\mathrm{D}t} \hat{\mathbf{k}} \tag{4.3}$$

Denoting the zonal, azimuthal (or meridional) and radial components of velocity by u, v and w respectively, then we have

$$u\hat{\mathbf{i}} + v\hat{\mathbf{j}} + w\hat{\mathbf{k}} = r\cos(\phi)\frac{\mathrm{D}\lambda}{\mathrm{D}t} \hat{\mathbf{i}} + r\frac{\mathrm{D}\phi}{\mathrm{D}t} \hat{\mathbf{j}} + \frac{\mathrm{D}r}{\mathrm{D}t} \hat{\mathbf{k}} \tag{4.4}$$

A second application is to calculate the gradient of any scalar, $S(\mathbf{r})$:

$$\nabla S = \frac{1}{h_\lambda} \frac{\partial S}{\partial \lambda} \hat{\mathbf{i}} + \frac{1}{h_\phi} \frac{\partial S}{\partial \phi} \hat{\mathbf{j}} + \frac{1}{h_r} \frac{\partial S}{\partial r} \hat{\mathbf{k}}$$

$$= \frac{1}{r\cos(\phi)} \frac{\partial S}{\partial \lambda} \hat{\mathbf{i}} + \frac{1}{r} \frac{\partial S}{\partial \phi} \hat{\mathbf{j}} + \frac{\partial S}{\partial r} \hat{\mathbf{k}}$$

(4.5)

4.2 Scalar equations

The thermodynamic equation was derived in Chapter 2. It simply states that the scalar potential temperature is advected by the flow and modified by a source or sink term which is in general a function of time and space:

$$\frac{D\theta}{Dt} = S(\mathbf{r},t)$$

The Lagrangian rate of change of a scalar in spherical coordinates follows from Equation 4.5, and so the thermodynamic equation can be written as

$$\frac{\partial \theta}{\partial t} + \frac{u}{r\cos(\phi)} \frac{\partial \theta}{\partial \lambda} + \frac{v}{r} \frac{\partial \theta}{\partial \phi} + w \frac{\partial \theta}{\partial z} = S(\mathbf{r},t)$$

(4.6)

The continuity equation can be written in several forms. Take the form

$$\frac{\partial \rho}{\partial t} + \mathbf{u} \cdot \nabla \rho + \rho \nabla \cdot \mathbf{u} = 0$$

The first two terms, the Lagrangian rate of change, take the same form as for the thermodynamic equation. The $\nabla \cdot \mathbf{u}$ term is more complicated. In general curvilinear coordinates, the divergence can easily be constructed from the definitions of the metrics given in the last section. Given that $\nabla \cdot \mathbf{u}$ is the net emergent flux of $\mathbf{u}$ per unit volume from an infinitesimal parallelepiped, it follows that

$$\nabla \cdot \mathbf{u} = \frac{1}{h_\lambda h_\phi h_r} \left(\frac{\partial}{\partial \lambda} (h_\phi h_r u) + \frac{\partial}{\partial \phi} (h_\lambda h_r v) + \frac{\partial}{\partial z} (h_\lambda h_\phi w) \right)$$

(4.7)

Using this result, the continuity equation in spherical coordinates becomes

$$\frac{\partial \rho}{\partial t} + \frac{u}{r\cos(\phi)} \frac{\partial \rho}{\partial \lambda} + \frac{v}{r} \frac{\partial \rho}{\partial \phi} + w \frac{\partial \rho}{\partial z}$$

$$+ \rho \left(\frac{1}{r\cos(\phi)} \left(\frac{\partial u}{\partial \lambda} + \frac{\partial}{\partial \phi} (\cos(\phi)v) \right) + \frac{1}{r^2} \frac{\partial}{\partial r} (r^2 w) \right) = 0$$

(4.8)

4.3 The momentum equations

We must now recognize explicitly the variation of the directions of the unit vectors $\hat{\mathbf{i}}, \hat{\mathbf{j}}$ and $\hat{\mathbf{k}}$ as a fluid element moves around. Figure 4.1 shows the geometry envisaged. The vectors $\hat{\mathbf{i}}, \hat{\mathbf{j}}$ and $\hat{\mathbf{k}}$ point in the eastward, northward and upward directions respectively. At the same time, it is helpful for the moment to refer to a set of unit vectors fixed in the solid Earth; they are denoted $\mathbf{a}, \mathbf{b}$ and $\mathbf{c}$. Unit vector $\mathbf{c}$ is parallel to the rotation axis, while $\mathbf{a}$ and $\mathbf{b}$ are in the plane of the equator, emerging from the Earth's surface at the Greenwich meridian and at 90°E respectively. Finally, unit vector $\mathbf{l}$ is perpendicular to the rotation axis and to the unit vector $\hat{\mathbf{i}}$. In terms of the constant unit vectors $\mathbf{a}, \mathbf{b}$ and $\mathbf{c}, \hat{\mathbf{i}}, \hat{\mathbf{j}}$ and $\hat{\mathbf{k}}$ are

$$
\begin{aligned}
\hat{\mathbf{i}} &= -\sin(\lambda)\mathbf{a} + \cos(\lambda)\mathbf{b} \\
\hat{\mathbf{j}} &= \cos(\phi)\mathbf{c} - \sin(\phi)\mathbf{l} \\
\hat{\mathbf{k}} &= \sin(\phi)\mathbf{c} + \cos(\phi)\mathbf{l}
\end{aligned}
\tag{4.9}
$$

Also,

$$
\mathbf{l} = \cos(\lambda)\mathbf{a} + \sin(\lambda)\mathbf{b} = -\sin(\phi)\hat{\mathbf{j}} + \cos\phi\hat{\mathbf{k}}
\tag{4.10}
$$

We are now in a position to work out the Lagrangian rates of change of the unit vectors $\hat{\mathbf{i}}, \hat{\mathbf{j}}$ and $\hat{\mathbf{k}}$. For example,

$$
\frac{D\hat{\mathbf{i}}}{Dt} = -(\cos(\lambda)\mathbf{a} + \sin(\lambda)\mathbf{b})\frac{D\lambda}{Dt} = \frac{D\lambda}{Dt}\mathbf{l}
$$

since the vectors $\mathbf{a}, \mathbf{b}$ and $\mathbf{c}$ are fixed. But $u = r\cos(\phi)D\lambda/Dt$ and so

$$
\begin{aligned}
\frac{D\hat{\mathbf{i}}}{Dt} &= -\frac{u}{r\cos(\phi)}\mathbf{l} = -\frac{u}{r\cos(\phi)}\left(-\sin(\phi)\hat{\mathbf{j}} + \cos(\phi)\hat{\mathbf{k}}\right) \\
&= \frac{u}{r}\left(\tan(\phi)\hat{\mathbf{j}} - \hat{\mathbf{k}}\right)
\end{aligned}
\tag{4.11}
$$

A similar argument leads to

$$
\frac{D\mathbf{l}}{Dt} = (-\sin(\lambda)\mathbf{a} + \cos(\lambda)\mathbf{b})\frac{D\lambda}{Dt} = \frac{u}{r\cos(\phi)}\hat{\mathbf{i}}
$$

and so we obtain

$$\frac{D\hat{\mathbf{j}}}{Dt} = -(\sin(\phi)\mathbf{c} + \cos(\phi)\mathbf{l})\frac{D(\phi)}{Dt} - \sin(\phi)\frac{D\mathbf{l}}{Dt}$$

$$= -\frac{v}{r}\hat{\mathbf{k}} - \frac{u\tan(\phi)}{r}\hat{\mathbf{i}}$$

(4.12)

and

$$\frac{D\hat{\mathbf{k}}}{Dt} = (\cos(\phi)\mathbf{c} - \sin(\phi)\mathbf{l})\frac{D\phi}{Dt} + \cos(\phi)\frac{D\mathbf{l}}{Dt}$$

$$= \frac{v}{r}\hat{\mathbf{j}} + \frac{u}{r}\hat{\mathbf{i}}$$

(4.13)

Referring back to the discussion of rotation as one moves about the sphere at the beginning of the chapter, the v/r terms in Equations 4.12 and 4.13 represent the movement poleward and the u/r terms in (4.11), (4.12) and (4.13) represent the movement around a latitude circle.

The acceleration in spherical coordinates is

$$\frac{D\mathbf{u}}{Dt} = \frac{D}{Dt}\left(u\hat{\mathbf{i}} + v\hat{\mathbf{j}} + w\hat{\mathbf{k}}\right)$$

$$= \frac{Du}{Dt}\hat{\mathbf{i}} + u\frac{D\hat{\mathbf{i}}}{Dt} + \frac{Dv}{Dt}\hat{\mathbf{j}} + v\frac{D\hat{\mathbf{j}}}{Dt} + \frac{Dw}{Dt}\hat{\mathbf{k}} + w\frac{D\hat{\mathbf{k}}}{Dt}$$

Substituting for the Lagrangian rates of change of the unit vectors, and gathering up the terms in each direction separately, yields a final expression for the acceleration in spherical coordinates:

$$\frac{D\mathbf{u}}{Dt} = \left\{\frac{Du}{Dt} - uv\frac{\tan(\phi)}{r} + \frac{uw}{r}\right\}\hat{\mathbf{i}} + \left\{\frac{Dv}{Dt} + u^2\frac{\tan(\phi)}{r} + \frac{vw}{r}\right\}\hat{\mathbf{j}} + \left\{\frac{Dw}{Dt} - \frac{u^2+v^2}{r}\right\}\hat{\mathbf{k}}$$

(4.14)

The remaining terms in the momentum equations are more straightforward. The pressure gradient term follows from Equation 4.5:

$$-\frac{1}{\rho}\nabla p = -\left\{\frac{1}{\rho r\cos(\phi)}\frac{\partial p}{\partial \lambda}\right\}\hat{\mathbf{i}} - \left\{\frac{1}{\rho r}\frac{\partial p}{\partial \phi}\right\}\hat{\mathbf{j}} - \left\{\frac{1}{\rho}\frac{\partial p}{\partial r}\right\}\hat{\mathbf{k}}$$

(4.15)

The Coriolis term is

$$2\boldsymbol{\Omega} \times \mathbf{u} = 2\left(0\hat{\mathbf{i}} + 2\Omega\cos(\phi)\hat{\mathbf{j}} + 2\Omega\sin(\phi)\hat{\mathbf{k}}\right) \times \left(u\hat{\mathbf{i}} + v\hat{\mathbf{j}} + w\hat{\mathbf{k}}\right) \qquad (4.16)$$
$$= \{2\Omega\cos(\phi)w - 2\Omega\sin(\phi)w\}\hat{\mathbf{i}} + \{2\Omega\sin(\phi)u\}\hat{\mathbf{j}} - \{2\Omega\cos(\phi)u\}\hat{\mathbf{k}}$$

The gravitational and friction terms complete the set:

$$\mathbf{g} = \nabla\Phi_e = -g_e\hat{\mathbf{k}}; \quad \dot{\mathbf{u}} = \dot{u}\hat{\mathbf{i}} + \dot{v}\hat{\mathbf{j}} + \dot{w}\hat{\mathbf{k}} \qquad (4.17)$$

Combining all these results leads to the three components of the momentum equation in polar spherical coordinates:

$$\frac{Du}{Dt} - \frac{uv}{r}\tan(\phi) + \frac{uw}{r} = 2\Omega\sin(\phi)v - 2\Omega\cos(\phi)w - \frac{1}{\rho r\cos(\phi)}\frac{\partial p}{\partial\lambda} + \dot{u} \qquad (4.18)$$

$$\frac{Dv}{Dt} + \frac{u^2}{r}\tan(\phi) + \frac{vw}{r} = -2\Omega\sin(\phi)u - \frac{1}{\rho r}\frac{\partial p}{\partial\phi} + \dot{v} \qquad (4.19)$$

$$\frac{Dw}{Dt} - \frac{u^2 + v^2}{r} = 2\Omega\cos(\phi)u - g_e - \frac{1}{\rho}\frac{\partial p}{\partial r} + \dot{w} \qquad (4.20)$$

where the Lagrangian (material) time derivative is

$$\frac{D}{Dt} = \frac{\partial}{\partial t} + \frac{u}{r\cos(\phi)}\frac{\partial}{\partial\lambda} + \frac{v}{r}\frac{\partial}{\partial\phi} + w\frac{\partial}{\partial r} \qquad (4.21)$$

The new terms involving products of wind components owe their origin to the rotation of the coordinate system used as air moves around the sphere. They are sometimes referred to as 'metric terms'. Terms involving Ω originate with the rotation of axes embedded in the sphere and the consequent Coriolis term, and reflect the variations in the angle between the rotation axis and the coordinate system from place to place.

These equations represent something of a climax of complexity. In later sections, we shall explore various ways of simplifying the equations in particular circumstances.

4.4 Energy and angular momentum

A cursory inspection of the metric and Coriolis terms suggests that the various additional terms may have very different magnitudes, raising the possibility that the equation set might be simplified without significant loss of accuracy.

However, care needs to be taken in carrying out such a simplification. By omitting even small terms, the fundamental character of the equation set might be changed in an undesirable way. Among the most important properties of the equation set to be preserved when they are simplified are their conservation properties.

The kinetic energy per unit mass of the motion relative to the rotating coordinate system is

$$K = \frac{1}{2}\left(u^2 + v^2 + w^2\right) \tag{4.22}$$

From Equations 4.18 to 4.20, K evolves according to

$$\frac{DK}{Dt} = u\frac{Du}{Dt} + v\frac{Dv}{Dt} + w\frac{Dw}{Dt}$$
$$= -\frac{1}{\rho}\mathbf{u}\cdot\nabla p - wg_e + \mathbf{u}\cdot\mathbf{F} \tag{4.23}$$

All the contributions to changes of kinetic energy due to the metric terms have cancelled out. Similarly, contributions from the Coriolis force, which operates at right angles to the motion, lead to no change of the kinetic energy. According to Equation 4.23, changes of kinetic energy can only arise from motion across the pressure contours so that the pressure gradient force does work, from vertical motion so the gravitational force does work or from motion with a component parallel to the friction force so the friction force does work. When simplifying the equations, it will be necessary to ensure that these properties are preserved. Consequently, arbitrary deletion of apparently small terms may not be consistent with energy conservation.

Now consider the potential energy of a fluid parcel. The rate of change of potential energy per unit mass is

$$\frac{D\Phi}{Dt} = wg_e \tag{4.24}$$

so that the equation for the sum of kinetic and potential energy becomes

$$\frac{D}{Dt}(K + \Phi) = -\frac{1}{\rho}\mathbf{u}\cdot\nabla p + \mathbf{u}\cdot\mathbf{F} \tag{4.25}$$

The first term on the right-hand side can be rewritten as

$$-\frac{1}{\rho}\mathbf{u}\cdot\nabla p = -\frac{1}{\rho}\nabla\cdot(\mathbf{u}p) + \frac{p}{\rho}\nabla\cdot\mathbf{u} \tag{4.26}$$

From the continuity equation, the last term in this equation can be expressed as:

$$\frac{p}{\rho}\nabla\cdot\mathbf{u} = p\frac{\mathrm{D}}{\mathrm{D}t}\left(\rho^{-1}\right) \qquad (4.27)$$

Following Equation 2.22, the first law of thermodynamics can be written in the following form:

$$c_v\frac{\mathrm{D}T}{\mathrm{D}t} = -p\frac{\mathrm{D}}{\mathrm{D}t}\left(\rho^{-1}\right) + \dot{q}$$

This equation relates the rate of change of internal energy, on the left-hand side, to the work done by the pressure exerted by the surrounding fluid and the effects of direct input of heat. The last term in Equation 4.26 is minus the pressure work term in the thermodynamic equation and therefore represents the conversion from internal to mechanical energy. Add the thermodynamic equation and the kinetic energy equation and thus obtain a total energy equation:

$$\frac{\mathrm{D}}{\mathrm{D}t}\left(K + \Phi + c_v T\right) = -\frac{1}{\rho}\nabla\cdot(\mathbf{u}p) + \mathbf{u}\cdot\dot{\mathbf{u}} + \dot{q} \qquad (4.28)$$

Therefore, the sum of the kinetic, potential and internal energies can change following the fluid only through the terms on the right-hand side. The first term on the right-hand side, when multiplied by ρ and integrated over a volume using the Gauss divergence theorem, gives $\oint_A p\mathbf{u}\cdot\mathbf{n}\,\mathrm{d}A$. It is simply the rate of working of the pressure force on the volume. The second and third terms in the energy equation represent generation of energy by frictional and heating processes. All contributions from the various metric terms and Coriolis terms have cancelled out. Again, any approximation to the governing equations must preserve this property if the approximated equations are to retain realistic conservation properties.

Angular momentum is another important quantity that has conservation properties. The absolute angular momentum per unit mass is the eastward component of the absolute velocity multiplied by the perpendicular distance to the rotation axis:

$$m = (u + \Omega r\cos(\phi))r\cos(\phi) \qquad (4.29)$$

As in the case of the energy equation, an equation for the rate of change of m can be derived from the momentum equations; the various metric and Coriolis terms cancel out, leaving

$$\frac{\mathrm{D}m}{\mathrm{D}t} = -\frac{1}{\rho}\frac{\partial p}{\partial \lambda} + r\cos(\phi)F_\lambda \qquad (4.30)$$

This equation reveals that m can only change moving with the fluid if there is an eastward pressure gradient or an east-west frictional force. Again, any approximated form of the equations should retain this property.

4.5 The shallow atmosphere approximation

We have already stated, and in the next chapter will deduce, that the vertical scale of the troposphere is of order 10 km. The horizontal scale of atmospheric motions is much larger and may be as large as a scale comparable with the radius of the Earth a. The thin atmosphere approximation recognizes this asymmetry between the vertical and horizontal scales and is based upon the inequality $|D| \ll a$.

For example, the zonal component of the momentum equation, Equation 4.18, contains a term uw/r. The ratio of this term to the vertical advection term, $w\partial u/\partial z$, will be of the order of D/a and so wu/r should be negligible. But dropping this term alone would lead to an inconsistency in deriving the principle of angular momentum conservation, Equation 4.30. The term uw/r is related to the use of r in the definition of specific angular momentum m given by Equation 4.29. If we redefine m as

$$m' = (u + a\Omega\cos(\phi))a\cos(\phi) \qquad (4.31)$$

then Equation 4.30 is maintained only if

1. The term $2\Omega\cos(\phi)w$ is simultaneously neglected in Equation 4.18

2. r is replaced by a in the $uv\tan(\phi)$ term in the same equation

3. r is replaced by a in the metrics h_i and h_f

Similar remarks apply to the derivation of the mechanical energy equation, Equation 4.25. If wu/r is neglected in Equation 4.18, then the corresponding terms vw/r and $(u^2 + v^2)/r$ in Equations 4.19 and 4.20 must also be neglected. Again, if the small term $2\Omega\cos(\phi)w$ is neglected in Equation 4.18, then so must the term $2\Omega\cos(\phi)w$ in Equation 4.19. In the mass conservation equation, r must be replaced by a everywhere.

Each of these extra approximations can be justified by detailed scale analysis. However, the arguments in this section show that either all or none of these approximations must be made. So, making these various approximations and using $z = r - a$, a consistent set of 'shallow atmosphere' equations results:

$$\frac{Du}{Dt} - \frac{uv\tan(\phi)}{a} = fv - \frac{1}{\rho}\frac{1}{a\cos(\phi)}\frac{\partial p}{\partial \lambda} + \dot{u} \qquad (4.32)$$

$$\frac{Dv}{Dt} - \frac{u^2\tan(\phi)}{a} = -fu - \frac{1}{\rho}\frac{1}{a}\frac{\partial p}{\partial \phi} + \dot{v} \qquad (4.33)$$

$$\frac{Dw}{Dt} = -\frac{1}{\rho}\frac{\partial p}{\partial z} - g_e + \dot{w} \qquad (4.34)$$

$$\frac{D\rho}{Dt} + \rho\left[\frac{1}{a\cos(\phi)}\left(\frac{\partial u}{\partial \lambda} + \frac{\partial}{\partial \phi}(\cos(\phi)v)\right) + \frac{\partial w}{\partial z}\right] = 0 \qquad (4.35)$$

where

$$\frac{D}{Dt} = \frac{\partial}{\partial t} + \frac{u}{a\cos(\phi)}\frac{\partial}{\partial \lambda} + \frac{v}{a}\frac{\partial}{\partial \phi} + w\frac{\partial}{\partial z} \qquad (4.36)$$

Together with the thermodynamic energy equation and the equation of state, these equations form a complete set for the atmosphere and serve as a starting point for further analysis. They are sometimes referred to as the 'primitive equations'. The only explicit reference to the rotation of the planet is through the 'Coriolis parameter'.

$$f = 2\Omega\sin(\phi) \qquad (4.37)$$

a measure of the component of the Earth's rotation about the local vertical. From this point on, the suffix e will be dropped from the geopotential Φ and the gravitational acceleration, g.

The shallow atmosphere approximation is adequate for the terrestrial planets and the oceans. It may not be satisfactory for the gas giants Jupiter and Saturn whose atmospheres are much deeper and probably extend to a significant fraction of the planetary radius. In any case, these rapidly rotating bodies are very distorted by the centrifugal force, and so the use of spherical coordinates may be inappropriate.

4.6   The beta effect and the spherical Earth

Replacing Cartesian coordinates with spherical polar coordinates greatly complicates the equations of fluid dynamics. It has added new nonlinear terms to the equations and has added variable coefficients to other terms that simply had constant coefficients in the Cartesian set. One may ask how important are these complications. The answer is usually not very important. With a couple of exceptions, the new terms are simply minor corrections which add no substantial new dynamical processes to the equation sets. In fact, for many purposes, it will prove helpful to revert to 'local Cartesian coordinates', with an origin at some reference latitude. All the metric terms associated with the Earth's curvature can be neglected, but two of the Coriolis terms must be retained.

These two terms are the terms involving f, the Coriolis parameter, in Equations 4.32 and 4.33. Based on a reference latitude ϕ_0, f can be written as
These equations can be written in local Cartesian coordinates:

$$\frac{Du}{Dt} = fv - \frac{1}{\rho}\frac{\partial p}{\partial x} + \dot{u}$$

$$\frac{Dv}{Dt} = -fu - \frac{1}{\rho}\frac{\partial p}{\partial y} + \dot{v}$$

or in vector form

$$\frac{D\mathbf{v}}{Dt} = f\mathbf{k} \times \mathbf{v} - \frac{1}{\rho}\nabla p + \dot{\mathbf{v}} \qquad (4.38)$$

$$f = 2\Omega \sin\left(\phi_0 + \frac{y}{a}\right)$$

Use the double angle formula

$$f = 2\Omega \sin(\phi_0)\cos\left(\frac{y}{a}\right) + 2\Omega \cos(\phi_0)\sin\left(\frac{y}{a}\right)$$

and assume that $|y/a| \ll 1$:

$$f = 2\Omega \sin(\phi_0) + \frac{2\Omega}{a}\cos(\phi_0)y \qquad (4.39)$$

$$= f_0 + \beta y$$

This is called the 'β-plane approximation', and it is the simplest modification of the equations which permits an important class of wave motions called Rossby waves. These will be introduced in Chapter 9. Their properties are relevant for much of the discussion in the latter part of this book. If meridional displacements are sufficiently small, the βy term can be neglected, and one then has the 'f-plane' approximation. The f-plane approximation is often appropriate for discussing local mesoscale phenomena.

5

Scale analysis and its applications

5.1 Principles of scaling methods

In the previous three chapters, we have assembled and elaborated the equations of fluid dynamics, particularly as they apply to the Earth's atmosphere. In many branches of physics, that would be the end of the interesting science. Once the governing equations are complete and appropriate boundary conditions have been specified, the problem is done, apart from a certain amount of routine algebra and computation. With the equations of fluid flow, however, this is far from the case. The equations are very complex, highly nonlinear and with many couplings between them. They have no known analytical solutions in general, and indeed only a handful of exact solutions for highly simplified and contrived situations exist. In general, exploring their solution requires recourse to numerical simulation, that is, to solving some discretized analogue to the full equations. Such an approach requires the use of large computers; solutions are only generated for specific cases. Numerical simulation raises questions of how representative the chosen cases might be, as well as difficulties in proving that the numerical results have converged onto the hypothetical exact solution to the continuous equations.

However, considerable progress can be made simply by estimating the typical magnitude of terms in the governing equations and using the results to examine the various balances between the terms that hold. The estimates can also suggest ways in which the equations can be approximated and simplified in particular circumstances. The technique is called scale analysis. The basic principle is to consider typical fluctuations in various state and dynamical variables and the time or space scales over which these fluctuations occur. For example, if any flow quantity Q has a typical fluctuation ΔQ over a distance of order L, then a term in the equations such as ∇Q will have a magnitude

Fluid Dynamics of the Midlatitude Atmosphere, First Edition. Brian J. Hoskins & Ian N. James.
© 2014 John Wiley & Sons, Ltd. Published 2014 by John Wiley & Sons, Ltd.

$$\nabla Q \sim \frac{\Delta Q}{L} \tag{5.1}$$

Here, the relation '~' should be read as 'of the same order of magnitude as'. Factors of 2 or 3 or less will be regarded as order-1 factors. Similarly, geometric factors such as $\sin(\alpha)$ or $\cos(\alpha)$ will generally be regarded as order 1. For a sinusoidal distribution of Q, $Q = Q_0 + A\sin(kx)$, the amplitude A is the natural choice for ΔQ, and for L, the natural choice is k^{-1}, which is the wavelength of the fluctuation divided by 2π. In this case, Equation 5.1 is exact. For a weather system with wavelength 3000 km, that is, about 1500 km across a low-pressure centre, then $L \sim 500$ km.

In this chapter, we will attempt to be as general as possible by analysing each equation in terms of various geometric and flow variables which will take different values in different systems, such as the midlatitude troposphere, the ocean or a laboratory experiment. Table 5.1 lists the principal such variables, but also suggests values relevant to a typical midlatitude weather system such as a depression.

As a very simple example of scale analysis, consider the continuity equation for incompressible flow:

$$\nabla_H \cdot \mathbf{v} + \frac{\partial w}{\partial z} = 0 \tag{5.2}$$
$$\quad (1) \quad (2)$$

Denote the typical fluctuation of horizontal velocity as U over a length scale L, and the typical fluctuation of vertical velocity W over a vertical length scale of D. Then, the orders of magnitude of each term can be estimated as

Table 5.1 Order of magnitude values of various constants and flow variables appropriate to a midlatitude weather system

Quantity	Value
Horizontal wind U	$10\,\mathrm{m\,s^{-1}}$
Horizontal length scale L	$5 \times 10^5\,\mathrm{m}$
System depth D	$\leq 5 \times 10^3\,\mathrm{m}$
Coriolis parameter $f = 2\Omega\sin(\phi)$	$10^{-4}\,\mathrm{s^{-1}}$
Gravitational acceleration g	$10\,\mathrm{m\,s^{-2}}$
Density ρ	$1 - 0.3\,\mathrm{kg\,m^{-3}}$
Radius of Earth a	$6.37 \times 10^6\,\mathrm{m} \sim 10^7\,\mathrm{m}$
Depth mean potential temperature q	$300\,\mathrm{K}$
Brunt–Väisälä frequency N	$10^{-2}\,\mathrm{s^{-1}}$

$$\nabla_H \cdot \mathbf{v} + \frac{\partial w}{\partial z} = 0$$

$$\frac{U}{L} \qquad \frac{W}{D}$$

$$(1) \qquad (2)$$

(5.3)

Since there are only two terms in the equation, they must balance, and so

$$\frac{U}{L} \sim \frac{W}{D} \text{ or } W \sim U\left(\frac{D}{L}\right)$$

(5.4)

The dimensionless factor (D/L) is called the aspect ratio; it is about 1 for a laboratory experiment but nearer to 0.01 for a synoptic weather system. For such a system, therefore, this scale analysis predicts typical vertical velocities of around $10\,\text{cm}\,\text{s}^{-1}$. In fact, this is at least an order of magnitude larger than observed on the scale of a weather system. What has gone wrong with our analysis is that term (1) is in fact the sum of two terms $\partial u/\partial x$ and $\partial v/\partial y$. They are usually comparable in magnitude but opposite in sign. Term (1) is therefore the small residual between two larger terms of opposite sign, and our estimate of W is very much an upper bound. We shall derive a more realistic estimate of W in Section 5.3.

5.2   The use of a reference atmosphere

We will write the various thermodynamic variables in terms of their deviation from some reference atmosphere relevant to the region under study. The reference thermodynamic variables are functions of z only:

$$p_R = p_R(z), \quad \rho_R = \rho_R(z), \quad T_R = T_R(z), \quad \theta_R = \theta_R(z)$$

This choice of reference atmosphere is such that it obeys the equation of state and the associated definition of potential temperature:

$$p_R = R\rho_R T_R, \quad \theta_R = T_R\left(\frac{p_R}{p_0}\right)^{-\kappa}$$

As for any system with no motion, the reference atmosphere is also in hydrostatic balance:

$$\frac{dp_R}{dz} = -\rho_R g \tag{5.5}$$

Use the equation of state to write the hydrostatic relationship as

$$\frac{dp_R}{dz} = -\frac{g}{RT_R} p_R \tag{5.6}$$

Assume g is constant with z. Then, if T_R is constant as well (i.e., an isothermal atmosphere), the equation is easily integrated to give

$$p_R = p_0 e^{-z/H} \tag{5.7}$$

where p_0 is the pressure at $z=0$ and $H=RT_r/g$ is called the 'pressure scale height'; it has a value of 7.5 km for the troposphere. Also, for an isothermal atmosphere,

$$\rho_R = \rho_0 e^{-z/H}$$

Although the assumption of constant temperature is rather crude in the troposphere, it is more accurate in the stratosphere. However, the roughly exponential decrease in pressure and density with height is quite robust. The hydrostatic equation is straightforward to integrate if the more realistic assumption of a constant lapse rate is made:

$$T_R(z) = T_0 - \Gamma z$$

A representative value of Γ is around $6\,K\,km^{-1}$.

Now write all thermodynamic variables in terms of their deviation from the reference atmosphere so that

$$p = p_R(z) + p'(x,y,z,t) \tag{5.8}$$

and similarly for ρ, T and θ. It is assumed that $|p'| \ll p_r$ and so on. We will scale the deviations from the reference atmosphere by Δp, $\Delta \rho$, ΔT and $\Delta \theta$.

The logarithm of the ideal gas equation is

$$\ln(p) = \ln(\rho) + \ln(T) + \ln(R)$$

Then differentiating gives

$$\frac{\delta p}{p_R} = \frac{\delta \rho}{\rho_R} + \frac{\delta T}{T_R} \tag{5.9}$$

For the scaling of the thermodynamic variables, this implies:

$$\frac{\Delta p}{\overline{p}_R} \sim \frac{\Delta \rho}{\overline{\rho}_R} + \frac{\Delta T}{\overline{T}_R} \tag{5.10}$$

The definition of potential temperature leads to a similar relationship:

$$\frac{\Delta \theta}{\overline{\theta}_R} \sim \frac{\Delta T}{\overline{T}_R} - \kappa \frac{\Delta p}{\overline{p}_R} \tag{5.11}$$

Here and subsequently the overbar for quantities such as $\overline{p}_R$, $\overline{\rho}_R$ and $\overline{T}_R$ denotes a typical value of p_R, ρ_R and T_R for the range of heights being considered. They may, for example, represent the relevant vertical average of these quantities.

5.3 The horizontal momentum equations

This and subsequent equations involve the Lagrangian rate of change of one of the variables. In general, propagating or stationary midlatitude weather systems are observed to change relatively slowly in structure in the time it takes for air parcels to move through them. Consequently, we will assume that the Eulerian rate of change is not larger in magnitude than the advective rate of change U/L, and we will therefore scale it as $\mathbf{u} \cdot \nabla \sim U/L$. With this assumption, the Lagrangian rate of change scales like the advective terms, U/L.

Consider first the zonal component of the momentum equation in spherical coordinates, Equation 4.32:

$$\frac{Du}{Dt} = \frac{uv}{a}\tan\phi - \quad 2\Omega\sin\phi v - \frac{1}{\rho a \cos\phi}\frac{\partial p'}{\partial \lambda}$$

$$(1) \qquad (2) \qquad (3) \qquad (4)$$

$$\frac{U^2}{L} \qquad \frac{U^2}{a} \qquad fU \qquad \frac{\Delta p}{\overline{\rho}_R L} \tag{5.12}$$

$$\mathrm{Ro} \qquad \mathrm{Ro}\left(\frac{L}{a}\right) \qquad 1 \qquad \frac{\Delta p}{\overline{\rho}_R fUL}$$

In the last line, the result has been made more general by dividing by the magnitude of term (3), so that the magnitude of each term can be represented by products of dimensionless ratios. The most significant of these is the Rossby number, which we introduced in Section 3.3:

$$\mathrm{Ro} = \frac{U}{fL} \qquad (5.13)$$

The Rossby number is the most important dimensionless number in geophysical fluid dynamics. It represents the ratio of the characteristic net acceleration of fluid elements to the Coriolis acceleration and so measures the importance of planetary rotation. The smaller the Rossby number, the more dominant is the Coriolis acceleration in the dynamics. For a typical midlatitude weather system, Ro is around 0.2, while in the ocean, where U is much smaller and L is rather smaller than in the atmosphere, the Rossby number may be closer to 10^{-2}.

Thus, for a midlatitude synoptic-scale weather system, for which $L/a \ll 1$, term (1) is relatively small and term (2) is even smaller. The meridional component of the momentum equation, Equation 4.33, scales similarly:

$$\frac{Dv}{Dt} = \quad -\frac{u^2}{a}\tan\phi - \quad 2\Omega\sin\phi u - \frac{1}{\rho a}\frac{\partial p'}{\partial \phi}$$

$$\quad (1) \qquad (2) \qquad (3) \qquad (4)$$

$$\frac{U^2}{L} \qquad \frac{U^2}{a} \qquad fU \qquad \frac{\Delta p}{\bar{\rho}_R L} \qquad (5.14)$$

$$\mathrm{Ro} \qquad \mathrm{Ro}\!\left(\frac{L}{a}\right) \qquad 1 \qquad \frac{\Delta p}{\bar{\rho}_R fUL}$$

Therefore, the metric terms (2) in both horizontal momentum equations are sufficiently small that they may generally be dropped compared to the remaining terms. In both equations, terms (3) and (4) have to provide the dominant balance. Therefore, the scaling gives

$$\Delta p \sim \bar{\rho}_R fUL \qquad (5.15)$$

For a typical midlatitude weather system, Equation 5.15 suggests that the typical horizontal pressure fluctuation will be of order $5 \times 10^2\,\mathrm{Pa}$. This is an Eulerian pressure fluctuation, the typical pressure fluctuation observed across the weather system at a given time and height. The Lagrangian pressure fluctuation, the pressure fluctuation in time to which an individual fluid parcel is subject to as it circulates through the weather system, is usually a good deal larger.

In Equations 5.12 and 5.14, the leading order balance is between the Coriolis terms and the pressure gradient terms. In these terms, provided the length scale L is small compared to the planetary radius a, the Coriolis parameter f can be approximated by a constant value f_0; the density ρ can also be approximated by its reference atmosphere value ρ_R. Then,

$$f_0 v = \frac{1}{\rho_R} \frac{1}{a \cos(\phi)} \frac{\partial p}{\partial \lambda}; \quad f_0 u = -\frac{1}{\rho_R} \frac{1}{a} \frac{\partial p}{\partial \phi} \qquad (5.16)$$

or, in vector notation,

$$\mathbf{v} = \frac{1}{\rho_R f_0} \mathbf{k} \times \nabla_{\mathbf{H}} p \qquad (5.17)$$

This result is called 'geostrophic balance', and it provides a direct relationship between the pressure field and the velocity field. In fact, it states that streamfunction and pressure are related by

$$\psi = \frac{p}{\rho_R f_0}$$

Under conditions of geostrophic balance, the flow is parallel to isobars, with low pressure to the left in the Northern Hemisphere (right in the Southern Hemisphere). The scaling for pressure, Δp, given by Equation 5.15 follows directly from geostrophic balance.

However, such a drastic simplification of the momentum equation as Equation 5.17 has thrown the baby out with the bath water. Equation 5.17 expresses a balance between the pressure and velocity fields, but it gives no way of calculating either. Indeed, the loss of the time derivative term means that we cannot predict the evolution of the flow at all at this level of approximation. We will return to this matter in Section 5.5, where account will also be taken of small yet crucial deviations from geostrophic balance. First, the next section elaborates the concept of balanced flow.

5.4 Natural coordinates, geostrophic and gradient wind balance

Geostrophic balance is the simplest balance condition. A parcel of air in exact geostrophic balance with the local pressure field would move at a constant speed in a straight line. If the parcel is accelerating in some way, for example, if it is moving along a curved trajectory, then the balance condition becomes more complicated. A straightforward way of exploring this is to rewrite the Euler equations in so-called natural coordinates, which are illustrated in Figure 5.1.

We use the momentum equation on an f- or beta- plane, Equation 4.38, and denote the local horizontal direction of the flow by a unit vector $\mathbf{s}$ and distance along the trajectory by s. Similarly, $\mathbf{n}$ is a unit vector perpendicular to the horizontal trajectory rotated anticlockwise from $\mathbf{s}$, while n denotes distance in the $\mathbf{n}$ direction. Then, the

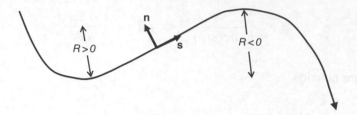

Figure 5.1 Illustrating natural coordinates. The thick curve represents the trajectory of a fluid parcel

horizontal component of the flow velocity is $\mathbf{v} = V\mathbf{s}$, where V is the speed and is always positive or zero. Since the Coriolis force always acts at right angles to the velocity vector, it plays no part in the equation of motion in the $\mathbf{s}$ direction, which is simply

$$\frac{DV}{Dt} = -\frac{1}{\rho}\frac{\partial p}{\partial s} \qquad (5.18)$$

This component of the Euler equation shows that the flow speed can only change in response to changes in pressure in the direction of motion.

The equation of motion in the $\mathbf{n}$-direction is

$$\frac{V^2}{R} + fV = -\frac{1}{\rho}\frac{\partial p}{\partial n} \qquad (5.19)$$

Here, R is the radius of curvature of the trajectory. The first term on the left-hand side is the centripetal acceleration associated with the fluid parcel moving in a curved trajectory. The rate of change of the direction of motion is

$$\frac{D\mathbf{s}}{Dt} = \frac{V}{R}\mathbf{n} \qquad (5.20)$$

so that the radius of curvature R must be taken as positive when the trajectory is curved in an anticlockwise direction, that is, cyclonically in the Northern Hemisphere, and as negative when it is curved clockwise, that is, anticyclonically in the Northern Hemisphere.

If we define V_g to be the component of the geostrophic wind in the direction of motion, then the pressure gradient force term in Equation 5.19 may be written as fV_g and the equation itself as

$$\frac{V^2}{R} + fV - fV_g = 0 \qquad (5.21)$$

From Equation 5.20, the equation of motion in the **n**-direction in the form Equation 5.19 or 5.21 is actually an equation for the curvature of the trajectory R and the material rate of change of **s**. Equation 5.21 can be rearranged to give

$$\frac{1}{R} = \frac{f(V_g - V)}{V^2} \tag{5.22}$$

Quite generally, it follows that the curvature of the trajectory is cyclonic, that is, $R > 0$, when $V < V_g$, that is, the flow is sub-geostrophic. Conversely, when $V > V_g$, that is, when the flow is super-geostrophic, trajectories are anticyclonically curved.

Equation 5.21 represents a balance of forces in the negative **n**-direction acting on a fluid parcel: the first term is the centrifugal force, the second term is the Coriolis force and the third is the pressure gradient force. Dividing this equation by the Coriolis force fV leads to

$$\frac{V}{fR} + 1 - \frac{V_g}{V} = 0$$

or   (5.23)

$$y + 1 - x = 0$$

Here, the first term, denoted y, is the rate of change of the direction of motion normalized by f, and the third term, denoted x, measures the relative strength of the component of the geostrophic wind in the direction of motion to the actual flow speed. Figure 5.2 shows the line described by Equation 5.23. When y is positive, the trajectory curvature is cyclonic, and when it is negative, the curvature is anticyclonic. When x is greater than 1, the flow is sub-geostrophic; between 0 and 1, it is super-geostrophic; and when x is less than 0, it may be called anti-geostrophic.

Near point A, where $x = 1$ and $y = 0$, the trajectories are nearly straight, $V \simeq V_g$ and the flow speed is close to the geostrophic wind speed in the direction of motion. The condition for this to be the case is that

$$\frac{V}{fR} \ll 1$$

The parameter V/fR is a Rossby number which uses the radius of curvature of the trajectory as a length scale. So geostrophic balance is a good approximation if $R \gg R_I = V/f$. For typical midlatitude weather systems, R_I is around 100 km. For an alternative perspective, Equation 5.20 shows that the condition is that the rate of change of the direction **s** for a fluid parcel must be much smaller than f. This means that the timescale for change in the direction of motion of a fluid parcel must be much longer than f^{-1}, which is around 3 hour in middle latitudes.

In Figure 5.2, moving from point A to larger x, the flow becomes sub-geostrophic. At point B, $x = 2$ and $y = 1$ and so $R = R_1$. Here the centrifugal and Coriolis forces are equal and together balance the pressure gradient force. At very large x, the Coriolis force is negligible compared with the centrifugal force. So the centrifugal force must balance the pressure gradient force alone. The flow is then said to be in 'cyclos-trophic balance'. Cyclostrophic balance is a feature of a very tight vortex such as a tornado. It also characterizes the large-scale flow on a slowly rotating planet such as Venus.

Moving from point A to smaller values of x, the flow becomes super-geostrophic and anticyclonic. At point C, where $x = 0.5$ and $y = -0.5$, $R = -2R_1$, the centrifugal and pressure gradient forces are equal and together they balance the Coriolis force. At point D, where $x = 0$ and $y = -1$, the pressure gradient force is 0 and the Coriolis and centrifugal forces must balance. In this case, the anticyclonic trajectory has the radius of curvature $R = -R_1$. When x is negative, the flow is anti-geostrophic, that is, the trajectories have anticyclonic curvature around a low-pressure centre. Although this configuration represents an equilibrium balance of forces, it would be highly unstable and is never observed.

In the aforementioned analysis, V_g is the component of the geostrophic wind in the direction of the parcel motion. If the geostrophic wind is nearly parallel to the parcel motion, or, equivalently, if the fluid parcel moves nearly parallel to isobars, then V_g is close to the magnitude of the total geostrophic wind. The balance

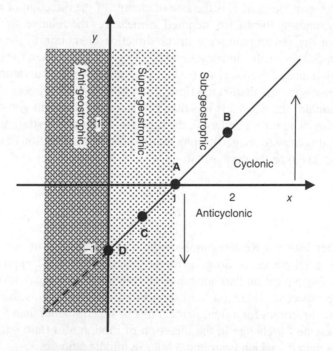

Figure 5.2   A plot of V_g/fR versus V/V_g

represented by Equation 5.22 is then called 'gradient wind balance'. The condition for this is that $|\partial p/\partial s|$ should be much smaller than $|\partial p/\partial n|$. Scaling these terms from the equations in the **s**- and **n**-directions, Equations 5.18 and 5.19 respectively, suggests that $|DV/Dt|$ must be much smaller than $|fV|$. Therefore, the condition for gradient wind balance is that the timescale of the change in speed following the motion must be much longer than f^{-1}.

The condition for the flow to be approximately equal to the component of the geostrophic wind in the direction of the motion discussed earlier is that the direction of movement of a parcel must also not change much in this time. Thus, the wind is approximately equal to the vector geostrophic wind if the Lagrangian timescale for change in both the speed and direction of the motion is much longer than f^{-1}.

A special case of gradient wind balance is when the analysis refers to steady flow around a circle of radius R. Then, the flow is parallel to the isobars, and so the gradient wind balance condition V_g nearly parallel to V is exactly satisfied. Many authors then treat Equation 5.21 as a quadratic equation for V given R and V_g. However, the approach outlined here, in which Equation 5.21 leads to R given V and V_g, is more consistent with the physical basis for the equation. For circular motion, the particular case when V_g is zero (point D) is anticyclonic motion around circles of radius R_I and is referred to as motion in inertial circles.

This discussion illustrates the point that geophysical flows are never exactly in geostrophic balance if air parcels experience any acceleration. If the Coriolis and pressure gradient accelerations balanced exactly, and friction were negligible, then no net force would act on air parcels. They would simply move in straight lines at a constant speed equal to the geostrophic speed. Whenever air parcels are accelerating, either because their speed changes or because their direction of motion changes, there must be a degree of imbalance between the Coriolis and pressure gradient forces acting, and the parcel will move with a velocity different from the geostrophic velocity. In the examples of gradient wind balance just discussed, cyclonically curved trajectories and a speed less than geostrophic are associated. In many important circumstances, the difference between the actual flow velocity and the geostrophic velocity is small. Nevertheless, this small difference is crucial in determining the evolution of the flow.

5.5 Vertical motion

In Section 5.3, we remarked that neglecting all $O(\text{Ro})$ terms in the momentum equation led to a diagnostic set of equations which gave no information about the flow evolution. In order to recover the central representation of flow evolution, we return to the geostrophic relationship, Equation 5.17. While this will generally give a fair approximation to the observed wind, at least outside the atmospheric boundary layer, it will rarely be exact. Instead, use Equation 5.17 to define an ideal 'geostrophic wind', close to but not equal to the actual wind:

$$\mathbf{v}_g = \frac{1}{\rho_R f_0} \mathbf{k} \times \nabla_H p \qquad (5.24)$$

The vector difference between the actual wind and the geostrophic wind is called the 'ageostrophic' wind:

$$\mathbf{v}_a = \mathbf{v} - \mathbf{v}_g$$

In the midlatitude troposphere, the magnitude of the geostrophic wind is typically 10–20 m s^{-1}, possibly more in upper tropospheric jet streams, while the magnitude of the ageostrophic wind is an order of magnitude smaller, often only 1–2 m s^{-1}, which is comparable to the measurement uncertainty in the wind observations. Take the f-plane horizontal momentum equation as in Equation 4.38; using Equation 5.24 to substitute for the pressure gradient gives

$$\frac{D\mathbf{v}}{Dt} = f_0 \mathbf{k} \times \mathbf{v}_g - f_0 \mathbf{k} \times \mathbf{v} = -f_0 \mathbf{k} \times \mathbf{v}_a \qquad (5.25)$$

This remarkable result says that there is a direct link between the ageostrophic part of the wind and the evolution or development of the flow. If a typical magnitude of the ageostrophic wind is denoted by U_a, then scale analysis of Equation 5.25 leads to

$$\frac{U^2}{L} \sim f_0 U_a, \quad \text{i.e.,} \, U_a \sim U\left(\frac{U}{f_0 L}\right) \sim \mathrm{Ro} U \qquad (5.26)$$

So here is another interpretation of the Rossby number. It is the ratio of the magnitude of the ageostrophic wind to the magnitude of the total wind.

Now the geostrophic wind is non-divergent. This is easily demonstrated by taking the horizontal divergence of Equation 5.24:

$$\nabla_H \cdot \mathbf{v}_g = 0 \qquad (5.27)$$

Then, the continuity equation for an incompressible fluid, Equation 5.2, can be written in terms of the ageostrophic wind and the vertical motion:

$$\nabla_H \cdot \mathbf{v}_a = -\frac{\partial w}{\partial z} \qquad (5.28)$$

Not only is the evolution of the flow directly related to the ageostrophic flow, but the vertical motion is also a proxy for the ageostrophic motion. It follows that vertical motion, or rather more precisely the rate of change of vertical motion in the vertical direction, is directly related to the evolution of the horizontal component of flow. The actual mechanism responsible is made clear by a consideration of vorticity dynamics, the subject of Chapter 8. But for now, a simple scaling analysis of Equation 5.28 leads to an estimate of the typical magnitude W of the vertical motion:

$$W \sim \left(\frac{D}{L}\right)\mathrm{Ro}U \qquad (5.29)$$

This analysis reveals that the estimate of vertical motion in Equation 5.4 is really no more than an upper bound on the vertical motion. In general, W is considerably less than the aspect ratio multiplied by the horizontal flow speed. In the limit $\mathrm{Ro} \to 0$, Equation 5.29 reduces to the Taylor–Proudman result that vertical velocity is suppressed in a rapidly rotating system.

5.6 The vertical momentum equation

The vertical form of the momentum equation using spherical coordinates and incorporating the shallow atmosphere approximation was given in Equations 4.34 and may be written as

$$\frac{Dw}{Dt} = -g - \frac{1}{\rho}\frac{\partial p}{\partial z}$$

where the viscous term has been ignored for the present. The centrifugal force has been absorbed into the effective gravitational acceleration. Setting $p = p_R + p'$ in the final term and using the hydrostatic relation, Equation 5.5, the last two terms can be rewritten as

$$-g - \frac{1}{\rho}\frac{\partial p}{\partial z} = -g\frac{\rho'}{\rho} - \frac{1}{\rho}\frac{\partial p'}{\partial z}$$

The characteristic vertical velocity is given by Equation 5.29. Then, a term by term scale analysis gives

$$\frac{Dw}{Dt} = -g\frac{\rho'}{\rho} - \frac{1}{\rho}\frac{\partial p'}{\partial z}$$

$$(1) \qquad (2) \qquad (3) \qquad\qquad (5.30)$$

$$\frac{UW}{L} \qquad g\frac{\Delta\rho}{\bar{\rho}_R} \qquad \frac{\Delta p}{\bar{\rho}_R D}$$

We already have the scaling for W and Δp. If these are included, then the terms scale as

$$
\begin{array}{ccc}
(1) & (2) & (3) \\[6pt]
\mathrm{Ro}\!\left(\dfrac{U^2}{L^2}\right)D & g\dfrac{\Delta\rho}{\rho} & \dfrac{fUL}{D} \\[14pt]
\mathrm{Ro}^2\!\left(\dfrac{D^2}{L^2}\right) & \left(\dfrac{\Delta\rho}{\rho}\right)\dfrac{gD}{fUL} & 1
\end{array}
$$

The last line compares the magnitude of each of the terms to the magnitude of term (3), the pressure gradient term. Term (1) is extremely small, and so in all but the most extreme circumstances, term (3) must balance term (2). To a good approximation, then, the vertical component of the momentum equation can be reduced to the diagnostic relationship:

$$
\frac{\partial p'}{\partial z} = -\rho' g \tag{5.31}
$$

Thus, the hydrostatic relationship also applies to the deviation from the reference atmosphere as well as to the reference atmosphere itself. The hydrostatic relationship is equivalent to the statement that the pressure at any point in the fluid is equal to the weight of overlying fluid. So, at the Earth's surface, with a pressure of around $10^5\,\mathrm{Pa}$, the mass of gas overlying each square metre is p_0/g, which is around $10\,\mathrm{t\,m^{-2}}$. As will be discussed later, hydrostatic balance can become inaccurate when vertical accelerations become comparable with g. From the scale analysis, this requires D to be comparable to L. But for the examples discussed in this book, hydrostatic balance is a very good approximation.

The balance between terms (3) and (4) in Equation 5.30 gives a scaling for the density fluctuations:

$$
\frac{\Delta\rho}{\bar{\rho}_R} \sim \frac{fUL}{gD} \equiv \mathrm{Ro}^{-1}\mathrm{Fr} \tag{5.32}
$$

Here, the dimensionless number $\mathrm{Fr}=U^2/gD$ is called the Froude number. One interpretation of the Froude number is that it is the ratio of the kinetic energy of fluid parcels to the change of potential energy over the depth of the system. A typical value of Fr for a midlatitude weather system is 2×10^{-3}, so that $\Delta\rho\,/\,\bar{\rho}_R \sim 10^{-2}$. Also, it follows directly from the scaling of terms (4) and (5) in Equation 5.30 that the ratio

$$
\frac{\left(\Delta p\,/\,\bar{p}_R\right)}{\left(\Delta\rho\,/\,\bar{\rho}_R\right)} \sim \frac{\bar{\rho}_R gD}{\bar{p}_R} \sim \frac{D}{\bar{H}}
$$

This number is close to, but is probably rather less than, 1. Consistent with this and Equation 5.11, then

$$\frac{\Delta\theta}{\bar{\theta}_R} \sim \frac{\Delta T}{\bar{T}_R} \sim \frac{\Delta\rho}{\bar{\rho}_R} \sim \mathrm{Ro}^{-1}\mathrm{Fr} \tag{5.33}$$

It follows that the typical temperature or potential temperature fluctuation across a synoptic weather system will be of the order of 3 K.

The Lagrangian pressure fluctuation was mentioned in Section 5.4. We are now in a position to estimate a scale for this. The pressure changes on an individual fluid element are dominated by vertical motions. To a first approximation,

$$\frac{Dp}{Dt} \simeq -w\frac{dp_R}{dz} = \rho_R g w$$

So, assuming the pressure acting on the parcel varies on the usual advective time-scale L/U, the scaling for Lagrangian pressure fluctuations is

$$\Delta p_L \sim \rho_R g \frac{W}{U} L$$

Comparing this with the Eulerian pressure fluctuation Δp, Equation 5.15, it follows that

$$\frac{\Delta p_L}{\Delta p} \sim \mathrm{Ro}^2 \mathrm{Fr}^{-1}$$

For a midlatitude synoptic-scale weather system, this ratio is around 20, confirming the dominance of vertical motion in changing the pressure of individual fluid elements.

5.7 The mass continuity equation

For compressible flow, the continuity equation has the following form:

$$\frac{1}{\rho}\frac{D\rho}{Dt} = -\nabla \cdot \mathbf{u}$$

which, on substituting $\rho = \rho_r + \rho'$, can be written as

$$\frac{1}{\rho}\frac{D\rho'}{Dt} + \frac{w}{\rho}\frac{d\rho_R}{dz} = -\nabla \cdot \mathbf{v} - \frac{\partial w}{\partial z}$$

$$\begin{array}{cccc} (1) & (2) & (3) & (4) \end{array}$$

$$\frac{U\Delta\rho}{L\bar{\rho}_r} \quad RoU\frac{D}{L}\frac{1}{H} \quad Ro\frac{U}{L} \quad Ro\frac{U}{L} \tag{5.34}$$

$$Ro^{-2}Fr \qquad \frac{D}{H} \qquad 1 \qquad 1$$

Here, Equation 5.32 gives the scaling for $\Delta\rho / \bar{\rho}_R$, and Equation 5.29 gives the scaling for W. For a midlatitude synoptic-scale weather system, term (1) is typically of order 0.05 and so is small compared with terms (3) and (4). Term (2) is negligible for a shallow system, but for a deep system could be almost as large as terms (3) and (4). Terms (2) and (3) will generally be comparable. In term (2), the denominator can be approximated by ρ_R. Then, the continuity equation may have the following form:

$$\nabla \cdot \mathbf{v} + \frac{\partial w}{\partial z} + \frac{w}{\rho_R}\left(\frac{d\rho_r}{dz}\right) = 0, \quad \text{that is, } \nabla \cdot \left(\rho_R \mathbf{u}\right) = 0 \tag{5.35}$$

This form of the continuity equation is the so-called anelastic approximation. For a shallow system, the motion is approximately non-divergent, so then the continuity equation for incompressible flow $\nabla \cdot \mathbf{u} = 0$ is adequate. Chapter 7 will contain further discussion of the forms of the continuity equation for a stratified atmosphere.

5.8   The thermodynamic energy equation

The analysis of the thermodynamic equation hinges upon an appropriate scaling for temperature or potential temperature fluctuations associated with a weather system. Such a scaling was deduced in Section 5.4 and given by Equation 5.33. Typical midlatitude values suggest that the horizontal fluctuations of potential temperature in a synoptic-scale weather system will be around 3 K.

Figure 1.2 is a useful preliminary to discussing the scaling of the thermodynamic equation. It shows the seasonal and zonal mean potential temperature, θ, for the December–January–February season. Since, to a first approximation, motion in the atmosphere is adiabatic for short periods, this picture gives a good idea of the mean state about which the atmosphere is continually fluctuating. Note first that the troposphere is stably stratified, with θ increasing upwards. The stratosphere is much more stably stratified, with potential temperature increasing upwards very rapidly. Secondly, at all tropospheric levels, the potential temperature is lower at the pole than in the tropics. As a result, isentropes slope downwards towards the tropics, with a typical gradient in the midlatitudes of around 10^{-3}. A typical potential temperature difference between the pole and the equator at any level is around $40\,\mathrm{K}$. The typical potential temperature difference between the surface and the tropopause in the midlatitudes is rather similar, around $50\,\mathrm{K}$. The change of potential temperature in the vertical direction describes the stratification of the atmosphere, about which more will be said in Section 7.3. The stratification is conveniently measured by a parameter N, the Brunt–Väisälä frequency, defined from

$$N^2 = \frac{g}{\theta}\frac{\partial\theta}{\partial z} \tag{5.36}$$

If H denotes the depth of the troposphere, and θ_0 a depth average potential temperature, then a typical value of N^2 is

$$N^2 \sim \frac{g}{\theta_0}\frac{\Delta\theta_V}{H}$$

Substituting in representative values, a typical N^2 is around $10^{-4}\,\mathrm{s}^{-2}$.

Setting $\theta = \theta_R + \theta'$, the thermodynamic equation may be rewritten as

$$\frac{\partial\theta'}{\partial t} + \mathbf{v}\cdot\nabla\theta' + w\frac{\partial\theta'}{\partial z} + w\frac{d\theta_R}{dz} = S \tag{5.37}$$

A scale analysis gives for the following various terms:

$$
\begin{array}{ccccc}
\dfrac{\partial\theta'}{\partial t} & +\mathbf{v}\cdot\nabla\theta' + w\dfrac{\partial\theta'}{\partial z} & +w\dfrac{d\theta_R}{dz} = & S \\[2ex]
\dfrac{U\Delta\theta}{L} & \dfrac{U\Delta\theta}{L} & \dfrac{W\Delta\theta}{D} & W\dfrac{\theta_0 N^2}{g} & S \\[2ex]
1 & 1 & \mathrm{Ro} & \dfrac{N^2 D^2}{f^2 L^2}\dfrac{S}{f\overline{\theta}_R}\left(\dfrac{gD}{U^2}\right) \\[2ex]
(1) & (2) & (3) & (4) & (5)
\end{array}
\tag{5.38}
$$

In the last line, the scale of each term has been divided by term (1) to express them in dimensionless form. The scaling of term (4) is a combination which will appear many times in subsequent discussion and is called the Burger number, denoted by Bu:

$$\mathrm{Bu} = \frac{N^2 D^2}{f^2 L^2} \tag{5.39}$$

From Equation 5.38, a physical interpretation of this number is that it is the ratio of the vertical to horizontal advection of θ. The result of the scaling of the thermodynamic equation is less simple than that of the earlier equations. For midlatitude synoptic-scale weather systems, $Bu \sim 1$, although the dependence of Bu on the squares of the horizontal and vertical scales means that it can deviate quite substantially from 1 in particular cases. Terms (1), (2) and (3) in Equation 5.38 are therefore generally comparable in magnitude, and so the balance of terms on the left-hand side of the equation cannot be simplified further. The discussion can be turned around to say that in the absence of heating, if θ is not just to be advected horizontally, then Bu must be of order 1. Equivalently, this implies that the ratio of vertical to horizontal length scales is

$$\frac{D}{L} \sim \frac{f}{N} \sim 0.01$$

This 'natural' ratio of scales is characteristic of midlatitude weather systems.

In the clear troposphere, a typical value of S, largely due to infrared radiative cooling, is $1\text{--}2\,\mathrm{K\,day^{-1}}$. The non-dimensional scaling for term (5) can be written as $S/(f\theta_R \mathrm{Fr})$. With typical midlatitude synoptic-scale values for the other parameters in the scaling of term (5), the typical magnitude of term (5) turns out to be around 0.3, smaller than terms (1–3), but not much smaller.

The tropics are rather different. There, with the reduced value of the Coriolis parameter, the scale analysis shown in Equation 5.38 is not appropriate. But as Figure 5.1 shows, the isentropes are extremely flat in the tropics, and the horizontal Eulerian variations of θ are therefore at least an order of magnitude smaller than in the midlatitudes. The vertical stratification is similar in magnitude to its midlatitude value. Repeating the scale analysis with $\Delta\theta$ of 0.5 K, we now find that terms (1) and (2) are an order of magnitude smaller than terms (4) and (5), which must approximately balance. It follows that in the tropics, the thermodynamic equation can be written, to a first degree of approximation, as

$$w\frac{\theta_R N^2}{g} \simeq S \text{ or } w = \frac{gS}{\theta_R N^2} \tag{5.40}$$

That is, the thermodynamic equation becomes simply a diagnostic equation for the vertical velocity. Within the deep tropics, $\partial\theta/\partial z$ is closely related to the saturated adiabatic lapse rate, which in turn depends principally on the level in the troposphere and the surface temperature.

The scaling analysis of the thermodynamic equation given by Equation 5.38 has split the vertical advection term into contributions from the advection of deviations from the reference atmosphere, term (3), and from the advection of the reference atmosphere, term (4). The ratio of terms (3–4) scales as RoBu. So, if $Bu \sim 1$ and Ro is small, the thermodynamic equation can be approximated by neglecting the impact of the perturbation on the stratification. These conditions just about hold through the

Table 5.2 Various parameters and variables emerging from the scale analysis of this chapter, together with characteristic estimates for a midlatitude weather system

Quantity	Midlatitude Estimate
Rossby number Ro $= U/fL$	0.2
Froude number Fr $= U^2/gD$	2×10^{-3}
Burger number Bu $= N^2 D^2/f^2 L^2$	1
Aspect ratio D/L	10^{-2}
Vertical velocity scale $W \sim \text{Ro} \left(\dfrac{D}{L} \right) U$	$0.02\,\text{m s}^{-1}$
Horizontal pressure fluctuation (Eulerian) $\Delta p \sim \rho f U L$	$10^3\,\text{Pa}$
Potential temperature fluctuation $\Delta \theta \sim \dfrac{fUL}{gD} \theta_0$	$3\,\text{K}$

Earth's troposphere. But dropping the term $w(\partial\theta'/\partial z)$ is not always a particularly good approximation. Nevertheless, rewriting the thermodynamic equation in the form

$$\frac{\partial \theta}{\partial t} + \mathbf{v} \cdot \nabla \theta + w \frac{d\theta_r}{dz} = S \tag{5.41}$$

is a basic part of what will be referred to as the 'quasi-geostrophic' approximation. This equation set is widely used for analytical developments even if it is too inexact an equation set for detailed numerical work. We shall explore the implications of Equation 5.41 further in Chapter 12.

To summarize the results of this chapter, Table 5.2 sets out a number of parameters and flow variables, of which our scale analysis gives estimates. Also shown are values appropriate to a midlatitude weather system. The reader may like to repeat these estimates for different circumstances, such as flow in an oceanic gyre or eddies in a rotating laboratory apparatus. Subsequent chapters will make regular reference to the summary in Table 5.2.

5.9   Scalings for Rossby numbers that are not small

For horizontal length scales significantly smaller than the synoptic-scale examples which have formed the basis of the discussions in the preceding sections, the Rossby number may be of order 1 or larger. In this case, the metric terms in the momentum Equations 5.12 and 5.14 are again negligible, but now the pressure gradient term will be of the order of the momentum advection, term (1), so that

$$\Delta p \sim \bar{\rho}_R U^2 \tag{5.42}$$

The scaling of the vertical momentum equation, Equation 5.30, when the Rossby number is of order 1 suggests that the acceleration term be of order (D^2/L^2), the square of the aspect ratio, compared with the perturbation pressure gradient term. The validity of the hydrostatic approximation depends upon this aspect ratio being small. It will not be valid for phenomena such as deep convection or flow over steep mountains, in which the vertical and horizontal scales are comparable. However, the buoyancy term, term (4), and the pressure gradient term, term (5), will still be of comparable magnitude. This leads to the scaling

$$\frac{\Delta\rho}{\bar{\rho}_R} \sim \frac{U^2}{gD} \equiv \mathrm{Fr} \qquad (5.43)$$

For a phenomenon with $U \sim 5\,\mathrm{m\,s^{-1}}$ and $D \sim 1\,\mathrm{km}$, $\mathrm{Fr} \sim 2.5 \times 10^{-3}$. As before, the ratio

$$\frac{\delta p/\bar{p}_R}{\delta\rho/\bar{\rho}_R} \sim \bar{\rho}_R \frac{gD}{\bar{p}_R} \equiv \frac{D}{\bar{H}_R}$$

So, for systems with vertical scale much smaller than $7\,\mathrm{km}$, the $\delta p/\bar{p}_R$ term in the equation of state, Equation 5.9, is small and so to a first approximation,

$$\frac{\delta\rho}{\rho_R} \sim \frac{\delta T}{T_R}$$

In the continuity equation, Equation 5.34, the first term on the left-hand side, term (1), scales as Fr compared with the remaining terms. So the anelastic approximation is again applicable, reducing to incompressibility for a shallow system.

Finally, consider the thermodynamic equation when $\mathrm{Ro} \sim 1$. The ratio of the vertical to horizontal advection scales as

$$\mathrm{Ri} = \frac{N^2 D^2}{U^2} \qquad (5.44)$$

where Ri is the Richardson number. For midlatitude scaling, it has a value around 25. However, in the atmospheric boundary layer, where a wind shear of $5\,\mathrm{m\,s^{-1}}$ over a depth of $500\,\mathrm{m}$ would be typical, it is around 1. The Richardson number can be written in terms of the Rossby and Burger numbers:

$$\mathrm{Ri} = \mathrm{Ro}^{-2}\mathrm{Bu} \qquad (5.45)$$

Therefore, small Rossby number and Burger number of order 1 imply a very large Richardson number. An alternative scaling of the horizontal momentum equations reaches geostrophic balance on the equivalent basis of large Richardson number rather than the small Rossby number basis used here.

6

Alternative vertical coordinates

6.1 A general vertical coordinate

The basic equations of atmospheric motion derived in the previous chapters use geometrical height z as vertical coordinate. While this might seem a natural and obvious choice, it leads to a number of undesirable features. Among them are:

1. The pressure gradient force term is nonlinear, involving the product of $1/\rho$ and ∇p.

2. The mass conservation equation is complicated if the flow extends over such a range of heights that the mean density at each level changes significantly.

3. All three components of velocity turn out to be equally important.

4. In general, the lower boundary is not a coordinate surface. This makes it particularly difficult to apply numerical methods to the lower boundary.

In this chapter and the next, we consider some ways of avoiding these problems. One way, which is the subject of this chapter, is to use variables other than z as the vertical coordinate. In the next chapter, we shall consider ways of retaining z as vertical coordinate while linearizing the pressure gradient term. Any variable which varies monotonically with height may be used to relabel the vertical axis. Such a general variable is denoted ξ; it may either increase or decrease with height, and in Figure 6.1, $\delta\xi$ may be either positive or negative.

In what follows, we shall neglect the metric terms and simply use local Cartesian coordinates. The adjustments that will be introduced all carry over straightforwardly to the spherical case. The hydrostatic approximation is made. This is not strictly necessary, but many of the advantages of the coordinate transform are lost if the hydrostatic approximation is relaxed.

Fluid Dynamics of the Midlatitude Atmosphere, First Edition. Brian J. Hoskins & Ian N. James.
© 2014 John Wiley & Sons, Ltd. Published 2014 by John Wiley & Sons, Ltd.

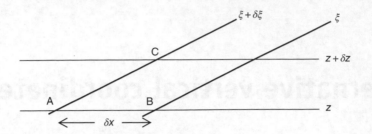

Figure 6.1   A general vertical coordinate

From the definition of geopotential,

$$g\delta z = \delta\Phi \tag{6.1}$$

and so the hydrostatic relation may be written as

$$\delta\Phi = -\frac{1}{\rho}\delta p \tag{6.2}$$

or, in terms of the new vertical coordinate ξ,

$$\frac{\partial\Phi}{\partial\xi} = -\frac{1}{\rho}\frac{\partial p}{\partial\xi} \tag{6.3}$$

The surfaces of constant ξ in Figure 6.1 are tilted with respect to level surfaces on which z is a constant. Their slope is $-\delta z/\delta x$, and this must be allowed for when transforming the gradient of pressure on a horizontal surface into the gradient of pressure along a $\xi = $ constant surface. Referring to Figure 6.1,

$$p_C - p_A = (p_B - p_A) + (p_C - p_B)$$

and so, in terms of the pressure gradient term:

$$\frac{1}{\rho}\frac{\partial p}{\partial x}\bigg|_\xi = \frac{1}{\rho}\frac{\partial p}{\partial x}\bigg|_x + \frac{1}{\rho}\frac{\partial p}{\partial z}\bigg|_x\frac{\partial z}{\partial x}\bigg|_\xi$$

Using the hydrostatic relation, this may be written as

$$\frac{1}{\rho}\frac{\partial p}{\partial x}\bigg|_z = \frac{1}{\rho}\frac{\partial p}{\partial x}\bigg|_\xi + \frac{\partial\Phi}{\partial x}\bigg|_\xi \tag{6.4}$$

A similar result holds in the y-direction. Notice that if $\xi = \xi(p)$ only, the first term on the right-hand side of Equation 6.4 is 0, and we are left simply with the linear term $\partial\Phi/\partial x$.

In height coordinates, the mass of a fluid element is $\delta m = \rho \delta x \delta y \delta z$. In the new general coordinate, this will become $r \delta x \delta y \delta \xi$, where the 'pseudo-density' r is defined so that $\rho \delta z = r \delta \xi$. Using the hydrostatic relation,

$$r = \rho \frac{\partial z}{\partial \xi} = -\frac{1}{g} \frac{\partial p}{\partial \xi} \qquad (6.5)$$

Following a fluid element, its pseudo-volume in a general vertical coordinate system $\delta \tau = \delta x \delta y \delta \xi$ will change as a result of velocity divergence in that coordinate system. We may write:

$$\frac{1}{\tau} \frac{D\tau}{Dt} = \nabla_\xi \cdot \mathbf{u} \qquad (6.6)$$

Conservation of mass $\delta m = r \delta \tau$ following the fluid element implies that:

$$\frac{1}{r} \frac{Dr}{Dt} + \frac{1}{\tau} \frac{D\tau}{Dt} = 0 \qquad (6.7)$$

Combining this with Equation 6.6, mass conservation becomes:

$$\frac{Dr}{Dt} + r \nabla_\xi \cdot \mathbf{u} = 0 \qquad (6.8)$$

Vertical velocity in the new coordinate system is replaced by the rate of change of ξ following the fluid motion. The new 'velocity' vector has components

$$\mathbf{u} = (u, v, \omega) \quad \text{where } \omega = \frac{D\xi}{Dt} = \frac{\rho}{r} w \qquad (6.9)$$

In terms of this new 'vertical velocity', the Lagrangian time derivative may be written as:

$$\frac{D}{Dt} = \frac{\partial}{\partial t} + u \frac{\partial}{\partial x}\Big|_\xi + v \frac{\partial}{\partial y}\Big|_\xi + \omega \frac{\partial}{\partial \xi} \qquad (6.10)$$

Introducing the notation

$$\mathbf{u} = \mathbf{v} + \omega \mathbf{k}, \quad \frac{D}{Dt} = \frac{\partial}{\partial t} + \mathbf{v} \cdot \nabla_H + \omega \frac{\partial}{\partial \xi}$$

provides a convenient separation between the horizontal and vertical components of flow and of advection. Note that we have introduced a relabelling of the vertical coordinate: $\mathbf{v}$ is the horizontal wind, but 'horizontal' derivatives are calculated keeping ξ constant.

The relationships derived in this section are sufficient to transform the various equations into the new vertical coordinate system. However, to specify the problem completely, boundary conditions, and in particular, a boundary condition at the planetary surface, must be derived. The surface of the planet is defined by its surface geopotential $\Phi^* = \Phi^*(x, y)$ or equivalently by the height of its surface above a reference geoid $z^* = z^*(x, y)$. If the surface is rigid and impenetrable to atmospheric flow, then fluid at the surface must share the same geopotential as the boundary:

$$\frac{D\Phi}{Dt} = \frac{D\Phi^*}{Dt} \quad \text{at } z = z^*$$

or since the planetary surface deforms only very slowly if at all,

$$\frac{D\Phi}{Dt} = \mathbf{v} \cdot \nabla \Phi^* \quad \text{at } z = z^* \tag{6.11}$$

If one writes $\Phi^* = \Phi_0 + gz^*$, then this boundary condition reduces to

$$\frac{D\Phi}{Dt} = gw^* \quad \text{at } z = z^* \tag{6.12}$$

Here, w^* is the vertical velocity of fluid elements at the surface, by virtue of their up- or downslope component of motion.

All the elements needed to transform the primitive equations from one vertical coordinate to another have now been assembled. They include the hydrostatic equation (Equation 6.3); the pressure gradient force (Equation 6.4); the mass conservation equation (Equation 6.8); the Lagrangian time derivative (Equation 6.10); and the lower boundary condition (Equation 6.11).

6.2   Isobaric coordinates

A popular and convenient choice for ξ is the pressure p. From the hydrostatic relationship in the form of Equation 4.9, it is clear that pressure must decrease monotonically with height, and so it makes a suitable alternative vertical coordinate. Note also from Equation 4.9 that the mass per unit area of a slab of atmosphere is related to the pressure drop across it. Integrating Equation 4.9 with respect to z gives:

$$m = \int_z^{z+\Delta z} \rho \, dz = -\frac{\Delta p}{g} \tag{6.13}$$

The minus sign results from the decrease of pressure with height, so that if Δz is positive, Δp must be negative.

From Equation 6.3, the hydrostatic equation in pressure coordinates is:

$$\frac{\partial \Phi}{\partial p} = -\frac{1}{\rho} \qquad (6.14)$$

Using the equation of state, Equation 2.17, and the definition of potential temperature, Equation 2.18, it is conveniently rewritten as:

$$\frac{\partial \Phi}{\partial p} = -\hat{R}(p)\theta \quad \text{where } \hat{R} = \frac{R}{p}\left(\frac{p}{p_0}\right)^{\kappa} \qquad (6.15)$$

Recall that p_0 is an arbitrary reference pressure, often but not necessarily taken to be 100 kPa.

The pressure gradient term takes a particularly simple form in pressure coordinates, since of course pressure does not vary on the coordinate surfaces. From Equation 6.4,

$$\frac{1}{\rho}\nabla_z p = \nabla_p \Phi \qquad (6.16)$$

This form of the pressure gradient term has an intuitive physical interpretation. Writing $\Phi = \Phi_0 + gz$, the pressure gradient force reduces to $-g\nabla_p z$, the component of the gravitational acceleration parallel to the sloping pressure surface.

From Equation 6.5, the pseudo-density is simply

$$r = -\frac{1}{g}$$

So the pseudo-density, and equivalently the pseudo-volume of a fluid element in pressure coordinates, is constant. The mass of a fluid element is proportional to its pseudo-volume, and so conservation of mass is simply a statement of conservation of volume in the pressure coordinates:

$$\nabla_p \cdot \mathbf{v} + \frac{\partial \omega}{\partial p} = 0 \qquad (6.17)$$

To summarize, the basic equation set in pressure coordinates is as follows:

$$\frac{D\mathbf{v}}{Dt} + f\mathbf{k} \times \mathbf{v} + \nabla\Phi = \mathbf{F}$$

$$\frac{\partial \Phi}{\partial p} = -\hat{R}\theta$$

$$\nabla \cdot \mathbf{v} + \frac{\partial \omega}{\partial p} = 0 \qquad (6.18)$$

$$\frac{D\theta}{Dt} = S$$

The use of pressure as a vertical coordinate has simplified and linearized the pressure gradient term. It has also simplified the continuity equation which even for a compressible fluid has the same simple form as for an incompressible fluid. However, these simplifications come at a cost. First, all three velocity components are equally important for the dynamics of a fluid element, including the virtually unmeasurable pressure vertical velocity ω. More seriously, the lower boundary condition has not been simplified. The lower boundary is no longer a coordinate surface, even for a flat boundary with no orography, and in general it varies with time.

The momentum equation in Equation 6.18 implies that the geostrophic wind is

$$\mathbf{v}_g = \frac{1}{f}\mathbf{k} \times \nabla \Phi \tag{6.19}$$

Taking a p derivative and eliminating Φ using the hydrostatic equation give the following thermal wind equation:

$$f\frac{\partial \mathbf{v}_g}{\partial p} = -\hat{R}\mathbf{k} \times \nabla \theta \tag{6.20}$$

The increase in geostrophic wind with p is in the horizontal direction of θ contours and proportional to the gradient of θ.

The lower boundary condition is obtained from Equation 6.12, using the hydrostatic equation in the form Equation 6.14:

$$\frac{\partial \Phi}{\partial t} + \mathbf{v} \cdot \nabla \Phi - \frac{\omega}{\rho} = \mathbf{v} \cdot \nabla \Phi^* \tag{6.21}$$

Note that each side of this equation is equal to wg at the surface. The vertical velocity at the surface must scale as $w \sim UH/L$ where H is the typical orography height. For high frequency, non-geostrophic motion, Equation 6.18, suggests

$$\frac{U}{T} \sim \frac{\Delta \Phi}{L}$$

Then, the ratio of the first term on the left-hand side of Equation 6.21 to the right-hand side is

$$\frac{LU}{T^2} \bigg/ g\frac{UH}{L} \sim \frac{L^2}{T^2}\bigg/(gH)$$

From this result, it is clear that the first term on the left-hand side of Equation 6.21 is only important for extremely fast motions, of order hundreds of metres per second.

The second term on the left-hand side is 0 for geostrophic motion. The ageostrophic velocity scales as URo, and consequently the ratio of this second term to wg is:

$$\max\left(\frac{U^2}{gH}, \text{Ro}\frac{f^2L^2}{gH}\right)$$

So the first two terms in Equation 6.21 are usually neglected. For a flat surface, the deviation Δp of the surface pressure from the mean surface pressure p_0 is generally sufficiently small that it suffices to apply the boundary condition at $p = p_0$. So for a flat surface, the lower boundary condition can often be approximated reasonably by

$$\omega = 0 \ \text{ at } p = p_0 \tag{6.22}$$

The other aspect of pressure coordinates that can be more difficult to deal with in pressure than in height coordinates, particularly in analytical work, is the background vertical gradient of potential temperature. Whereas $\partial\theta_R/\partial z$ is roughly uniform in the troposphere, $\partial\theta_R/\partial p$ increases with decreasing p, varying approximately as ρ^{-1}.

6.3   Other pressure-based vertical coordinates

A closely similar vertical coordinate is based upon the logarithm of pressure:

$$Z = -H\ln\left(\frac{p}{p_0}\right) \quad \text{where } H = \frac{RT_0}{g} \tag{6.23}$$

Here T_0 is some constant reference temperature and p_0 some constant reference pressure. The latter is often but arbitrarily taken as $100\,\text{kPa}$. For an isothermal atmosphere of temperature T_0, Z would be the geometric height. In the real atmosphere, in which temperature varies with position, it is similar to, but not identical with, the geometric height. The use of $\ln(p)$ as a vertical coordinate combines the intuitive appeal of height coordinates while retaining the advantages of p-coordinates in simplifying the pressure gradient term. The vertical coordinate (6.23) is often used in middle atmosphere studies. Since the atmosphere is more nearly isothermal in the stratosphere than in the troposphere, the coordinate is more height-like in the stratosphere.

From the equation of state, Equation 2.17, it is easily shown that the hydrostatic relation becomes:

$$\frac{\partial\Phi}{\partial Z} = g\frac{T}{T_0}$$

However, the pseudo-density is:

$$r = \left(\frac{p_0}{gH}\right)e^{-Z/H}$$

and so the mass conservation equation may be written as

$$\nabla \cdot \mathbf{v} + \frac{\partial W}{\partial Z} - \frac{W}{H} = 0$$

Here, $W = DZ/Dt$ denotes vertical velocity in this coordinate. The extra complexity compared with p-coordinates is offset by the vertical gradient of θ_R being nearly uniform because of the height-like nature of $\log(p)$ coordinates. The geostrophic velocity has the same form as in p-coordinates, and from the hydrostatic relation, the thermal wind equation is

$$f\frac{\partial \mathbf{v}_g}{\partial Z} = -\frac{g}{T_0}\mathbf{k} \times \nabla T \tag{6.24}$$

Another particularly useful height-like, but pressure based, vertical coordinate is defined to be

$$z_a = \frac{c_p\theta_0}{g}\left[1 - \left(\frac{p}{p_0}\right)^{\kappa}\right] \tag{6.25}$$

Since this coordinate is a simple function of pressure, the pressure gradient term will have the same form as for simple pressure coordinates, Equation 6.16. An increment in z_a is related to an increment in pressure p by:

$$\delta z_a = -\frac{R\theta_0}{g}\left(\frac{p}{p_0}\right)^{\kappa}\frac{1}{p}\delta p = -\frac{\theta_0}{g\theta\rho}\delta p \tag{6.26}$$

From this result, the hydrostatic equation in the new coordinate system can be obtained as

$$\frac{\partial \Phi}{\partial z_a} = \frac{g}{\theta_0}\theta \tag{6.27}$$

from which it follows that

$$\frac{\partial z_a}{\partial z} = \frac{\theta}{\theta_0} \tag{6.28}$$

Consider an adiabatic atmospheric temperature profile, for which $\theta=\theta_0$ at all levels. Then, $z_a=0$ for $p=p_0$; at other levels, $z_a=z$. So z_a becomes identical with geometrical height for such an adiabatic atmosphere. For the real stratified troposphere, z_a differs from z, but not by much. For example, throughout the Earth's troposphere, z_a differs from z by less than $1\,km$. The analogue of density is:

$$r = \rho \frac{\partial z}{\partial z_a} = \rho \frac{\theta}{\theta_0} = \rho_0 \left(\frac{p}{p_0} \right)^{1-\kappa}$$

$$= \rho_0 \left(1 - \frac{gz_a}{c_p \theta_0} \right)^{\frac{1}{\kappa}-1}$$

(6.29)

Since this is a function of z_a only, the mass conservation relation, Equation 6.17, can be reduced to:

$$\nabla \cdot (r\mathbf{u}) = 0$$

(6.30)

Again the geostrophic wind has the same form as in p-coordinates, and from the hydrostatic relation, Equation 6.25, the thermal wind relation is

$$f \frac{\partial \mathbf{v}_g}{\partial z_a} = -\frac{g}{\theta_0} \mathbf{k} \times \nabla \theta$$

(6.31)

Using these results, the choice of z_a as vertical coordinate leads to equations which are identical in form to approximate height coordinate versions discussed in Chapter 7. The coordinate itself is similar to height, z, and the static stability measure, $(g/\theta_0)d\theta_R/dz_a$, is approximately uniform with height. It has a simpler form for the hydrostatic relation than other coordinates, having merely a constant times θ on the right-hand side. The height-like nature of the coordinate again means that the vertical gradient of θ_R is almost uniform. However, the mass conservation equation, Equation 6.30, is perhaps more complicated than for $\log(p)$ coordinates, save in the limit $H \ll H_p$, when the factor r can be omitted. Of course, this coordinate suffers from the same difficulties as p as far as the lower boundary is concerned; the lower boundary is not a coordinate surface, and the value of z_a on the boundary varies in time. In Chapter 7, we will specify a standard set of equations that will be used in many subsequent chapters. This set can be derived using z_a as vertical coordinate or using z but making an additional approximation in the equations. For the sake of brevity, and recognizing that z_a and z are very similar, we shall not include the subscript 'a'.

Any other monotonic function of p can be used as the basis of a vertical coordinate. A final choice which is popular for numerical models but which is rarely used for theoretical studies is the so-called σ-coordinate, defined as:

$$\sigma = \frac{p}{p^*}$$

(6.32)

where p^* is the actual surface pressure, that is, the surface pressure not corrected to mean sea level. The σ coordinate system has the advantage that the mountainous surface of the Earth is a coordinate surface, $\sigma = 1$. However, in other respects, it loses the advantages of pressure coordinate. Consider the pressure gradient term. The hydrostatic relation takes the form:

$$\frac{\partial \Phi}{\partial \sigma} = -\frac{RT}{\sigma}$$ (6.33)

From Equation 6.4, the pressure gradient force is

$$-\frac{1}{\rho}\sigma \nabla p^* - \nabla \Phi = -RT\nabla \ln\left(p^*\right) - \nabla \Phi$$ (6.34)

From Equation 6.5, the analogue of density is $r = -p^*/g_e$, leading to the mass conservation relationship:

$$\frac{D}{Dt}\ln\left(p^*\right) + \nabla \cdot \mathbf{u} = 0$$ (6.35)

The great disadvantage of this system is revealed by the pressure gradient force. It has two terms, one of which is nonlinear. For steep orography, when the slope of the σ-surfaces is much larger than the slope of the pressure surfaces, the cancellation between the two terms in the pressure gradient term is generally very large. Another disadvantage is that Equation 6.35 appears to be a prognostic equation for p^* but applicable at every level of the atmosphere. In fact, the $\nabla \cdot \mathbf{u}$ term dominates away from low levels near steep orography. Nevertheless, the system is widely used for numerical work.

6.4   Isentropic coordinates

In the absence of heating, parcels of fluid conserve their potential temperature θ, that is, they move on surfaces of constant potential temperature. Furthermore, in a stably stratified atmosphere, θ always increases with height. This suggests that the equations of fluid dynamics might usefully be reformulated, using potential temperature θ as the vertical coordinate. The main advantage of such a coordinate system is that advection by the 'vertical' component of velocity $\dot{\theta}$ is small compared to horizontal advection, unless the heating or cooling is strong.

To formulate the equations in θ- or 'isentropic' coordinates, it is helpful to introduce two functions, the Exner function:

$$\pi = c_p \left(\frac{p}{p_0}\right)^\kappa$$ (6.36)

and the Montgomery potential:

$$M = c_p T + \Phi = \pi\theta + \Phi \tag{6.37}$$

From Equation 6.36, it follows that:

$$\theta\delta\pi = T\left(\frac{p_0}{p}\right)^\kappa \frac{\kappa\pi}{R\rho T}\delta p = \frac{1}{\rho}\delta p$$

and consequently:

$$\delta M = \pi\delta\theta + \frac{1}{\rho}\delta p + \delta\Phi \tag{6.38}$$

The hydrostatic equation (6.3), becomes, for isentropic coordinates:

$$\frac{\partial\Phi}{\partial\theta} = -\frac{1}{\rho}\frac{\partial p}{\partial\theta}$$

which means that Equation 6.38 can be rewritten as:

$$\frac{\partial M}{\partial\theta} = \pi \tag{6.39}$$

Use Equation 6.4 to write the pressure gradient term as:

$$-\frac{1}{\rho}\nabla_H p - \nabla_H \Phi = -\nabla_H M \tag{6.40}$$

This means that the use of isentropic coordinates linearizes the pressure gradient term, just as does the use of pressure coordinates. It also means that the Montgomery potential M is a streamfunction for the geostrophic wind

$$\mathbf{v}_g = \frac{1}{f}\mathbf{k}\times\nabla_\theta M \tag{6.41}$$

From Equation 6.39, it follows that the thermal wind equation has the form:

$$f\frac{\partial\mathbf{v}_g}{\partial\theta} = \mathbf{k}\times\nabla\pi \tag{6.42}$$

The analogue of density, Equation 6.5, is $r = -(1/g)\partial p/\partial\theta$, and so the continuity equation is:

$$\frac{D}{Dt}\left(\frac{\partial p}{\partial\theta}\right) + \frac{\partial p}{\partial\theta}\nabla\cdot\mathbf{u} = 0 \tag{6.43}$$

In adiabatic conditions, flow is parallel to θ-surfaces. When there is strong heating or cooling, material crosses isentropes. The pressure gradient term is a simple linear gradient term. Against these advantages, the major reservation is that isentropes have a significant slope, generally between 10^{-2} and 10^{-3}. This means that the θ-surfaces have a significant angle of intersection with the lower boundary, even when the lower boundary is nearly flat. The lower boundary condition can be derived from Equation 6.12 as follows:

$$\frac{D}{Dt}\left(M - \pi\theta\right) = \mathbf{v} \cdot \nabla\Phi^*$$

(6.44)

In general, there is no simple approximation to this boundary condition. This means that isentropic coordinates are of limited value in qualitative or analytic discussion in the troposphere, at least in problems where the kinematic constraint imposed by the lower boundary is important. A further drawback is that the daytime atmospheric boundary layer is often close to isentropic.

7

Variations of density and the basic equations

7.1 Boussinesq approximation

This chapter contains a discussion of the role of density, particularly in the pressure gradient term. In some classical laboratory and engineering fluid dynamics, we simply assume that flow is incompressible with constant ρ so that the pressure gradient term remains a linear term. But for most geophysical flows, density variations are crucial elements in the flow. This is true of the ocean where even quite small variations of density give rise to important buoyancy forces. It is even more true of the atmosphere, which, being composed of compressible fluid, has very large variations of density with height and rather smaller, but nevertheless important, variations in the horizontal.

First, we consider the vertical structure of the atmosphere, building upon some of the concepts introduced in Chapter 1. Define a 'reference atmosphere' in which pressure, temperature, density and other thermodynamic variables are functions of height only. Such a reference variation of a thermodynamic variable is denoted by a subscript R. So $T_R(z)$ could be defined as the horizontal average of temperature T on a level surface at height z. Similarly, reference profiles of density, pressure, potential temperature and so on may all be defined. Some of these quantities vary very markedly with height, others less so. For any quantity Q, define a 'scale height' H_Q such that

$$H_Q \sim \left(\frac{1}{Q_R} \left| \frac{\partial Q_R}{\partial z} \right| \right)^{-1} \tag{7.1}$$

If H_Q is constant with height, Equation 7.1 implies that Q decreases or increases exponentially with height, with a characteristic vertical scale H_Q. For example, if Q is pressure and temperature is constant with height, then using the hydrostatic relationship in Equation 7.1 leads to

$$p_R = p_0 e^{-z/H_p} \tag{7.2}$$

Fluid Dynamics of the Midlatitude Atmosphere, First Edition. Brian J. Hoskins & Ian N. James.
© 2014 John Wiley & Sons, Ltd. Published 2014 by John Wiley & Sons, Ltd.

where

$$H_p = \frac{RT_R}{g} \tag{7.3}$$

is called the 'pressure scale height'. For a typical value of $T_R \sim 260$ K, Equation 7.3 leads to H_p around 7.6 km. Although T_R varies with height, in fact, the value of H_p does not change greatly, so 7.6 km is a typical vertical scale in the atmosphere. Similarly, scale heights for density, temperature and potential temperature are

$$H_\rho = -\left(\frac{1}{\rho_R}\frac{\partial \rho_R}{\partial z}\right)^{-1}, \quad H_T = -\left(\frac{1}{T_R}\frac{\partial T_R}{\partial z}\right)^{-1}, \quad H_\theta = \left(\frac{1}{\theta_R}\frac{\partial \theta_R}{\partial z}\right)^{-1} \tag{7.4}$$

These scale heights are interrelated through the equation of state. For example, taking the usual form of the ideal gas equation and differentiating logarithmically gives

$$\frac{1}{p_R}\frac{\partial p_R}{\partial z} = \frac{1}{\rho_R}\frac{\partial \rho_R}{\partial z} + \frac{1}{T_R}\frac{\partial T_R}{\partial z}$$

or

$$\frac{1}{H_p} = \frac{1}{H_\rho} + \frac{1}{H_T} \tag{7.5}$$

In the lower troposphere, temperature is observed to decline with height at about 6.5 K km^{-1}. Hence H_T is typically 40 km. It follows from Equation 7.5 that the density scale height is 9.4 km. The potential temperature scale height can be estimated from the equation of state in the form

$$\theta = p^{1-\kappa}\rho^{-1}Rp_0^\kappa \tag{7.6}$$

Noting that 'theta' increases with height, as before, differentiate logarithmically and find

$$\frac{1}{H_\theta} = -\frac{(1-\kappa)}{H_p} + \frac{1}{H_\rho} \tag{7.7}$$

Putting in values for H_p and H_ρ, a value for H_θ is around 80 km.

A secondary heading for this chapter might be 'ways of linearizing the pressure gradient term'. The pressure gradient term in the Navier–Stokes equation is nonlinear, since it involves the product of the pressure gradient with the specific volume $1/\rho$. In many circumstances, the variations of density are sufficiently small that the pressure gradient term can be simplified. Suppose density can be split into a large background part which is constant, together with a small fluctuation which varies in time and space:

$$\rho(\mathbf{r},t) = \rho_0 + \delta\rho(\mathbf{r},t) \tag{7.8}$$

A similar partitioning can also be applied to the pressure

$$p = p_0(z) + \delta p \quad \text{where} \quad \frac{\partial p_0}{\partial z} = -\rho_0 g \tag{7.9}$$

The Euler equations become

$$(\rho_0 + \delta\rho)\left(\frac{Du}{Dt} + 2\Omega \times u\right) = -\nabla\delta p - \frac{\partial p_0}{\partial z}k - (\rho_0 + \delta\rho)gk \tag{7.10}$$

Assume $|\delta\rho| \ll \rho_0$ and so neglect $\delta\rho$ compared with ρ_0 on the left-hand side of this equation. Then,

$$\frac{Du}{Dt} + 2\Omega \times u = -\frac{1}{\rho}\nabla\delta p + b_\rho k \tag{7.11}$$

where the only place that $\delta\rho$ still occurs is

$$b_\rho = -g\frac{\delta\rho}{\rho_0} \tag{7.12}$$

The quantity b_ρ is called the 'buoyancy'. It looks like a scaled-down gravitational acceleration and is sometimes called the 'reduced gravity'. But it is better thought of as a pressure gradient force. It results from the pressure field which is consistent with hydrostatic balance in a fluid of density ρ_0 acting upon an element of fluid with slightly different density $\rho_0 + \delta\rho$. In other words, it is that part of the pressure gradient force which is not balanced by the gravitational acceleration.

This approximation is called the 'Boussinesq approximation'. It is satisfactory in a wide variety of circumstances, wherever density fluctuations are small compared to the mean density. It is widely used in the oceans where salinity and temperature variations both lead to important changes in density, to laboratory systems using water as a working fluid in which small variations of temperature or salinity introduce density fluctuations, and in the atmosphere for circumstances where the flow is confined to a small range of heights. However, in many circumstances, the range of heights spanned by atmospheric circulation systems is comparable to or greater than the density scale height. In this case, the Boussinesq approximation may be inadequate.

7.2 Anelastic approximation

A development of the Boussinesq approximation to a system such as the atmosphere which spans heights comparable to or larger than the density scale height is called the anelastic approximation. It amounts to recognizing the larger variation of density across the domain, principally in the vertical, while assuming that the density fluctuations of an individual fluid element remain small. Let the density be partitioned, not into a constant and a perturbation, but into a reference profile that varies only in the vertical, and a fluctuating part which in general varies in both space and time:

$$\rho = \rho_R(z) + \rho'(\mathbf{r},t) \tag{7.13}$$

with the assumption that $|\rho'| \ll \rho_R$. For synoptic systems, in Equation 5.32 of Section 5.6, it was shown that $\rho'/\rho_R \sim fUL/gD = \mathrm{Fr}/\mathrm{Ro} \sim 10^{-2}$, and so this approximation is certainly valid. It is very similar to the Boussinesq approximation, save the background density varies with height. As before, we also partition the pressure:

$$p = p_R(z) + p'(\mathbf{r},t)$$

The reference density profiles are chosen to satisfy the hydrostatic equation, so that

$$\frac{\partial p_R}{\partial z} = -\rho_R g \tag{7.14}$$

First, consider the horizontal component of the pressure gradient force:

$$\frac{1}{\rho_R + \rho'} \nabla \left(p_R + p' \right)$$

Neglecting ρ' compared with ρ_R, this becomes

$$\frac{1}{\rho_R} \nabla p' = \nabla \left(\frac{p'}{\rho_R} \right)$$

With this approximation, the horizontal pressure gradient force can be written as a gradient as is the case for pressure-type vertical coordinates. Using the reference profile and deviation splitting also in the vertical component of the momentum equation gives

$$\left(\rho_R + \rho' \right) \frac{Dw}{Dt} = -\frac{\partial p_r}{\partial z} - \frac{\partial p'}{\partial z} - g\rho_R - g\rho' \tag{7.15}$$

$$= -\frac{\partial p'}{\partial z} - g\rho'$$

Again, neglect ρ' on the left-hand side to obtain

$$\frac{Dw}{Dt} = -\frac{1}{\rho_R} \frac{\partial p'}{\partial z} - g \frac{\rho'}{\rho_R} \tag{7.16}$$

$$= -\frac{\partial}{\partial z} \left(\frac{p'}{\rho_R} \right) - \frac{p'}{\rho_R^2} \frac{\partial \rho_r}{\partial z} - g \frac{\rho'}{\rho_R}$$

Finally, use Equation 7.6 to eliminate ρ'

$$\frac{Dw}{Dt} = b_\theta - \frac{\partial}{\partial z} \left(\frac{p'}{\rho_R} \right) + \frac{p'}{\rho_R H_\theta} \tag{7.17}$$

where $b_\theta = -g(\theta'/\theta_R)$ is closely similar to the buoyancy term in the Boussinesq equation, except that b_θ is defined in terms of potential temperature rather than density. We can approximate θ_R by a constant value θ_0 in b to give

$$b = -g\left(\frac{\theta'}{\theta_0}\right)$$

The last term in Equation 7.17 is small since the potential temperature scale height is so large compared to the pressure or density scale heights. The term is lost altogether if θ_R is an adiabatic profile, with θ_R constant.

It remains to discuss the continuity equation within the anelastic framework. The full continuity equation can be written in its Eulerian form as

$$\frac{\partial \rho}{\partial t} + \nabla \cdot (\rho \mathbf{u}) = 0 \qquad (7.18)$$

The scale analysis in Section 5.7 gave the magnitude of the first term relative to the second term to be

$$\mathrm{Ro}^{-2}\mathrm{Fr} = \frac{f^2 L^2}{gD} \sim 0.05$$

Dropping this first term is equivalent to filtering sound waves from the system so that the compressibility or 'elasticity' of an individual parcel of air is neglected. The second term is usefully separated into contributions from the horizontal and vertical components of the motion. The anelastic version of the continuity equation therefore becomes

$$\nabla_{\mathrm{H}} \cdot \mathbf{v} + \frac{1}{\rho_R}\frac{\partial}{\partial z}(\rho_R w) = 0 \qquad (7.19)$$

The principal qualitative result from the anelastic system is contained in this equation; it concerns the increasing asymmetry of circulations in the vertical plane as their depth increases. For example, in order to conserve mass flux, an ascending plume of air must rise more rapidly at upper levels, where the density is low, and less rapidly at low levels, where the density is large.

As will be discussed in Section 7.5, using height coordinates and the anelastic approximation is one way to derive the basic set of equations that will be used for much of the analysis of atmospheric motions in the rest of this book.

7.3   Stratification and gravity waves

Consider an initially motionless atmosphere in which parcels of air undergo small vertical displacements δz from their equilibrium level. The displacements are adiabatic so the parcels conserve their potential temperature. Assume pressure perturbations are negligible. Because parcels conserve their potential temperature and have the same potential temperature as their environment at their equilibrium level, it follows that

$$b_\theta = -g \frac{\theta'}{\theta_R} = -\frac{g}{\theta_R} \frac{\partial \theta_R}{\partial z} \delta z \qquad (7.20)$$

The Equation 7.15 may therefore be written as follows:

$$\frac{Dw}{Dt} = \frac{d^2 \delta z}{dt^2} = -\frac{g}{\theta_R} \frac{\partial \theta_R}{\partial z} \delta z = -N^2 \delta z \qquad (7.21)$$

The parameter $N = (g(\partial \theta_R/\partial z)/\theta_R)^{1/2}$ has the unit s^{-1} of frequency. It is the frequency with which vertically displaced parcels of air will oscillate about their home level, and it is called the Brunt–Väisälä frequency. An alternative expression for the Brunt–Väisälä frequency is $N = (g/H_\theta)^{1/2}$. It should be emphasized that this argument is rather crude: a more rigorous derivation of this result is given in Section 7.5. As discussed briefly in Section 1.1, provided $\partial \theta_R/\partial z$ is positive, N is real, and gently displaced parcels will remain in the neighbourhood of their original level, simply oscillating in the vertical. Such an atmosphere is said to be 'stably stratified'. If $\partial \theta_R/\partial z$ were negative, N would be imaginary, corresponding to even initially small parcel displacements increasing exponentially with time. In such a case, there would be very rapid exchange of parcels from different levels and mixing of potential temperature, leading eventually to a 'neutral' or 'adiabatic' profile with θ_R constant with height. Significantly unstable stratification, in which θ_R decreases with height, is rarely observed, apart from in shallow layers immediately above strongly heated surfaces. The bulk of the Earth's atmosphere is significantly stably stratified; given H_θ is around 80 km, the Brunt–Väisälä frequency works out to be around 1.1×10^{-2} s^{-1}. This translates into an oscillation period of around 10 minutes.

Two properties are crucial in determining the nature of the flow in a planet's atmosphere. The first is the planet's rotation rate Ω. The second is the characteristic Brunt–Väisälä frequency N of its atmosphere. The factors determining the mean value of N for an atmosphere are complicated and poorly understood. Indeed, for the giant planets Jupiter, Saturn, Uranus and Neptune, the large-scale values of N are more or less unknown, making any account of their large-scale atmospheric circulations extremely uncertain.

In the Earth's atmosphere, the rotation frequency is two orders of magnitude smaller than the Brunt–Väisälä frequency. This means that the atmosphere supports both low-frequency motion, with periods of order the day length or longer, and high-frequency motions, with periods of 10 minutes or so. A third, even higher, frequency is associated with sound waves. However, as said in Section 7.2, these are normally filtered out of the equations by approximating the continuity equation. Even with this simplification, the atmosphere exhibits multiple timescales. This is one major source of difficulty in setting up numerical models of the atmosphere, either for forecasting or in order to simulate the large-scale atmospheric circulation and its sensitivities.

7.4 Balance, gravity waves and Richardson number

In Section 5.3, scale analysis of the horizontal momentum equation revealed that for synoptic scales, there was a near-balance between the Coriolis acceleration and the pressure gradient acceleration. This balanced state has been elaborated in the previous sections. In subsequent chapters, the conditions for such a balance to exist will be shown to impose strong constraints on the flow and its evolution. The assumption of balance will lead to important insights into the nature of small Rossby number flow in the atmosphere and in other rotating fluid systems. In this section, we address the reasons why balanced states are established and persist. The role of density variations turns out to be central to the persistence of near-balanced states.

Consider a motionless stratified atmosphere, into which a small, localized disturbance is introduced. The disturbance involves perturbations to the velocity and pressure fields, not generally in a state of balance. Provided the perturbations are small, the various advection terms can be neglected as being even smaller. Furthermore, for simplicity, we shall assume that the disturbances vary only in the zonal or x-direction. With these assumptions, the two horizontal momentum equations are as follows:

$$\frac{\partial u}{\partial t} = fv - \frac{\partial \phi}{\partial x} \tag{7.22}$$

$$\frac{\partial v}{\partial t} = -fu \tag{7.23}$$

The vertical accelerations are retained; no assumption about hydrostatic balance is made. Then, the vertical component of the momentum equation is

$$\frac{\partial w}{\partial t} = -\frac{\partial \phi}{\partial z} + b' \tag{7.24}$$

Here, the buoyancy field has been split into two parts $b = b_0(z) + b'$, where b_0 reflects the variation of density with height in the unperturbed, motionless, stably stratified fluid. The Boussinesq approximation has been made in writing Equation 7.24. The evolution of the buoyancy field is given by the thermodynamic equation in the following form:

$$\frac{\partial b'}{\partial t} = -w \frac{\partial b_0}{\partial z} \equiv -wN^2 \tag{7.25}$$

Here, N is the Brunt–Väisälä frequency, introduced in Section 7.3. It is assumed constant with height. Finally, the continuity equation for slow motions and the Boussinesq approximation takes the form:

$$\frac{\partial u}{\partial x} + \frac{\partial w}{\partial z} = 0 \tag{7.26}$$

The equation set is now complete, with five equations in the five variables u, v, w, b' and ϕ.

The evolution of the flow field is revealed by eliminating variables from Equations 7.22–7.26 to form a single equation for w. The algebra is straightforward and left as an exercise for the reader. The essential steps are, firstly, to eliminate f between Equations 7.22 and 7.24, secondly to eliminate v using Equation 7.23, and finally to eliminate u using Equation 7.26. The second step involves taking a time derivative of the prognostic equation; effectively, this eradicates any balanced modes from the problem, because they cannot evolve in this idealized linear system. The result is

$$\left(w_{xx} + w_{zz}\right)_{tt} + N^2 w_{xx} + f^2 w_{zz} = 0 \tag{7.27}$$

where the subscripts denote differentiation.

Seek wave-like solutions of the following form:

$$w = W \exp\left(i\left(kx + mz - \omega t\right)\right) \tag{7.28}$$

Here, k and m are the horizontal and vertical wavenumbers respectively, and ω is the frequency of the wave. Substituting into Equation 7.27 gives a consistency condition for this form of solution to be valid. It is

$$\omega^2 = \frac{k^2 N^2 + m^2 f^2}{k^2 + m^2} \tag{7.29}$$

This consistency condition is called the 'dispersion relationship'. It is a relationship between the frequency of disturbances and their wavenumbers. The waves it describes, which depend both on the stratification N and the Coriolis parameter f, are called 'inertia–gravity waves'.

Equation 7.29 has a number of interesting properties and limiting forms which bear upon the matter of how flows achieve a balanced state. First, note the quadratic form of the dispersion relationship. This means that both positive and negative roots for ω exist, and therefore that the phase speed ω/k and group velocity $\partial\omega/\partial k$ can take either sign. In other words, a localized region of imbalance will radiate disturbances which can propagate equally happily in either direction. Secondly, consider the case of deep waves whose structure is vertically coherent and for which therefore $m \rightarrow 0$. In this limit, fluid parcels oscillate only in the vertical, and the dispersion relationship reduces to $\omega = \pm N$. This is the result derived less rigorously as Equation 7.21 in Section 7.3.

In order to explore the implications of Equation 7.29, some typical magnitudes of the various quantities involved are needed. In the Earth's troposphere, the Brunt–Väisälä frequency N is around $10^{-2}\,\mathrm{s}^{-1}$. The Coriolis parameter f is two orders of magnitude smaller, around $10^{-4}\,\mathrm{s}^{-1}$. For the purposes of the present discussion, consider disturbances whose vertical scale is comparable to the depth of the troposphere H, which we may take to be around $10\,\mathrm{km}$. In that case, the vertical wavenumber is $m \sim H^{-1}$. The dispersion relation is illustrated in Figure 7.1, which shows the variation of ω with k for the case $f = 10^{-4}\,\mathrm{s}^{-1}$, $N = 10^{-2}\,\mathrm{s}^{-1}$ and $H = 10^4\,\mathrm{m}$. For small k, the frequency

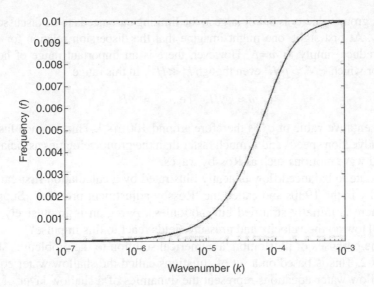

Figure 7.1 The dispersion relationship, Equation 7.29, for inertia gravity waves. Here, $f=10^{-4}\,s^{-1}$, $N=10^{-2}\,s^{-1}$ and $m=H^{-1}$ where $H=10^4\,m$

$\omega \rightarrow f$. At larger k, the frequency increases approximately linearly with k. before asymptoting to N.

The numerator of Equation 7.29 is made up of two terms, the first dependent upon the stratification and the second upon the rotation rate. These two terms are comparable when

$$\frac{k}{m} \sim \frac{f}{N} \sim \frac{1}{100}$$

This is referred to as the natural ratio of vertical and horizontal scales, D/L. For the particular case with $m=H^{-1}$,

$$k^{-1} \sim \frac{NH}{f} \equiv L_R$$

The horizontal length scale L_R is called the 'Rossby radius of deformation' or simply the 'Rossby radius'. It distinguishes between small scales, for which stable stratification acts to flatten pressure or density surfaces, and larger scales, for which the Coriolis effect tends to deform and tilt such surfaces. In later chapters, the Rossby radius will turn out to be a crucial horizontal scale. For example, it determines the synoptic scale, the typical scale of midlatitude weather systems. A representative midlatitude value of L_R is $10^6\,m$.

The analysis of inertia–gravity waves is relevant to the problem we are considering of the development of an isolated, imbalanced disturbance, because such a disturbance can be thought of as a super-position of inertia–gravity waves of various wavelengths. Such a wave complex propagates away from its source at the group velocity; the component of group velocity parallel to the x-axis is given by $c_{gx}=\partial\omega/\partial k$. For small k, with

$\omega \approx f$, the group velocity is 0. For large k, the limiting case needs to be discussed rather carefully. At first sight, one might imagine that the dispersion relation for $k \gg L_R^{-1}$ should reduce simply to $\omega = N$. However, there is an important range of horizontal scales for which $k^2 N^2 \gg f^2/H^2$ even though $k^2 \ll H^{-2}$. In this range,

$$\omega \approx kNH, \quad \text{i.e.,} \quad c_{gx} \approx NH$$

A representative value of c_{gx} is therefore around $100\,\text{m}\,\text{s}^{-1}$. This is much faster than typical advection speeds and is much faster than the group velocities associated with balanced wave motions such as Rossby waves.

The route to balanced flow is neatly illustrated by a calculation first carried out by Rossby in the 1930s and called the 'Rossby adjustment problem'. Suppose the atmosphere is initially stratified but motionless. A region is impulsively set into motion. How do the velocity and pressure fields react to this impulse?

For the purposes of illustration, a numerical solution to the problem is shown in Figure 7.2. This is based on a set of equations called the shallow water equations. The shallow water equations represent the dynamics of a shallow layer of fluid of constant density, depth H. Mathematically, the results are very similar to the results given in the discussion just given earlier but with a prescribed vertical scale H to the motions in place of a varying vertical wavenumber m. For simplicity, the initial disturbance is assumed to be of so small an amplitude that the equations can be linearized about a state of rest, meaning that all the nonlinear advection terms are dropped. The momentum equations may be written as follows:

$$\frac{\partial u}{\partial t} = fv - g\frac{\partial h}{\partial x}$$
$$\frac{\partial v}{\partial t} = -fu \tag{7.30}$$

Here, h is the small deviation of the fluid depth from H. We assume no variations in the y-direction. The continuity equation for this system is

$$\frac{\partial h}{\partial t} = -H\nabla \cdot \mathbf{v} = -H\frac{\partial u}{\partial x} \tag{7.31}$$

The results can be generalized by making these equations dimensionless. A characteristic time is f^{-1}, sometimes called the 'inertial timescale', and a characteristic length scale is the Rossby radius $L_R = (gH)^{1/2}/f$. Vertical displacements of the fluid surface are scaled by the mean depth H. The dimensionless form of the equation set is

$$\frac{\partial u^*}{\partial t^*} = v^* - \frac{\partial h^*}{\partial x^*}$$
$$\frac{\partial v^*}{\partial t^*} = -u^* \tag{7.32}$$
$$\frac{\partial h^*}{\partial t^*} = -\frac{\partial u^*}{\partial x^*}$$

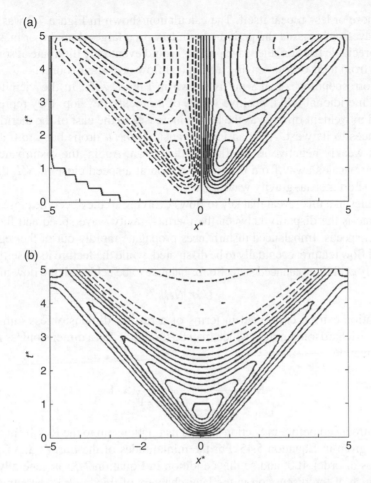

Figure 7.2 The Rossby adjustment problem: longitude-time plots showing (a) geopotential height and (b) zonal wind. Distances along the x-axis are in units of L_R while time is given in units of f^{-1}. Negative values are indicated by dashed contours

where the asterisk indicates dimensionless variables. Figure 7.2 was generated by solving these equations in a domain periodic in x from an initial state in which u and h were both 0, and v was 0 apart from a region around $x = 0$ where it was positive.

Figure 7.2a shows a longitude–time plot, or 'Hovmöller' plot, of the height field. Initially zero everywhere, the height develops a characteristic slope in the region of non-zero v, bringing it into geostrophic balance with the v field on a timescale of $O(f^{-1})$. At the same time, a packet of disturbances propagates away from the region of non-zero v towards both larger and smaller x. The speed of the leading edge of this packet is very close to the shortwave group velocity NH. In the rather artificial configuration described here, with its periodic boundary condition and virtual absence of any dissipation, the wave packets propagate right around the domain and would eventually meet again in the region of non-zero v. Thereafter, the sequence would more or less repeat itself. The calculation shown in Figure 7.2 was stopped as the wavefront crossed the periodic boundary. In a larger domain, or in a domain with appreciable friction, the wave packets would eventually dissipate at some large distance from their origin, leaving a balanced flow near their source.

The corresponding zonal wind is shown in Figure 7.2b. In the vicinity of the imposed meridional wind, positive values of u quickly develop: they represent the eastward movement of mass as the height increases to the east of the meridional jet and reduces to its west. After an initial transient, the u drops back to 0 and even becomes weakly negative as the flow settles down. Again, the disturbance in the u-field propagates away from the source region at a speed close to NH, the group speed of short inertia–gravity waves.

Examples like this reveal that key to flows evolving so they become close to a balanced state is the disparity between the inertia–gravity wave speed and the typical advection speeds. Imbalanced disturbances propagate rapidly out of the region of a localized flow feature, eventually to be dissipated, while the feature itself settles down to a slowly evolving balanced state. So for balance to be achieved, the flow must have

$$U \ll NH$$

This relation can be expressed in terms of a further dimensionless number, the so-called Richardson number, which may be thought of as a dimensionless measure of stratification:

$$\mathrm{Ri} = \frac{N^2 H^2}{U^2} \sim \frac{N^2}{\left(\partial u / \partial z\right)^2} \gg 1 \qquad (7.33)$$

Ri was introduced before in Section 5.9 and its relationship to the Rossby and Burger numbers given in Equation 5.45. For the midlatitudes of the Earth's atmosphere, it has values of order 100, and so this condition in Equation 7.33 is generally met. A key to much of the discussion in the later chapters of this book is that atmospheric flow evolves so that it remains close to a balanced state, although crucially, it is rarely exactly in a state of balance. This is the basis of 'quasi-geostrophic theory', which will be introduced in Chapter 12 and will be explained in subsequent chapters.

The Richardson number has been interpreted in terms of the ratio of the advection speeds to the speed of inertia–gravity waves. It can also be related to the thermal structure of the atmosphere. Note that

$$N^2 = \frac{g}{\theta}\frac{\partial \theta}{\partial z}$$

In Chapter 6, we have given the thermal wind relations in various coordinate systems, and thermal wind balance will be discussed in detail in Chapter 12. Taking characteristic scales in the relation as given in Equation 6.24 or 6.31 leads to

$$\frac{U}{H} \sim \left|\frac{\partial \mathbf{v}}{\partial z}\right| \sim \frac{g}{f\theta_0}\left|\nabla_{\mathrm{H}}\theta\right|$$

It follows that

$$\frac{1}{Ri} = \left(\frac{\partial\theta/\partial s}{\partial\theta/\partial z}\right)\frac{U}{fD} = \left(\frac{\partial\theta/\partial s}{\partial\theta/\partial z}\right)Ro\left(\frac{L}{D}\right)$$

(Here, the variable s is simply horizontal distance parallel to $\nabla_H\theta$). Now the first factor on the right-hand side of this relationship is simply the magnitude of the slope of the θ-surfaces. Rearranging gives

$$\left.\frac{\partial z}{\partial s}\right|_{\theta} \sim \left(\frac{D}{L}\right)\frac{1}{RiRo} \sim \left(\frac{D}{L}\right)\left(\frac{Ro}{Bu}\right) \tag{7.34}$$

Taking $H \sim 10^4$ m, $L \sim 10^6$ m, Ri ~ 100, Ro ~ 0.1 and Bu ~ 1 for the Earth's midlatitudes, the isentropic slope is around $|\partial z/\partial s|_{\theta} \sim 10^{-3}$. This slope indicates that isentropes are flat compared to the purely geometric slope (D/L). In turn, such flat isentropes mean that changes of θ over a depth H in the vertical are large compared to the changes of θ over a horizontal distance L. We shall return to this important result in Chapter 12, where it lies at the heart of the quasi-geostrophic approximation.

7.5 Summary of the basic equation sets

In this section, we gather together the basic equation sets that will be used for much of the analysis in the later chapters. From Section 7.2, the frictionless, adiabatic equations with the hydrostatic and anelastic approximations may be written as follows:

$$\frac{D\mathbf{v}}{Dt} + f\mathbf{k}\times\mathbf{v} + \nabla\Phi' = \dot{\mathbf{v}},$$

$$\frac{\partial\Phi'}{\partial z} = b,$$

$$\nabla\cdot(\rho_R\mathbf{u}) = 0, \tag{7.35}$$

$$\frac{Db}{Dt} + N^2 w = \dot{b},$$

Here, $\Phi' = p'/\rho_R$, ρ_R is a function of z only, the buoyancy $b = g\theta'/\theta_0$ and

$$N^2(z) = \frac{g}{\theta_R}\frac{d\theta_R}{dz}$$

For a flat bounding surface, the boundary condition is $w = 0$ on $z = 0$.

If the hydrostatic approximation is not made, then the corresponding vertical momentum equation is

$$\frac{Dw}{Dt} = b - \frac{\partial\Phi'}{\partial z}$$

If we set

$$\Phi = \int_0^z \left(\frac{g}{\theta_0}\right)\theta_R\left(z'\right)\mathrm{d}z' + \Phi'$$

then the equations are

$$\frac{D\mathbf{v}}{Dt} + f\mathbf{k}\times\mathbf{v} + \nabla\Phi = \dot{\mathbf{v}}$$

$$\frac{\partial\Phi}{\partial z} = \left(\frac{g}{\theta_0}\right)\theta$$

$$\nabla\cdot\left(\rho_R\mathbf{u}\right) = 0 \tag{7.36}$$

$$\frac{D\theta}{Dt} = \dot{\theta}$$

with $w=0$ at $z=0$. For completeness, a friction term $\dot{\mathbf{v}}$ and heating terms $\dot{b}$ or $\dot{\theta}$ have been retained in Equations 7.35 and 7.36. But for many of the applications in later chapters, where the relevant timescales are less than a few days, we shall assume these terms are small and so will drop them.

As shown in Section 6.3, the equation set, Equation 7.36 is obtained directly using the pressure-based coordinate

$$z = \frac{H_0}{\kappa}\left[1 - \left(\frac{p}{p_0}\right)^\kappa\right]$$

Here $H_0 = R\theta_0/g$. For this vertical coordinate, in Equation 7.36, Φ is the geopotential, ρ_R stands for the equivalent density r defined in Equation 6.25 and the vertical coordinate z is similar to geometrical height, provided θ is close to θ_0. In this case, no equivalent of the anelastic approximation that was necessary in height coordinates has been made. However, using this pressure-based vertical coordinate, the boundary condition $w=0$ at $z=0$ is an approximation. The equation set in the form Equation 7.35 can be derived by setting $\Phi=\Phi_R+\Phi'$ and $\theta=\theta_R+\theta'$, and then $b=g\theta'/\theta_0$, and $N^2(z)=(g/\theta_R)(\mathrm{d}\theta_R/\mathrm{d}z)$ as before.

These two basic equation sets will provide the starting point for many analyses in the subsequent chapters.

7.6   The energy of atmospheric motions

The atmosphere is in continual motion. By forming an energy equation from Equations 7.35 or 7.36, the mechanism for generating the kinetic energy of atmospheric motion becomes clear.

The kinetic energy per unit volume of the atmosphere is

$$K = \rho_R \frac{\mathbf{u} \cdot \mathbf{u}}{2} \tag{7.37}$$

There is a tiny contribution from the vertical velocity, but for synoptic-scale motion, it is several orders of magnitude smaller than the contribution from the horizontal components of the velocity. It is convenient to retain it here and use the non-hydro-static form of the momentum equations in Equation 7.35a, in the following form:

$$\frac{\partial \mathbf{u}}{\partial t} + \mathbf{u} \cdot \nabla \mathbf{u} + f \mathbf{k} \times \mathbf{u} + \nabla \Phi' - b\mathbf{k} = \dot{\mathbf{v}} \tag{7.38}$$

Here, $\dot{\mathbf{v}}$ denotes the acceleration due to friction. Use vector identities to write the advection term in the form

$$\mathbf{u} \cdot \nabla \mathbf{u} = \nabla \cdot \left(\frac{\mathbf{u} \cdot \mathbf{u}}{2} \right) + \mathbf{u} \times (\nabla \times \mathbf{u})$$

and take the scalar product of the momentum equation with $\rho_R \mathbf{v}$ to form a kinetic energy equation:

$$\frac{\partial K}{\partial t} + \rho_R \mathbf{u} \cdot \nabla K + \rho_R \mathbf{u} \cdot \nabla \Phi' - \rho_R w b = \rho_R \mathbf{v} \cdot \dot{\mathbf{v}}$$

Using the mass conservation equation, this may be written:

$$\frac{\partial K}{\partial t} + \nabla \cdot (\rho_R \mathbf{u} K) + \nabla \cdot (\rho_R \mathbf{u} \Phi') - \rho_R w b + \rho_R \mathbf{v} \cdot \dot{\mathbf{v}} \tag{7.39}$$

Finally, integrate over the whole domain, so that the divergence terms become integrals around the boundary which are zero with the boundary conditions that $\mathbf{v} = 0$ on vertical boundaries and $w=0$ on horizontal boundaries, to obtain an equation for the total kinetic energy

$$\frac{d}{dt} \langle K \rangle = \langle \rho_R w b \rangle + \langle \rho_R \mathbf{v} \cdot \dot{\mathbf{v}} \rangle \tag{7.40}$$

Here, the angle brackets represent the domain averaging operator

$$\langle \; \rangle = \int_V dx \, dy \, dz \Big/ \int_V dx \, dy \, dz$$

No simple exact form for the friction $\dot{\mathbf{v}}$ is realistic, but in nearly all circumstances, $\mathbf{v} \cdot \dot{\mathbf{v}} < 0$, that is, the friction will be in the opposite direction to the velocity $\mathbf{v}$. So the second term on the right-hand side of Equation 7.40 is negative; it will deplete kinetic energy and dissipate atmospheric motions. The first term on the right-hand side will generate kinetic energy if there is a positive correlation between ascent and buoyancy, that is, if on average warm air rises and cold air sinks. The existence of such an upward buoyancy flux is a necessary condition for any process which generates winds in the atmosphere. In the long-term average, the friction and buoyancy terms must balance.

A complementary energy equation results from a similar development from the thermodynamic equation, Equation 7.35d. Multiply by b/N^2 and integrate over the whole domain, applying appropriate boundary conditions, to give

$$\frac{\mathrm{d}}{\mathrm{d}t}\langle A \rangle = -\langle \rho_R wb \rangle + \left\langle \rho_R \frac{b\dot{b}}{N^2} \right\rangle \tag{7.41}$$

where

$$\langle A \rangle = \left\langle \frac{\rho_R b^2}{2N^2} \right\rangle \tag{7.42}$$

has the units of energy per unit volume and is called the 'available potential energy'. Available potential energy is generated if there is heating of warm air and cooling of colder air, so the buoyancy variance is increased. On the global scale, heating of the warm tropics and cooling of the cold polar regions generate global available potential energy. The first term on the right-hand side of Equation 7.41 depletes available potential energy if there is on average a positive correlation between ascent and positive buoyancy. It is equal in magnitude but opposite in sign to the first term on the right-hand side of Equation 7.40; it therefore represents a conversion from available potential energy to kinetic energy. Figure 7.3 summarizes this energy cycle.

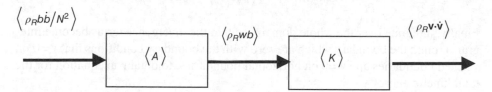

Figure 7.3 Energy cycle of atmospheric motion. Differential heating generates available potential energy $\langle A \rangle$. Upward buoyancy fluxes convert available potential energy to kinetic energy $\langle K \rangle$, which is dissipated by friction

Theme 2
Rotation in the atmosphere

8

Rotation in the atmosphere

8.1   The concept of vorticity

Geophysical flows are characterized by rotational components of the velocity field
which are large compared to the divergent components. This asymmetry is exploited
in this section by developing an alternative to the momentum equations, using a vorticity equation in place of the momentum equation to predict the evolution of the flow.
The rotational and divergent components of a flow were discussed in Section 2.2
in terms of two-dimensional flows. There, a measure of the spin of individual fluid
parcels, the 'vorticity', was introduced. For a two-dimensional flow in the x–y plane,

$$\xi = \frac{\partial v}{\partial x} - \frac{\partial u}{\partial y}$$

The concept of 'vorticity' or 'spin' is readily generalized to three-dimensional flow:

$$\begin{aligned}
\xi &= \nabla \times \mathbf{u} \\
&= \left(\frac{\partial w}{\partial y} - \frac{\partial v}{\partial z}\right)\mathbf{i} - \left(\frac{\partial w}{\partial x} - \frac{\partial u}{\partial z}\right)\mathbf{j} + \left(\frac{\partial v}{\partial x} - \frac{\partial u}{\partial y}\right)\mathbf{k}
\end{aligned} \tag{8.1}$$

In three dimensions, vorticity is a vector quantity. Its direction is parallel to the local
spin axes of fluid parcels, and its magnitude is twice the magnitude of the angular
velocity of the parcels.

Consider some closed loop L in the fluid, which encloses a surface A, as shown
in Figure 8.1. From Stokes's theorem,

$$C = \oint_L \mathbf{u} \cdot d\mathbf{l} = \int_A \nabla \times \mathbf{u} \cdot \mathbf{n} dA = \int_A \xi \cdot \mathbf{n} dA \tag{8.2}$$

The quantity C is called the 'circulation' of the fluid around L. This relation gives
a simple physical interpretation of vorticity. If the loop is small so that the unit

Fluid Dynamics of the Midlatitude Atmosphere, First Edition. Brian J. Hoskins & Ian N. James.
© 2014 John Wiley & Sons, Ltd. Published 2014 by John Wiley & Sons, Ltd.

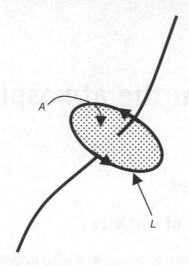

Figure 8.1   A vortex line and its associated circulation

vector **n** is more or less constant across A, then the average component of vorticity perpendicular to A is

$$|\xi \cdot \mathbf{n}| = \frac{C}{A},$$

that is, the magnitude of the vorticity is simply the circulation per unit area. For example, suppose that L is a circle of radius R and that a constant tangential wind of magnitude V blows around it. Then, the circulation around L is $2\pi RV$. Dividing by the area A enclosed by the loop tells us that the magnitude of the average vorticity perpendicular to A is $\zeta = 2V/R$. Circulation is a directly measurable quantity, referring to a finite region of a flow. Vorticity is a mathematical idealization of that concept, the circulation per unit area, as the loop becomes infinitesimally small. While the continuum hypothesis presumes that ξ tends towards some well-defined limiting value as $A \to 0$, in practice, this is not usually so, and there should really be a fixed minimum scale in mind when discussing vorticity. We shall discuss one example in Chapter 9 where the pointwise concept of vorticity appears to break down in a way which the concept of circulation does not.

An aid to visualizing the vorticity of a flow is analogous to the concept of a streamline. Imagine a line through the fluid whose direction is always parallel to the local vorticity. Such an imaginary line is called a 'vortex line' or a 'vortex filament', and it indicates the local rotation axis of spinning fluid elements. Notice that such a vortex line, like a streamline, must either be a closed loop or must extend to the boundaries of the fluid. Vortex lines may not be broken.

Now imagine a bundle of such vortex lines. These constitute a 'vortex tube'. Like individual vortex lines, vortex tubes must either be closed or must extend to the fluid boundaries. The vortex tube may be stretched and narrowed, but it cannot be broken and disconnected. We shall see in later sections that for inviscid flow with constant density, the vortex tube is also a material tube. It will be advected with the flow field and deformed by it. This leads to a very helpful conceptual model of the evolution of the vorticity field in terms of the twisting and stretching of vortex tubes.

In atmospheric applications, it is useful to separate the contribution of the horizontal and vertical components of velocity to the vorticity separately. Writing

$$\mathbf{u} = \mathbf{v} + w\mathbf{k}$$

and taking its curl gives

$$\xi = \nabla \times \mathbf{v} - \mathbf{k} \times \nabla_H w \qquad (8.3)$$

The first term contributes to both the horizontal and vertical components of vorticity. The second term contributes only to the horizontal components of vorticity. Consider the scaling of the first term. Its components can be deduced from Equation 8.1. Its horizontal components are of order U/D, while its vertical terms are of order U/L. For typical synoptic scale motions with

$$U \sim 10 \text{ ms}^{-1}, \quad L \sim 500 \text{ km}, \quad D \sim 5 \text{ km}$$

the horizontal components will have a typical magnitude of $2 \times 10^{-3} \text{s}^{-1}$, while the typical vertical components will be perhaps 10^{-5}s^{-1}, some two orders of magnitude smaller. In the lowest kilometre or so of the atmosphere, through the atmospheric boundary layer, the typical horizontal components of vorticity are another order of magnitude larger still, at around 10^{-2}s^{-1}. Following the scale analysis of Section 4.6, the contribution of the second term will be of order $\mathrm{Ro} U D/L^2$ to the horizontal components of vorticity and is of order $\mathrm{Ro}(D^2/L^2)$ smaller than the contribution from the first term. We conclude that on the synoptic scales, the second term in Equation 8.3 can be neglected, so that

$$\xi = \nabla \times \mathbf{v} \qquad (8.4)$$

to an adequate level of approximation. In the following text, we shall find that the hydrostatic approximation implies that this approximation should be made. One might also jump to the conclusion that the vertical components of vorticity could be neglected compared to the horizontal components. This turns out to be incorrect. We shall see in the next chapter that it is the vertical component of vorticity which is dynamically active, even though it is numerically smaller than the horizontal components. Indeed, meteorologists, when referring to the atmospheric vorticity, often mean the vertical component of vorticity only.

8.2   The vorticity equation

We start with the momentum equation in an absolute frame of reference, written as

$$\frac{\partial \mathbf{u}_A}{\partial t} = -\mathbf{u}_A \cdot \nabla \mathbf{u}_A - \frac{1}{\rho} \nabla p - \nabla \Phi + \dot{\mathbf{u}}$$

Here, $\dot{\mathbf{u}}$ represents acceleration due to friction and will generally be small away from the ground. To form a vorticity equation, take the curl of this equation to obtain

$$\frac{\partial \boldsymbol{\zeta}}{\partial t} = -\nabla \times \left(\mathbf{u}_A \cdot \nabla \mathbf{u}_A\right) - \nabla\left(\frac{1}{\rho}\right) \times \nabla p + \nabla \times \dot{\mathbf{u}} \tag{8.5}$$

where $\boldsymbol{\zeta} = \nabla \times \mathbf{u}_A$ is the absolute vorticity, the spin in the absolute frame of reference. The term involving the gravitational potential has vanished since the gravitational force can be expressed as the gradient of a scalar, and the curl of any gradient must be 0. The advection term needs further simplification, as in Section 7.6. Using the vector identity,

$$\nabla\left(\mathbf{A} \cdot \mathbf{B}\right) = \left(\mathbf{B} \cdot \nabla\right)\mathbf{A} + \left(\mathbf{A} \cdot \nabla\right)\mathbf{B} + \mathbf{B} \times \left(\nabla \times \mathbf{A}\right) + \mathbf{A} \times \left(\nabla \times \mathbf{B}\right)$$

with $\mathbf{A} = \mathbf{B} = \mathbf{u}_A$ gives

$$\mathbf{u}_A \cdot \nabla \mathbf{u}_A = \nabla\left(\frac{1}{2}\mathbf{u}_A \cdot \mathbf{u}_A\right) - \mathbf{u}_A \times \left(\nabla \times \mathbf{u}_A\right) \tag{8.6}$$

Hence, the vorticity equation becomes

$$\frac{\partial \boldsymbol{\zeta}}{\partial t} = \nabla \times \left(\mathbf{u}_A \times \boldsymbol{\zeta}\right) - \nabla\left(\frac{1}{\rho}\right) \times \nabla p + \nabla \times \dot{\mathbf{u}} \tag{8.7}$$

Use the vector identity

$$\nabla \times \left(\mathbf{A} \times \mathbf{B}\right) = \left(\mathbf{B} \cdot \nabla\right)\mathbf{A} - \mathbf{B}\left(\nabla \cdot \mathbf{A}\right) - \left(\mathbf{A} \cdot \nabla\right)\mathbf{B} + \mathbf{A}\left(\nabla \cdot \mathbf{B}\right)$$

to re-write the first term on the right-hand side:

$$-\nabla \times \left(\mathbf{u}_A \times \boldsymbol{\zeta}\right) = -\mathbf{u}_A \cdot \nabla \boldsymbol{\zeta} + \mathbf{u}_A\left(\nabla \cdot \boldsymbol{\zeta}\right) + \boldsymbol{\zeta} \cdot \nabla \mathbf{u}_A - \boldsymbol{\zeta}\left(\nabla \cdot \mathbf{u}_A\right) \tag{8.8}$$

The second term on the right-hand side is 0, and the last term is 0 for incompressible flow. For the moment assuming compressible flow, the final form of the vorticity equation is

$$\frac{D_A \boldsymbol{\zeta}}{Dt} = -\nabla\left(\frac{1}{\rho}\right) \times \nabla p + \left(\boldsymbol{\zeta} \cdot \nabla\right)\mathbf{u}_A - \boldsymbol{\zeta}\left(\nabla \cdot \mathbf{u}_A\right) + \nabla \times \dot{\mathbf{u}} \tag{8.9}$$

The vorticity equation is in the form of a 'material conservation relationship', in which the rate of change of absolute vorticity moving with the fluid is equal to the sum of a number of source or sink terms. If the source or sink terms are 0, then absolute vorticity is conserved moving with the fluid. But this is not in general the case.

Equation 8.9 can be transformed into a rotating frame of reference, using the relationships developed in Chapter 2, Equation 2.8 *et seq.*

$$\frac{D_A \boldsymbol{\zeta}}{Dt} = \frac{D_R \boldsymbol{\zeta}}{Dt} + \boldsymbol{\Omega} \times \boldsymbol{\zeta}$$

Also note that

$$(\zeta \cdot \nabla) \mathbf{u}_A = (\zeta \cdot \nabla)(\mathbf{u}_R + \Omega \times \mathbf{r}) = (\zeta \cdot \nabla) \mathbf{u}_R + \Omega \times \zeta$$

and

$$\nabla \cdot \mathbf{u}_A = \nabla \cdot \mathbf{u}_R$$

With these relationships, the vorticity equation in a rotating frame of reference is

$$\frac{D_R \zeta}{Dt} = (\zeta \cdot \nabla) \mathbf{u}_R - \zeta (\nabla \cdot \mathbf{u}_R) - \nabla \left(\frac{1}{\rho} \right) \times \nabla p + \nabla \times \dot{\mathbf{u}}, \qquad (8.10)$$

an equation which has the same form as Equation 8.9. Note however that it involves the absolute vorticity but the velocity relative to a rotating frame of reference. From now on, the A and R subscripts will be dropped. Equation 8.10 could have been obtained directly by taking the curl of the momentum equation in a rotating frame of reference, Equation 3.12, and recalling that the absolute and relative vorticities are related by $\zeta = \xi + 2\Omega$.

8.3   The vorticity equation for approximate sets of equations

When the shallow atmosphere approximation applies, the momentum equations (Equations 4.32, 4.33 and 4.34), yield the full vorticity equation:

$$\frac{D\zeta}{Dt} = (\zeta \cdot \nabla) \mathbf{u} - \zeta (\nabla \cdot \mathbf{u}) - \nabla \left(\frac{1}{\rho} \right) \times \nabla p + \nabla \times \dot{\mathbf{u}} \qquad (8.11)$$

This will again take the form of Equation 8.10, with the various metric terms approximated using a in place of r, and with the local horizontal component of the Earth's rotation neglected, so that the absolute vorticity is written as

$$\zeta = \xi + f\mathbf{k} \qquad (8.12)$$

In rectangular Cartesian coordinates on an f- or β-plane,

$$\xi = \left(\frac{\partial w}{\partial y} - \frac{\partial v}{\partial z} \right) \mathbf{i} + \left(\frac{\partial u}{\partial z} - \frac{\partial w}{\partial x} \right) \mathbf{j} + \left(\frac{\partial v}{\partial x} - \frac{\partial u}{\partial y} \right) \mathbf{k}. \qquad (8.13)$$

If in addition, the hydrostatic approximation is also made, the same vorticity equation again results, provided the relative vorticity is redefined to exclude the tiny contribution from the vertical velocity:

$$\xi = \nabla \times \mathbf{v} \qquad (8.14)$$

so that on an f- or β-plane,

$$\xi = -\frac{\partial v}{\partial z}\mathbf{i} + \frac{\partial u}{\partial z}\mathbf{j} + \left(\frac{\partial v}{\partial x} - \frac{\partial u}{\partial y}\right)\mathbf{k}. \tag{8.15}$$

With the anelastic approximation, the third term on the right-hand side of Equation 8.10 takes a simpler form, and the vorticity equation becomes

$$\frac{D\zeta}{Dt} = (\zeta \cdot \nabla)\mathbf{u} - \zeta(\nabla \cdot \mathbf{u}) - \mathbf{k} \times \nabla b + \nabla \times \dot{\mathbf{u}}. \tag{8.16}$$

This form applies to the basic equation sets (Equations 7.35 and 7.36), the latter with $g\theta/\theta_0$ replacing b. Neglecting friction term and using the mass conservation equation, the vertical component of the vorticity equation is

$$\frac{D\xi}{Dt} = (\xi_H \cdot \nabla w) + \zeta\left(\frac{\partial w}{\partial z} + \frac{w}{\rho_R}\frac{d\rho_R}{dz}\right) = (\xi_H \cdot \nabla w) + \frac{\zeta}{\rho_R}\frac{\partial(\rho_R w)}{\partial z} \tag{8.17}$$

where ξ_H denotes the horizontal components of the vorticity.

8.4 The solenoidal term

The third term on the right-hand side of Equation 8.10 is called the 'solenoidal' term or the 'baroclinic' generation term. Its magnitude depends upon the angle between surfaces of constant density and surfaces of constant pressure, and it is zero if these surfaces are parallel, that is, if the fluid is barotropic with $\rho = \rho(p)$ only. When this is not so, in other words, when the pressure and density surfaces are not parallel, the fluid is said to be 'baroclinic'.

The equation of state enables the solenoidal term to be re-written in terms of temperature and pressure rather than density and pressure or indeed in terms of potential temperature:

$$\nabla\left(\frac{1}{\rho}\right) \times \nabla p = \frac{R}{p}\nabla T \times \nabla p = \frac{R}{p}\left(\frac{T}{\theta}\right)\nabla\theta \times \nabla p \tag{8.18}$$

In the midlatitudes, the slope of the temperature or potential temperature surfaces is generally a great deal larger than the slope of the pressure surfaces. Assuming that geostrophic balance gives at least a rough estimate of the flow speeds, the slope of the pressure surfaces is of order $(\partial p/\partial y)/(\partial p/\partial z) \sim fU/g \sim 10^{-4}$. The slope of a $\theta = $ constant surface is of order $(\partial\theta/\partial y)/(\partial\theta/\partial z) \sim fU/N^2 D$, which is around 5×10^{-3}. So, if the pressure surfaces are effectively horizontal, by using the hydrostatic relationship, the solenoidal term can be written as follows:

$$\nabla\left(\frac{1}{\rho}\right) \times \nabla p \simeq \frac{g}{T}\mathbf{k} \times \nabla T \equiv \frac{g}{\theta}\mathbf{k} \times \nabla\theta \tag{8.19}$$

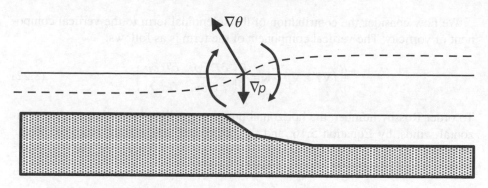

Figure 8.2 A sea breeze circulation set up by relatively cold water adjacent to a warm land mass, a circulation which can be set up during the daytime when the sun warms the land surface

This form of the solenoidal term is very similar to that given with the anelastic approximation as in Equation 8.16, and it is easy to see that the term results in the generation of a horizontal component of vorticity whenever there are horizontal temperature gradients. The sign of the resulting circulation is such that warmer air rises. An example of the operation of the solenoidal term is a sea breeze circulation. Here a temperature contrast between the land and sea sets up an overturning circulation with the warmer air rising. Figure 8.2 illustrates such a sea breeze circulation.

In examples such as that in Figure 8.2, the solenoidal term acts principally in the horizontal directions. So we split the vorticity into its horizontal and vertical parts:

$$\zeta = -\frac{\partial v}{\partial z}\mathbf{i} + \frac{\partial u}{\partial z}\mathbf{j} + (f + \xi)\mathbf{k} \equiv \zeta_H + (f + \xi)\mathbf{k}$$

Then, using the continuity equation, the horizontal components of the vorticity equation can be written as follows:

$$\rho\frac{D}{Dt}\left(\frac{\zeta_H}{\rho}\right) = \zeta \cdot \nabla \mathbf{v} + \left(\frac{g}{T}\right)\mathbf{k} \times \nabla T \tag{8.20}$$

To order Rossby number, this equation is dominated by the terms on its right-hand side, implying a near balance between those terms. So, to $O(\text{Ro})$, the horizontal component of the vorticity equation is simply

$$0 = f\frac{\partial \mathbf{v}}{\partial z} + \left(\frac{g}{T}\right)\mathbf{k} \times \nabla T \equiv f\frac{\partial \mathbf{v}}{\partial z} + \left(\frac{g}{\theta}\right)\mathbf{k} \times \nabla \theta \tag{8.21}$$

This is the thermal wind equation, relating the vertical shear of the horizontal components of the wind to the horizontal gradients of temperature that was introduced in Chapter 6. We shall return to thermal wind balance in Chapter 12.

We now consider the contribution of the solenoidal term to the vertical component of vorticity. The vertical component of the term is as follows:

$$\frac{R}{p}\left(\nabla T \times \nabla p\right) \cdot \mathbf{k} = \frac{R}{p}\left(\frac{\partial T}{\partial x}\frac{\partial p}{\partial y} - \frac{\partial T}{\partial x}\frac{\partial p}{\partial y}\right) \tag{8.22}$$

To order Rossby number, the horizontal pressure gradients are related to the horizontal winds, by Equation 5.16, and the horizontal temperature gradients to the vertical shear of the horizontal winds, by Equation 8.21. Therefore, the magnitude of the vertical component of the solenoidal term is

$$\left|\frac{R}{p}\left(\nabla T \times \nabla p\right) \cdot \mathbf{k}\right| \sim \frac{f^2 U^2}{gD}$$

Compare this with the horizontal advection of the vertical component of vorticity, which scales as

$$\frac{D\xi}{Dt} \sim \frac{U^2}{L^2}$$

The ratio of these terms is therefore

$$\frac{\text{Solenoidal}}{\text{Advection}} \sim \frac{f^2 L^2}{gD} \sim \text{Ro}^{-2}\text{Fr} \tag{8.23}$$

For a typical synoptic scale system, this ratio is around 0.05, and so we conclude that the solenoidal term does not make an appreciable contribution to the evolution of the vertical component of vorticity, at least on the synoptic and larger scales.

8.5   The expansion/contraction term

The second term on the right-hand side of Equation 8.10 relates the change of vorticity to the change of volume of a fluid element. Isolating this term

$$\frac{D\zeta}{Dt} = -\zeta\left(\nabla \cdot \mathbf{u}\right) + \text{other terms} \tag{8.24}$$

We shall designate this term as the 'expansion/contraction' term. It acts to increase or decrease the existing absolute vorticity; it cannot generate vorticity in an initially non-rotating fluid. The term is zero if the flow is incompressible. If the flow is compressible, so that the volume V of a fluid element changes according to

$$\nabla \cdot \mathbf{u} = \frac{1}{V}\frac{DV}{Dt},$$

then divergence, which increases the volume of fluid elements, reduces the absolute vorticity while convergence increases it. That is, the rotation rate of an expanding fluid element will slow down, and the rotation rate of a contracting volume will increase consistent with ideas of angular momentum conservation.

This term is closely related to the 'stretching term' to be described in the next section. Indeed, a little manipulation of the first two terms on the right-hand side of Equation 8.10 enables the two terms to be combined. For clarity, we shall treat them separately in this chapter.

8.6 The stretching and tilting terms

In order to interpret the effects of the first term upon the right-hand side of Equation 8.10, consider a single vortex line, shown in Figure 8.3, and set up Cartesian coordinates, x_l, y_l, z_l, the directions of whose axes are defined by the unit vectors $\mathbf{i}_l, \mathbf{j}_l, \mathbf{k}_l$. The unit vector $\mathbf{k}_l$ is parallel to the vortex line at the point of interest while both $\mathbf{i}_l$ and $\mathbf{j}_l$ are perpendicular to it. The local absolute vorticity is

$$\zeta = 0\mathbf{i}_l + 0\mathbf{j}_l + \zeta\mathbf{k}_l$$

Then the term can be written as follows:

$$\zeta \cdot \nabla \mathbf{u} = \zeta \left(\frac{\partial u_l}{\partial z_l} + \frac{\partial v_l}{\partial z_l} \right) + \zeta \frac{\partial w_l}{\partial z_l} \mathbf{k}_l$$
$$= \zeta \frac{\partial \mathbf{v}_l}{\partial z_l} + \zeta \frac{\partial w_l}{\partial z_l} \mathbf{k}_l$$

(8.25)

where $\mathbf{v}_l$ represents the components of fluid flow perpendicular to the vortex line.

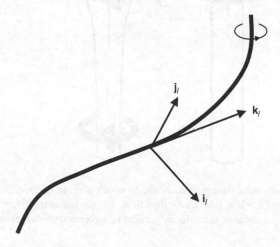

Figure 8.3 Local coordinates on a vortex line

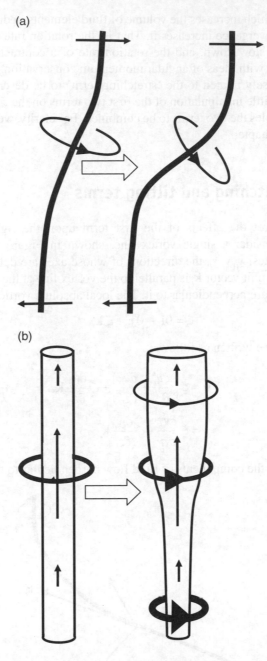

Figure 8.4 (a) The vortex tilting mechanism, in which a shear of the velocity perpendicular to a vortex line twists the line into a new direction, and (b) the vortex stretching or squashing mechanism, whereby a shear of the velocity parallel to a vortex tube causes thickening and thinning of the tube

The first term of Equation 8.25 is called 'vortex tilting'. It represents a local change of the (vector) vorticity resulting from shears of the velocity perpendicular to the vortex line. These tilt the line, changing the direction in which the vortex line points. The second term in Equation 8.25 represents a process called 'vortex stretching'. It may be thought of as a local change of the vorticity due to difluence in the component of flow parallel to the vortex tube, lengthening or shortening the tube. Equivalently, if there is net horizontal convergence, a vortex tube will narrow, and because it must conserve its angular momentum, its spin or vorticity must increase. The process reverses if there is net divergence ('vortex squashing'). Both processes are illustrated in Figure 8.4.

Like the expansion/contraction term, both these terms act on the existing absolute vorticity to change it. If there is no absolute vorticity initially, then these terms cannot create any. All they can do is to change the magnitude or direction of the pre-existing vorticity. The baroclinic term, on the other hand, can create vorticity from an initial situation of zero vorticity. In a rotating system, this is sometimes obscured, and it appears that vorticity is created in a barotropic flow with zero vorticity initially. Of course, in a rotating system, there is always copious planetary vorticity to hand. Creation of relative vorticity can be at the expense of planetary vorticity. We will return to the topic of generating vorticity in an initially irrotational flow in Section 11.1.

Which of these two processes will be most important in the Earth's atmosphere? We have already seen that the horizontal component of vorticity is large compared to the vertical, although we have asserted that it is the vertical component which is of greater dynamical significance. Let us first of all estimate the rate of generation of vertical vorticity by the tilting of horizontal vortex tubes. Tilting of a horizontal tube requires variations of the vertical velocity in the horizontal. Using the usual scalings, the horizontal component of vorticity will be of order U/D. Then, the tilting terms will have the following magnitude:

$$\left| \zeta \frac{\partial \mathbf{v}_l}{\partial z_l} \right| \sim \left| \zeta \frac{\partial w}{\partial x} \right| \sim \frac{U}{D} \frac{W}{L} \sim \mathrm{Ro} \frac{U^2}{L^2} \sim \mathrm{Ro}^3 f^2$$

where Equation 5.29 for the scale of vertical motion has been used to relate W to U. Now consider the stretching term, applied to the vertical component of vorticity. A typical magnitude of the vertical component of the absolute vorticity is simply f, the planetary vorticity. Then,

$$\left| \zeta \left(\frac{\partial u_l}{\partial x_l} + \frac{\partial v_l}{\partial y_l} \right) \right| \sim \mathrm{Ro} \frac{fU}{L} \sim \mathrm{Ro}^2 f^2$$

This result follows because since the divergence of the geostrophic wind is zero, so that the divergence in the expression for the stretching term must be the divergence of the ageostrophic wind, which scales as $\mathrm{Ro}U$. So, comparing the effectiveness of the tilting term and the stretching term at increasing the vertical component of vorticity, we find

$$\frac{\text{Stretching}}{\text{Tilting}} \sim \frac{1}{\text{Ro}} \tag{8.26}$$

This is an important result. For synoptic and large-scale flow, it indicates that the tilting of the large horizontal component of vorticity into the vertical is a minor contribution to the vertical vorticity budget. It implies that the large reservoir of horizontal vorticity is effectively decoupled from the vertical component. With this approximation, the vertical component of the vorticity equation (Equation 8.17), becomes

$$\frac{D\xi}{Dt} = \frac{\varsigma}{\rho_R} \frac{\partial (\rho_R w)}{\partial z} \tag{8.27}$$

where both expansion and stretching are represented by the term on the right-hand side.

8.7   Friction and vorticity

Up till this point, the friction has simply been represented schematically as an acceleration $\dot{\mathbf{v}}$ in the momentum equation. In this section, some more specific forms of $\dot{\mathbf{v}}$ are considered, and some general implications for the vorticity will be drawn. One difficulty is that the surface stress results from turbulent flow over the ground, and it is communicated to neighbouring levels in the atmosphere by turbulent momentum transports. Simple formulae for $\dot{\mathbf{v}}$ are very approximate and are generally replaced by complicated empirical relationships in all but the simplest atmospheric models.

Figure 8.5 illustrates the general principle schematically. In the free atmosphere, only the pressure gradient and Coriolis forces act on a fluid parcel. So if isobars are straight and uniformly spaced, in equilibrium, the parcel moves parallel to isobars at speed U where

$$F_c = fU = \frac{1}{\rho}|\nabla p| = F_p$$

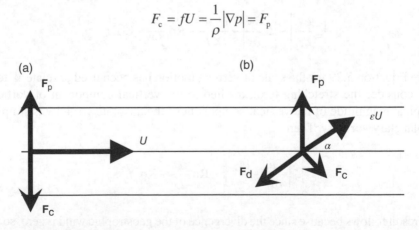

Figure 8.5   Schematic diagram of the horizontal forces acting on a fluid parcel in the northern hemisphere, with lower pressure to the north (a) in the free atmosphere (b) in the boundary layer. Air in the boundary layer is slowed by a factor ε, and its velocity vector is turned through an angle α towards low pressure

Close to the surface, a drag force $\mathbf{F}_d$ acts, and it is supposed that this acts in the opposite direction to the velocity. In this case, the three forces can balance only if the velocity vector is turned through an angle α towards low pressure and the parcel speed is reduced by a factor ε. A balance of forces can still exist, with the air parcel moving at constant speed at an angle α to the isobars.

Suppose that the drag force takes the idealized form of 'Rayleigh friction' in which the drag is simply proportional to flow speed and is directed in the opposite direction:

$$\mathbf{F}_d = -\frac{\mathbf{v}}{\tau_D} \qquad (8.28)$$

Then, balancing the forces acting perpendicular to the velocity leads to

$$F_c = F_p \cos(\alpha) \text{ or } \varepsilon = \cos(\alpha) \qquad (8.29)$$

while balancing the forces acting parallel to the isobars gives

$$F_d \cos(\alpha) = F_c \sin(\alpha) \text{ or } \tan(\alpha) = \frac{1}{f\tau_d} \qquad (8.30)$$

This calculation is very crude, and in particular, it does not determine the depth of the boundary layer, which becomes a free parameter of the system. What the model does show is that a mass flux from high pressure towards low pressure will develop within the boundary layer. This mass flux has important implications for the vorticity budget of the atmosphere.

Now take this flow and imagine bending the isobars into circles. Figure 8.6 shows a cross section through the system. Within the boundary layer, the Coriolis force on the retarded parcel does not balance the pressure gradient force, and as in the uniform flow case, there must be a component of ageostrophic flow from high pressure towards low pressure. In these circular systems, this ageostrophic flow converges into the centre of cyclones and diverges out of the centre of anticyclones. By continuity, this boundary layer flow must be balanced by vertical motion at the top of the boundary layer. This vertical motion may be thought of a lower boundary condition for the flow in the free atmosphere, and it is the basis of a series of secondary circulations wherever the flow is rotational. This vertical motion, exchanging boundary layer air and free atmosphere air at the top of the boundary layer, is called 'Ekman pumping'. The consequences of Ekman pumping in the region of the cyclone are vortex stretching in the boundary layer that offsets the frictional damping of the cyclone there and vortex shrinking above the boundary layer. The vortex shrinking above the boundary layer produces a spin-down of the cyclone above the boundary layer. In this way, secondary ageostrophic circulations act to distribute the boundary layer friction through a deeper layer of the atmosphere. A similar process, with reversal of the appropriate signs, acts to attenuate anticyclones. The impact of Ekman pumping is generally more significant than the direct effect of friction on the

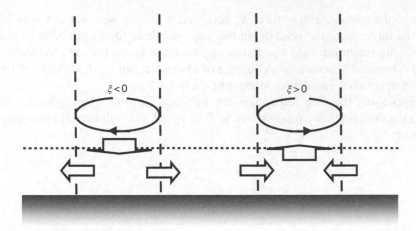

Figure 8.6 Cross section through a cyclone–anticyclone pair, showing the ageostrophic flow in the boundary layer and the resulting Ekman pumping

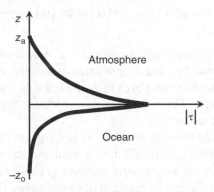

Figure 8.7 Schematic tangential stress profile over the oceans

vorticity field and is a much more important mechanism for dissipating relative vorticity in the atmosphere and ocean.

The conceptual model shown in Figure 8.6 is a plausible qualitative picture of the interaction between the boundary layer and the interior flow, but without more detail, it does not provide quantitative estimates of, for example, the magnitude of the Ekman pumping or the time needed for the mechanism to dissipate vorticity. Two further calculations which make the arguments more quantitative are now given.

Around two-thirds of the Earth's surface is fluid: it consists of ocean. So Figure 8.7 illustrates the boundary layer at the interface of the atmosphere and ocean. It is characterized by tangential stresses which fall off with distance from the interface and which are continuous across that fluid interface. For steady flow in either the atmosphere or ocean, there must be a three-way balance between the Coriolis force, the pressure gradient force, and the gradient of tangential stress:

$$f\mathbf{k} \times \mathbf{v} + \frac{1}{\rho}\nabla p = \frac{1}{\rho}\frac{\partial \tau}{\partial z} \qquad (8.31)$$

Break the flow into its geostrophic and ageostrophic parts $\mathbf{v} = \mathbf{v}_g + \mathbf{v}_a$ where

$$f\mathbf{k} \times \mathbf{v}_g = -\frac{1}{\rho}\nabla p$$

and so

$$\rho f\mathbf{k} \times \mathbf{v}_a = \frac{\partial \tau}{\partial z}.$$

This can be re-arranged to give

$$\rho \mathbf{v}_a = -\frac{\partial}{\partial z}\left(\frac{1}{f}\mathbf{k} \times \tau\right). \qquad (8.32)$$

There will therefore be an Ekman drift of matter between any two levels z_1 and z_2, with a mass flux of

$$\int_{z_1}^{z_2}(\rho \mathbf{v}_a)\,dz = -\frac{1}{f}\mathbf{k} \times \tau\Big|_{z_1}^{z_2} \qquad (8.33)$$

If the mass continuity equation is written in the form

$$\nabla \cdot (\rho \mathbf{v}_a) + \frac{\partial}{\partial z}(\rho w) = 0$$

then the Ekman pumping velocity is related to the Ekman drift by

$$\rho w\Big|_{z_1}^{z_2} = -\nabla \cdot \left(\int_{z_1}^{z_2}\rho \mathbf{v}_a\,dz\right) \qquad (8.34)$$

Finally, combining Equations 8.33 and 8.34 gives the relationship between the Ekman pumping velocity and the curl of the wind stress:

$$\rho w\Big|_{z_1}^{z_2} = -\frac{1}{f}\mathbf{k} \cdot (\nabla \times \tau)\Big|_{z_1}^{z_2} \qquad (8.35)$$

From this relationship, two results follow. First, integrating Equation 8.33 separately through the atmospheric and oceanic boundary layers, the Ekman mass fluxes through either are found to be equal but opposite in sign:

$$\mathbf{M}_{EA} = \int_{0}^{z_A} \rho_A \mathbf{v}_a dz = \frac{1}{f} \mathbf{k} \times \tau_s$$

$$\mathbf{M}_{EO} = \int_{-z_O}^{0} \rho_O \mathbf{v}_a dz = -\frac{1}{f} \mathbf{k} \times \tau_s$$

(8.36)

In retrospect, the result that the total mass flux produced by internal stresses in the two systems is zero is a simple example of Newton's third law of motion. Since the ocean is much denser than the air, the actual velocities are much smaller in the ocean, but the mass fluxes are the same in magnitude. Finally, using the continuity equations, it follows that

$$w_{EA} = \frac{1}{\rho_A f} \mathbf{k} \cdot \left(\nabla \times \tau_s \right)$$

$$w_{EO} = \frac{1}{\rho_O f} \mathbf{k} \cdot \left(\nabla \times \tau_s \right)$$

(8.37)

Again, the mass fluxes are identical, but this time, they are in the same direction. In the atmosphere, a typical midlatitude value for w_{EA} is around $1–2\,\mathrm{cm\,s^{-1}}$. In the ocean, the Ekman pumping is smaller by a factor of 10^3 because of the greater density of seawater. The coupling between the atmosphere and ocean by means of the Ekman layer is the primary mechanism for the wind-driven circulation in the ocean, generated by vortex stretching and squashing as a result of the Ekman pumping induced by the atmospheric vorticity.

In the laboratory, an explicit expression for the Ekman boundary layer can be deduced since the tangential stresses are given by Newton's law of viscosity $\tau = \mu \partial \mathbf{v}/\partial z$. Suppose that outside the boundary layer, flow is steady, geostrophic, and parallel to the x-axis. Then, balancing the viscous and Coriolis terms gives an estimate of the boundary layer thickness:

$$fU \sim v\frac{U}{\delta^2} \text{ or } \delta \sim \left(\frac{v}{f}\right)^{1/2}$$

For laboratory scales, a typical f might be $1\,\mathrm{rad\,s^{-1}}$ and a typical v around $10^{-6}\,\mathrm{m^2\,s^{-1}}$, suggesting a boundary layer thickness of around $1\,\mathrm{mm}$. The momentum equations for straight steady flow reduce to

$$v\frac{\partial^2 u}{\partial z^2} + fv = 0$$

$$v\frac{\partial^2 v}{\partial z^2} + f\left(u - u_g\right) = 0$$

(8.38)

A neat trick for solving these equations is to define a complex variable $U=u+iv$ so the pair of equations reduces to

$$\nu \frac{\partial^2 U}{\partial z^2} - ifU = -ifU_g$$

subject to the boundary conditions $U=0$ at $z=0$ and $U \to U_g = u_g + i0$ as $z \to \infty$. Noting that $i^{1/2} = (1+i)/\sqrt{2}$, the solution is

$$u = u_g \left(1 - \exp\left(-\frac{z}{H_E} \right) \cos\left(\frac{z}{H_E} \right) \right)$$

$$v = u_g \exp\left(-\frac{z}{H_E} \right) \sin\left(\frac{z}{H_E} \right)$$

(8.39)

where $H_E = (2\nu/f)^{1/2}$. The Ekman pumping velocity can be calculated using these expressions to calculate the horizontal divergence in the boundary layer and then integrating from the surface to the top of the boundary layer:

$$w_E = \frac{1}{2} H_E \xi_g$$

(8.40)

Here, ξ_g is the geostrophic relative vorticity above the boundary layer. Taking $w = w_E$ to be the lower boundary condition for the interior flow, the rate of change of vorticity due to vortex stretching is, from Equation 8.25,

$$\frac{D\xi}{Dt} = \zeta \frac{\partial w}{\partial z} = f \frac{w_{top} - w_E}{D} = -\frac{fw_E}{D}$$

From this expression and formula (8.40), the time taken for the Ekman pumping to erode significantly the vortex above the boundary layer can be calculated. This timescale is called the spin-down time, τ_E:

$$\tau_E = \left(\frac{1}{\xi} \frac{D\xi}{Dt} \right)^{-1} = \left(\frac{(f+\xi)w_E}{\xi D} \right) \sim \left(\frac{2D^2}{f\nu} \right)^{1/2}$$

(8.41)

We have assumed that $|\xi| \ll f$, which is equivalent to small Rossby number flow. For a typical laboratory experiment using water as a working fluid, with f around $1\,\mathrm{rad\,s^{-1}}$, D around $10\,\mathrm{cm}$, and ν around $10^{-6}\,\mathrm{m^2\,s^{-1}}$, the spin-down time is about $300\,\mathrm{s}$. A timescale based on vertical diffusion to erode the interior vorticity would be much longer, around

$$\tau_D \sim \frac{D^2}{\nu}$$

which for the same parameters is some $10^4\,\mathrm{s}$.

8.8   The vorticity equation in alternative vertical coordinates

An analogue to the vorticity can be defined in pressure-based or in entropy-based vertical coordinates. However, some care is needed in the use of the words 'vertical' and 'horizontal', for we are now dealing with non-orthogonal coordinates. In general, p-surfaces and θ-surfaces make an angle with surfaces of constant geopotential. In the midlatitudes, the angle between p and Φ surfaces is typically around 10^{-3} rad. The angle between θ- and Φ surfaces is typically 10^{-2} rad. However, in the following analysis, the 'vertical' component of vorticity refers to that component which is perpendicular to geopotential surfaces. Similarly, expressions for the vorticity will be given in terms of the wind $\mathbf{v}=u\mathbf{i}+v\mathbf{j}$, which blows parallel to Φ-surfaces. However, these components are differentiated with p (or θ) constant, not with Φ constant. Thus, the magnitudes of ζ, ζ_p and ζ_θ may differ, but they are all in the same direction, namely, parallel to $\nabla\Phi$.

Using the operator $\nabla_p = \left(\dfrac{\partial}{\partial x}\bigg|_p , \dfrac{\partial}{\partial y}\bigg|_p , \dfrac{\partial}{\partial p} \right)$, the pressure co-ordinate analogue of vorticity may be written as follows:

$$\xi_p = \nabla_p \times \mathbf{v} = \left(-\frac{\partial v}{\partial p}, \frac{\partial u}{\partial p}, \frac{\partial v}{\partial x}\bigg|_p - \frac{\partial u}{\partial y}\bigg|_p \right) \tag{8.42}$$

The pressure co-ordinate version of absolute vorticity is $\zeta_p = \xi_p + f\mathbf{k}$. Then, taking the curl of Equation 6.18a and using the hydrostatic relation in the form Equation 6.18b gives

$$\frac{D\zeta_p}{Dt} = \left(\zeta_p \cdot \nabla \right)\mathbf{v} + \mathbf{k} \times \left(\hat{R}\nabla\theta' \right) + \nabla \times \dot{\mathbf{v}} \tag{8.43}$$

Each of the terms in this equation is interpreted analogously to the corresponding term in the basic vorticity equation (Equation 8.10). The first term represents stretching and tilting, just as before. The second term in Equation 8.10 is not present in Equation 8.43 because it is identically zero in pressure coordinates. The second term in Equation 8.43 is a p-coordinate version of the baroclinic (or 'solenoidal') term; note that, like the pressure gradient term in the momentum equations, without any further approximation, it is a linear term in pressure coordinates.

A similar development can be carried out in θ-coordinates. The θ-coordinate version of vorticity is

$$\zeta_\theta = f\mathbf{k} + \left(-\frac{\partial v}{\partial \theta}, \frac{\partial u}{\partial \theta}, \frac{\partial v}{\partial x}\bigg|_\theta - \frac{\partial u}{\partial y}\bigg|_\theta \right) \tag{8.44}$$

The vorticity equation becomes

$$\frac{D\zeta_\theta}{Dt} = \left(\zeta_\theta \cdot \nabla \right)\mathbf{u} - \zeta_\theta \left(\nabla \cdot \mathbf{u} \right) - \mathbf{k} \times \nabla\Pi + \nabla \times \dot{\mathbf{u}} \tag{8.45}$$

Note that the vertical component of ζ_θ can be written as follows:

$$\zeta_\theta = f + \left.\frac{\partial v}{\partial x}\right|_\theta - \left.\frac{\partial u}{\partial y}\right|_\theta$$

$$= f + \left.\frac{\partial v}{\partial x}\right|_z - \left(\frac{\partial\theta}{\partial x}\bigg/\frac{\partial\theta}{\partial z}\right)\frac{\partial v}{\partial z} - \left.\frac{\partial u}{\partial y}\right|_z + \left(\frac{\partial\theta}{\partial y}\bigg/\frac{\partial\theta}{\partial z}\right)\frac{\partial u}{\partial z} \qquad (8.46)$$

$$= (\boldsymbol{\zeta}\cdot\nabla\theta)\bigg/\frac{\partial\theta}{\partial z}$$

At the level of the hydrostatic approximation, $\partial\theta/\partial z = |\nabla\theta|$ so that ζ_θ approximates the component of absolute vorticity normal to an isentropic surface. In the same way,

$$\xi_p = f + \left.\frac{\partial v}{\partial x}\right|_p - \left.\frac{\partial u}{\partial y}\right|_p$$

is approximately the component of absolute vorticity perpendicular to a pressure surface.

8.9   Circulation

We noted in the opening of this chapter the close relationship between vorticity and circulation. Circulation is arguably the more basic quantity. It has meaning even when the vorticity fields fail to converge to a finite value at a point, and indeed, any attempt to measure vorticity ends up as an attempt to measure circulation around an appropriately small circuit.

In an absolute or inertial frame of reference, the circulation about some circuit L is

$$C_A(t) = \oint_L \mathbf{u}_A \cdot d\mathbf{l} \qquad (8.47)$$

The evolution of the circulation is described by

$$\frac{dC_A}{dt} = \oint_L \frac{D_A\mathbf{u}_A}{Dt}\cdot d\mathbf{l} + \oint_L \mathbf{u}_A \cdot \frac{D}{Dt}d\mathbf{l}.$$

Now

$$\frac{D_A\delta\mathbf{l}}{Dt} \equiv \delta\mathbf{u}_A \qquad (8.48)$$

while the first term can be expanded using the momentum equation (Equation 2.40), to give

$$\frac{dC_A}{dt} = \oint_L \left(-\frac{1}{\rho}\nabla p + \nabla\Phi + \dot{\mathbf{u}}\right)\cdot d\mathbf{l} + \oint_L \mathbf{u}_A \cdot d\mathbf{u}_A$$

$$= \oint_L \left(-\frac{1}{\rho}dp + d\Phi + d\left(\frac{1}{2}\mathbf{u}_A\cdot\mathbf{u}_A\right) + \dot{\mathbf{u}}\cdot d\mathbf{l}\right)$$

Since the circuit L is closed, the exact differentials all integrate to 0, so that the absolute circulation equation takes the following form:

$$\frac{dC_A}{dt} = \oint_L \left(-\frac{1}{\rho} dp + \dot{\mathbf{u}} \cdot d\mathbf{l} \right) \tag{8.49}$$

For an incompressible fluid, ρ is a constant so that it can be taken outside the integral of the first term, and this becomes zero. Therefore, if the fluid is also frictionless, this equation shows that the circulation can never change. This result is sometimes called 'Kelvin's circulation theorem'. In particular, if the initial absolute circulation is 0, it must remain zero at all subsequent times. This result appears paradoxical. Incompressible flow in the laboratory and in nature is full of circulating structures, even when the friction term is expected to be negligible. Where does all this circulation come from? Furthermore, at first sight, friction looks an unlikely candidate to generate circulation. Suppose the friction term takes the form of 'Rayleigh friction':

$$\dot{\mathbf{u}} = -\frac{\mathbf{u}_A}{\tau_D}$$

Then, from Equation 8.49, the absolute circulation evolves according to

$$\frac{dC_A}{dt} = \oint_L \left(-\frac{1}{\rho} dp \right) - \frac{C_A}{\tau_D}$$

If the flow is incompressible, so the first term is zero, then any initial absolute circulation will decay towards zero on a timescale τ_D. We shall return to the problems posed by the Kelvin circulation theorem in Chapter 11.

Now consider the effect of rotation of the system. The absolute circulation can be written in terms of the relative velocities:

$$C_A = \oint_L (\mathbf{u} + \mathbf{\Omega} \cdot \mathbf{r}) \cdot d\mathbf{l}$$

The right-hand side is equal to the relative circulation C plus a contribution which may be transformed to an integral over a surface S bounded by L:

$$C_A = C + \int_S \left(\nabla \times (\mathbf{\Omega} \times \mathbf{r}) \right) . d\mathbf{s} = C + \int_S 2\mathbf{\Omega} \cdot d\mathbf{s} \tag{8.50}$$

So, finally the absolute circulation may be written as follows:

$$C_A = C + 2\Omega S_e \tag{8.51}$$

where S_e is the area of the circuit S projected onto the equatorial plane. The relative circulation equation can be written as follows:

$$\frac{dC}{dt} + 2\Omega \frac{dS_e}{dt} = \oint_L \left(-\frac{1}{\rho} dp + \dot{\mathbf{u}} \cdot d\mathbf{l} \right) \tag{8.52}$$

This result could have been obtained directly from the momentum equation in rotating coordinates, Equation 3.15. Alternatively, both Equations 8.49 and 8.52 can be obtained by integrating the absolute vorticity equations, Equations 8.9 and 8.10 over a surface bounded by the circuit S.

Under the shallow atmosphere approximation, it is only the vertical component of Ω which is relevant. Equation 8.52 remains valid, but S_e now must be defined as the area of the projection onto the equatorial plane of S_h, where S_h is the projection of S onto a geopotential surface. Alternatively, write Equation 8.50 in the following form:

$$C_A = C + \oint_{S_h} f \, dS$$

If the hydrostatic approximation is assumed, then this equation and the circulation equation (8.49) remain valid, but involve only the horizontal velocities in the definition of C:

$$C = \oint_L \mathbf{v} \cdot d\mathbf{l}$$

Consider a circuit which is a line of latitude at a fixed height and a flow which is purely zonal. Then, the absolute circulation is

$$C_A = \left(u + r\Omega \cos\phi\right) 2\pi r \cos\phi$$

and for frictionless flow, this is conserved. This result is equivalent to the conservation of angular momentum per unit mass.

The close relationship between the vorticity and the circulation means that a similar parallelism exists between the circulation equation (Equation 8.52), and the vorticity equation (Equation 8.10). The stretching/tilting terms of the vorticity equation do not appear explicitly in the circulation equation. However, they are implicitly expressed by the motion of the circuit L. The friction term and the solenoidal terms are explicit in both equations.

Finally, let us examine the solenoidal term more closely. Different forms of it may be written, using the equation of state in its various guises. These are as follows:

$$-\oint_L \frac{1}{\rho} dp = -\oint_L RT d\ln p = -\oint_L c_p \theta d\left(\frac{p}{p_0}\right)^\kappa \tag{8.53}$$

Under certain circumstances, this term will be 0. In particular, it will be 0 if ρ can be expressed as a function of p alone. Equivalently, it will be 0 if either T or θ can be expressed as a function of p only. Such an atmosphere is said to be barotropic. A particular case is when one of ρ, T, or θ is constant on the circuit. Only one such condition can be expected to persist with time: if there are no heat sources or sinks, so that the motion is adiabatic, then a circuit initially on a θ-surface must remain on that surface and the solenoidal term will vanish identically. The absolute circulation must be conserved on all such isentropic circuits. This proves a very strong constraint on atmospheric motion and leads to the development of a concept called 'potential vorticity', a concept developed in detail in Chapter 10.

For a frictionless adiabatic atmosphere, the vertical component of the vorticity equation (8.45) in isentropic co-ordinates is

$$\frac{\mathrm{D}\zeta_\theta}{\mathrm{D}t} = -\zeta_\theta\left(\nabla_\theta \cdot \mathbf{v}\right) \tag{8.54}$$

The absolute circulation around a circuit on an isentropic circuit is

$$C_\mathrm{A} = \oint_L \mathbf{v}_\mathrm{A} \cdot \mathbf{dl} = \int_{S_\mathrm{h}}\left(\nabla_\theta \times \mathbf{v}_\mathrm{A}\right).\mathbf{dS} = \int_{S_\mathrm{h}}\zeta_\theta \mathrm{d}S \tag{8.55}$$

Note that the surface integral is over the horizontal surface S_h since the circulation is defined only in terms of the horizontal wind $\mathbf{v}$. The conservation of C_A on an isentropic circuit is then consistent with the vorticity equation (8.45), which states that ζ_θ only changes as a result of expansion or contraction of the areas of fluid parcels on isentropic surfaces.

9

Vorticity and the barotropic vorticity equation

9.1 The barotropic vorticity equation

The barotropic vorticity equation is an over-simplification of the full vorticity equation in most circumstances. Nevertheless, it is useful because it isolates some basic dynamical processes which are still represented by more complicated equation sets. The configuration envisaged is illustrated in Figure 9.1. Homogeneous fluid of constant density flows between two rigid horizontal surfaces, vertical separation D. Away from the boundaries, the velocity vector $\mathbf{v}$ is independent of height, so fluid moves as coherent columns. The vorticity has only a vertical component, which is non-zero. Ekman layers form on the top and bottom boundaries, and a small Ekman pumping vertical velocity provides the top and bottom boundary conditions for the flow in fluid interior. The top boundary is supposed to be horizontal, but the bottom boundary may have a topography $h(x, y)$ where $|h| \ll D$.

With these assumptions, the baroclinic and tilting terms are zero in the fluid interior. The curl of the tangential stress will be ignored, but the Ekman pumping velocity w_E will represent the effects of boundary layer friction. The vorticity equation, Equation 8.17, becomes

$$\frac{D\zeta}{Dt} = \zeta \frac{\partial w}{\partial z}. \tag{9.1}$$

Since $\zeta = f + \xi$ and

$$\frac{\xi}{f} \sim \frac{U}{fL} = \text{Ro},$$

we shall assume small Ro and write the right-hand side of Equation 9.1 in terms of f only. We shall also adopt the beta-plane approximation, so the vorticity equation becomes

Fluid Dynamics of the Midlatitude Atmosphere, First Edition. Brian J. Hoskins & Ian N. James.
© 2014 John Wiley & Sons, Ltd. Published 2014 by John Wiley & Sons, Ltd.

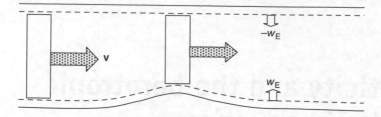

Figure 9.1 The configuration envisaged. The velocity $\mathbf{v}$ is independent of z outside the Ekman layers, and so fluid moves in coherent columns

$$\frac{D\xi}{Dt} + \beta v = f\frac{\partial w}{\partial z}.$$ (9.2)

At the upper boundary, $z = D$, the boundary condition is simply $w = -w_E$. At the lower boundary, the flow must also be parallel to the topography, so that

$$w = w_E + \mathbf{v}\cdot\nabla h \text{ at } z = 0.$$

Here we have assumed that $h \ll D$ so that the boundary condition can be applied at $z=0$ rather than $z=h$. For the motion to be independent of height, it follows from Equation 9.2 that $\partial w/\partial z$ must be uniform with height. So $\partial w/\partial z$ is determined by the values of w to at $z=0$ and D. Therefore, the vorticity equation becomes

$$\frac{D\xi}{Dt} + \beta v = -\frac{f}{D}\left(2w_E + \mathbf{v}\cdot\nabla h\right).$$ (9.3)

Finally, substituting from Equation 8.40 for w_E, the barotropic vorticity equation becomes

$$\frac{D\xi}{Dt} = -\beta v - f\mathbf{v}\cdot\nabla\left(\frac{h}{D}\right) - \frac{\xi}{\tau_E}$$ (9.4)

where

$$\tau_E = \frac{D}{\left(2f\nu\right)^{1/2}}$$ (9.5)

is called the 'Ekman spin-up time'. If the top boundary was frictionless, the factor 2 would be omitted in Equation 9.3 and the time τ_E in Equation 9.5 would be multiplied by 2. The different behaviours of this equation are made clearer by introducing dimensionless variables. Scaling horizontal distance by L, velocity by U, time by L/U, vorticity by U/L, and topographic height by

$$\frac{h_{max}}{\text{Ro}D},$$

Equation 9.4 can be rewritten in terms of dimensionless variables as

$$\frac{D\xi}{Dt} = -(\text{Be})v - (\text{Hi})\mathbf{v} \cdot \nabla(h) - \frac{(\text{Ek})^{1/2}}{(\text{Ro})}\xi. \qquad (9.6)$$

Here, various dimensionless numbers have been introduced:

$$\text{Be} = \frac{\beta L^2}{U}, \quad \text{Ek} = \frac{2v}{fD^2}, \quad \text{Ro} = \frac{U}{fL}, \quad \text{Hi} = \frac{h_{max}}{\text{Ro}D}. \qquad (9.7)$$

The first three of these are the β-number, the Ekman number, and the Rossby number, respectively. The importance of Hi was pointed out by Hide (1961) and will be designated as the 'Hide number' for present purposes.

9.2 Poisson's equation and vortex interactions

In order to solve either Equation 9.4 or 9.6, the velocity field must be related to the vorticity field. This is done by introducing a streamfunction ψ such that

$$\mathbf{v} = \left(-\frac{\partial\psi}{\partial y}\right)\mathbf{i} + \left(\frac{\partial\psi}{\partial x}\right)\mathbf{j}$$

which is possible because the flow is non-divergent. Then,

$$\xi = \nabla_H^2\psi \qquad (9.8)$$

This important relationship is called 'Poisson's equation'. It is an elliptic partial differential equation.

Elliptic equations will be met in several different contexts in this book, and so here, at their first appearance, it is worth digressing to discuss some of their basic properties. The discussion will focus on functions of two variables, x and y, but it is easily generalized to three variables. The general second-order partial differential equation

$$a\psi_{xx} + 2b\psi_{xy} + c\psi_{yy} = f(x, y, \psi, \psi_x, \psi_y)$$

is said to be elliptic if $ac - b^2 > 0$. This is certainly true for the Poisson equation, in which $a = c = 1$ and $b = 0$. The equation can be solved on some finite domain A in (x, y) space, which is enclosed by a closed circuit L, provided boundary conditions for ψ on L are specified. These boundary conditions may be one of a number of different kinds. Dirichlet boundary conditions mean that values of ψ at all points on L are specified.

For Neumann boundary conditions, the normal gradient of ψ rather than values of ψ itself is specified, that is, $\mathbf{n} \cdot \nabla \psi$ is specified on the boundary, $\mathbf{n}$ being a unit vector normal to L and directed out of A. Finally, of considerable utility in many geophysical problems, boundary conditions may be periodic, that is, values of ψ and $\mathbf{n} \cdot \nabla \psi$ on one part of L are the same as the values of ψ and $-\mathbf{n} \cdot \nabla \psi$ on another part of L. In a particular problem, it is possible that different types of boundary condition apply on different segments of L.

From now on, we focus on the Poisson Equation 9.8. Note that this is a linear equation, so that if ψ_1 and ψ_2 are both solutions to the equation, then $\alpha \psi_1 + \beta \psi_2$, α and β both being constants, is also a solution. This property leads to a little trick which makes it easier to deal with general boundary conditions. Suppose that ψ_1 is the solution to the Laplace equation (a special case of the Poisson equation with zero on the right-hand side) with appropriate boundary conditions on ψ_1 and that ψ_2 is a solution to the full Poisson equation subject to simple Dirichlet boundary conditions $\psi_2 = 0$ on L:

$$\nabla_H^2 \psi_1 = 0 \text{ with } \psi = \psi_L \text{ on } L$$
$$\nabla_H^2 \psi_2 = \xi \text{ with } \psi = 0 \text{ on } L$$

Then adding the two equations, it is clear that $\psi_1 + \psi_2$ is the solution to the Poisson equation with the boundary condition $\psi = \psi_L$. In other words, there is no loss of generality in just discussing solutions to the Poisson equation with a $\psi_L = 0$ boundary condition.

For an elliptic equation such as Poisson's equation, with boundary conditions on ψ prescribed on some closed loop in x–y space, values of ψ are defined at all interior points of that loop. In particular, suppose ξ and ψ vary sinusoidally:

$$\xi = Z \exp\left(i\left(kx + ly\right)\right), \quad \psi = \Psi \exp\left(i\left(kx + ly\right)\right)$$

Then, substituting this solution in Equation 9.8,

$$\Psi = -\frac{Z}{\left(k^2 + l^2\right)} \tag{9.9}$$

This reveals two points about the relationship between the vorticity and stream-function fields. First, they are anti-correlated, with maxima in ξ corresponding to minima in ψ and vice versa. Secondly, the ψ field is a smoothed version of the ξ field, with the larger scales in ξ emphasized and the small scales suppressed. The choice of these particular forms for ξ and ψ is not as restrictive as might be at first thought, for our sinusoidal forms may be but one term in a Fourier summation over all possible wavenumbers k and l.

It is worth exploring the relationship between ξ and ψ in some simple cases. Figure 9.2 shows a single isolated patch of vorticity. Suppose that for $r < r_0$, the relative vorticity is uniform with value ξ and that it is zero for larger values of r.

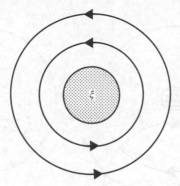

Figure 9.2 The flow around an isolated patch of vorticity

The mean vorticity is related to the tangential velocity and circulation around the edge of the vortex using Stokes' theorem:

$$\pi r_0^2 \xi = 2\pi r_0 V_0 = C_0$$

Also at larger radii,

$$\pi r^2 \xi = 2\pi r V = \pi r_0^2 \xi_0 = C_0 \tag{9.10}$$

Writing Poisson's equation in cylindrical coordinates and noting that there is no variation in the vertical or azimuthal directions, the streamfunction is

$$\psi = \psi_0 - \frac{C}{2\pi} \ln\left(\frac{r}{r_0}\right) \tag{9.11}$$

This reveals one of the most important consequences of Poisson's equation, Equation 9.8. A localized patch of vorticity induces a circulation which extends to infinity. Fluid remote from the centre of a vortex is in motion due to its presence. We may describe this principle as 'the principle of action at a distance'.

This principle of action at a distance means that two discrete separate vortices interact with one another. Assume for the moment that the radius of each vortex is small compared to their separation. In the case of two vortices of the same sign, shown in Figure 9.3a, the vortices move in opposite directions perpendicular to the line joining their centres. As a result, they will orbit around their mutual 'centre of circulation', which for vortices of the same strength will be the midpoint of the line joining their centres. In the case of vortices of equal strength but opposite sign, Figure 9.3b, they will both move along parallel trajectories, in the direction perpendicular to the line joining their centres.

In more realistic situations, the radius of the vortex may not be particularly small compared to the distance separating vortices. In this case, the circulation induced by the one vortex will vary across the second vortex, leading to shearing as well as simple

(a)   (b)

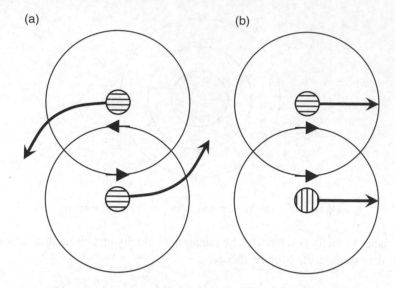

Figure 9.3   The interaction of two compact vortices. (a) Two vortices of the same sign orbit their centre of circulation. (b) Two vortices of opposite sign move in parallel tracks, perpendicular to the line joining them

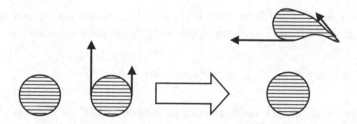

Figure 9.4   Schematic illustration of the shearing of one vortex by a neighbouring vortex which is taken to be so strong that it is not appreciably sheared itself

advection. In terms of the flow kinematics of Section 1.3, the vortex will be subject both to rotation and deformation. Such shearing is the first step leading to the formation of long vorticity streamers, a feature characteristic of many geophysical flows.

As well as action at a distance, the behaviour illustrated in Figure 9.4 relies on the other major characteristic of the solutions of Poisson's equation, namely, that the streamfunction is a highly smoothed version of the vorticity field. Thus, in Figure 9.4, the change of the orbiting vortex from nearly circular to drawn-out has hardly any effect on the induced flow far away from it, and the two vortices continue to circulate around each other. If two vortices are separated by a distance small compared to their radius, then their combined streamfunction is more or less a single extremum, with only small internal structure towards its centre. The result is that the component vortices circulate around one another, deforming and shearing, until they consist of two intertwining streamers of vorticity. These streamers continue to wrap ever more tightly around each other until there is effectively just a single vortex. Such a process is called 'vortex merging' in the literature since well outside the region of the two vortices, the circulation will depend only upon the integrated vorticity. It will be as if the two vortices have merged. However, in the vicinity of the two vortices, the process is perhaps better described as vortex 'intervolving'.[1] Instabilities and small-scale mixing processes may act to remove some of the detail of the intertwining streamers to produce a more homogeneous distribution of vorticity. Vortex merging/intervolving is a ubiquitous process in two-dimensional or quasi-two-dimensional flows. In the absence of processes to regenerate small-scale eddies, repeated inter-volving events ultimately result in a flow containing just one large positive and one large negative vortex.

These various examples give us a vocabulary with which to discuss the evolution of quasi-two-dimensional flows in terms of vortex dynamics. But the description is essentially quasi-linear, in that we have described the flow as if the vorticity is more or less passively advected by the streamfunction field. This is only roughly true, principally for weak isolated vortices. For stronger vortices, the vorticity field deter-mines the streamfunction field and vice versa, a strong nonlinear feedback which dominates most flows.

9.3   Flow over a shallow hill

According to the barotropic vorticity Equation 9.6, only the second term on the right-hand side can generate vorticity in a flow that has zero vorticity initially. That term represents the effect of orography on the boundaries on the vorticity distribu-tion. So in this section, we will discuss the operation of the orographic term and the subsequent evolution of the vorticity field it generates.

Equation 9.6 can be solved numerically for particular choices of parameter and for particular initial conditions. It is not the purpose of this book to discuss the appropriate numerical methods in detail, and so only a brief indication of the approach which may be used is included here. The domain is divided into a rectan-gular network of gridpoints, at each of which values of the relative vorticity ξ and the streamfunction ψ are defined. Then, the continuous differential Equation 9.6 may be replaced by a finite-difference analogue:

$$\frac{\xi_{x,y}^{t+1} - \xi_{x,y}^{t-1}}{2\Delta t} = T_{x,y} \qquad (9.12)$$

where $T_{x,y}$ is the Eulerian relative vorticity tendency at time t at each gridpoint, calculated by forming a finite-difference analogue to each of the terms in

Equation 9.6. Rearranging Equation 9.12, the vorticity at a new time level can be calculated from the current and previous time levels. The new streamfunction is then obtained by solving the Poisson equation with the new vorticity as right-hand side. The iteration is repeated as often as needed to take the solution to the time required. There are a number of subtleties. The desired solution to the continuous equation can be swamped by unwanted amplifying solutions which result from the finite-difference approximations. The various terms in $T_{x,y}$, particularly the nonlinear advection terms and the dissipative Ekman pumping terms, have to be formulated carefully to control such unwanted numerical solutions.

The flows described in this section are examples of 'inertial flow' in which $Ro \rightarrow 0$, $Ek \rightarrow 0$ and also for which $Ek^{1/2}/Ro \rightarrow 0$. This last condition means that the Coriolis force is dominant but that Ekman pumping is insignificant. Figure 9.5 shows a sequence of streamfunction and relative vorticity fields obtained by the numerical process outlined in the last paragraph. All variables have been rendered dimensionless using the scaling suggested for Equation 9.6. The initial state consisted of zero relative vorticity and uniform flow everywhere. After a short time, fluid initially over the hill has been advected downstream and replaced by fluid advected up the slope. The fluid downstream of the hill has been subjected to vortex stretching and so acquires positive relative vorticity. The fluid which is advected onto the hill has been subjected to vortex squashing and so acquires negative relative vorticity.

The positive vorticity which has been created as fluid was advected off the hill gradually dissipates as a result of Ekman pumping. Neglecting all terms but the Ekman pumping term, the barotropic vorticity equation reduces to

$$\frac{D\xi}{Dt} = -\frac{\xi}{\tau_E}$$

which has the solution

$$\xi = \xi_0 e^{-t/\tau_E}$$

following the fluid. Thus, non-zero relative vorticity decays downstream to zero on a timescale τ_E. In this example, the fixed negative vortex on top of the hill and the shed positive vortex interact while they remain close together. The shed eddy is deflected to the right before being advected out of range of the fixed eddy.

If the beta term and the Ekman pumping term are both ignored, the barotropic vorticity equation may be written as a conservation relationship:

$$\frac{D}{Dt}\{\xi + \mathrm{Hi}h\} = 0 \tag{9.13}$$

So, if the hill is a right circular cylinder of height 1 dimensionless unit, the steady-state relative vorticity at its summit will simply be Hi. This will induce a tangential flow of speed:

$$u' \sim \frac{1}{2}\mathrm{Hi} \tag{9.14}$$

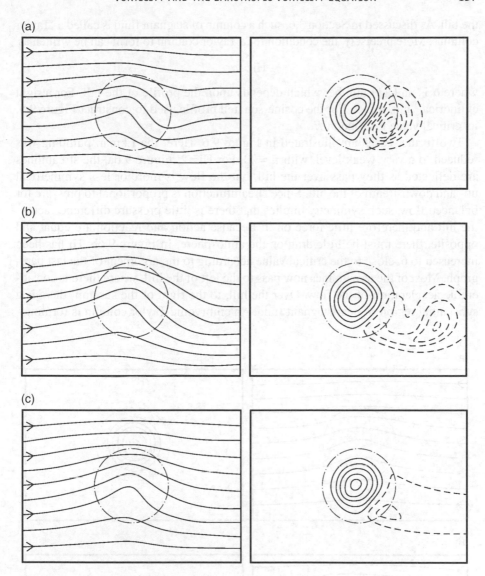

Figure 9.5 A time sequence of streamfunction (left) and vorticity (right) fields from a numerical integration of the barotropic vorticity equation, Equation 9.6, with flow from top to bottom of the plots. Negative vorticity is shown by solid contours and positive vorticity by dashed contours. Parameters Be$=0.0$, Hi$=4.0$, $\tau_E = 3.3$, contour interval$=0.680$. (a) Time $t=0.82$, (b) $t=2.41$ and (c) $t=16.07$. From James (1980)

around the edge of the hill. On the left-hand side of the hill, this flow is in the same direction as the far upstream flow and so the flow to the left of the hill is accelerated to $(1+\text{Hi}/2)$ dimensionless units. To the right of the hill, however, the opposite is true and the flow is slowed down to $(1 - \text{Hi}/2)$. If Hi>2, the flow will reverse so that a patch of stagnant fluid, which is never able to leave the hill, will form on the left-hand side of the hill. As discussed in Section 3.6, such a column of stagnant fluid is called a 'Taylor column'. More precisely, the condition for a Taylor column to form can be written as

$$Hi > \alpha$$

where α is an O(1) number which depends upon the profile of the hill. For a right cylindrical[2] hill, α is 2. For the cosine-squared profile used to construct Figure 9.5, its critical value is around 6.7.

Two further cases are illustrated in Figure 9.6. Here, the Ekman pumping was reduced to a very weak level, with $\tau_E = 20$. For Hi = 2, Figure 9.6a, the streamlines are deflected as they pass over the hill, but the flow is more or less symmetrical up- and downstream of the hill. Since streamfunction is proportional to pressure for balanced flow, such symmetry implies that there is little pressure difference across the hill and therefore little force on it. Because action and reaction are equal and opposite, there must be little drag on the atmosphere. In Figure 9.6b, Hi has been increased to 6, close to the critical value according to the argument of the last paragraph. Most of the streamlines now pass to the left of the hill, where there is considerable acceleration of the flow. Over the hill, to the right of the summit, there is a substantial region of near-stagnant fluid. An embryonic Taylor column is forming.

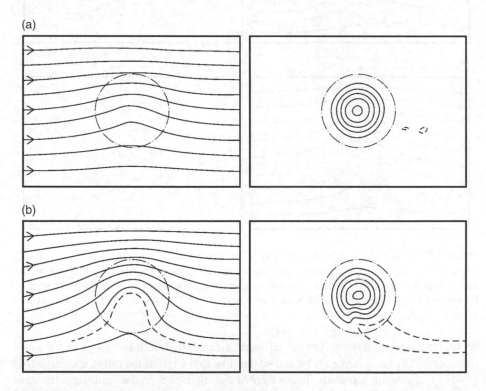

Figure 9.6 Showing the effect of obstacle height on the flow over a shallow hill. In both cases, Be = 0 and $\tau_E = 20$. (a) Hi = 2, contour interval = 0.390, (b) Hi = 6, contour interval = 1.15

A feature of Figure 9.6b is the 'wake' of weak positive vorticity downstream of the hill. The Ekman pumping, though weak, has time to significantly attenuate the negative vorticity of the near-stagnant fluid on the top of the hill. When eventually it is advected off the hill, so that vortex tubes return to their original length, it will then acquire weak positive relative vorticity. A result of this wake is the asymmetry of the streamlines up- and downstream of the hill. Consequently, the Ekman pumping increases the drag exerted on the flow by the hill.

9.4   Ekman pumping

In the example shown in Figure 9.6, the Ekman pumping was a small correction to the solution. An alternative limit has been explored theoretically and in fact is somewhat easier to reproduce in laboratory experiments. This is the limit $Ro \rightarrow 0$, $Ek \rightarrow 0$ but $Ek^{1/2} \gg Ro$. This means that the Coriolis force dominates the other accelerations to which a fluid element is subject, and that the Ekman pumping timescale is short compared to the advection timescale. In this case, a more meaningful scaling of the hill height is in terms of $Ek^{1/2}D$.

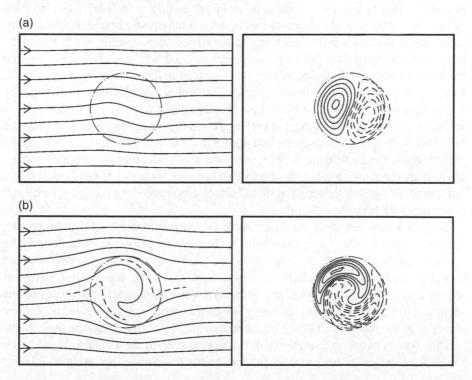

Figure 9.7   The viscous Taylor column. Two integrations of the barotropic vorticity equation with Be $= 0$, $\tau_E = 0.1$. (a) $h_{max}/Ek^{1/2}D = 1$, contour interval $= 0.286$, (b) $h_{max}/Ek^{1/2}D = 10$, contour interval $= 1.46$

Figure 9.7 shows two cases. The first is for a relatively low hill. The streamlines are scarcely deflected by the hill, but the vorticity field is very different from the inertial case, with negative vorticity on the upslope and positive vorticity on the downslope. The pattern is very nearly anti-symmetric up- and downstream of the hill. The two vortices are weak and tend to rotate around each other a small amount. The second case is a tall hill, with $h_{max} = 10\mathrm{Ek}^{1/2}D$. The positive and negative vortices now interact strongly; they spiral into each other and are trapped in the vicinity of the hill. The streamline pattern shows a region of near-stagnant fluid over the hill, while other streamlines pass either side of the hill. Away from the hill, the flow is remarkably similar to non-rotating 'potential flow' past a solid cylinder.

9.5   Rossby waves and the beta plane

The remaining term to be discussed in Equation 9.6 is the first term on the right-hand side, $-\mathrm{Be}v$. An integration with Be of 5.0 is shown in Figure 9.8. Here, the boundary conditions are periodic on the inflow and outflow boundaries. Any disturbances which are advected across the outflow boundary reappear at the inflow boundary. This of course is analogous to large-scale flow in the Earth's midlatitudes, since flow around latitude circles is subject to periodic boundary conditions. A shallow hill with Hi of 3.0 was retained to provide some disturbance to the otherwise uniform steady flow, and a rather short τ_E of 3.0 was specified to ensure that the wake of vorticity from the hill largely decayed before it interfered with itself. In the vicinity of the hill, a nearly circular patch of negative vorticity develops by vortex squashing. But downstream of the hill, an intense wake of wavy disturbances forms. A weaker wake of longer wavelength waves propagates upstream from the hill. The wake spreads both along and across the channel, eventually reflecting off the sidewalls. The disturbances which pass over the downstream boundary are starting to reappear at the upstream boundary, but because of the rather fast spin-up time, only weak vestiges of the original disturbances propagate right around the channel to interact with flow over the hill.

These waves when $\beta \neq 0$ are an example of 'Rossby waves', a most important category of wave motion in the midlatitude atmosphere. They dominate flow on the synoptic and larger scales. Rossby waves have some very curious properties which contrast sharply with the behaviour of sound waves, water waves, and waves on strings with which most people are familiar. We will explore these properties in this section. First, a mathematical derivation of the existence and properties of Rossby waves is given, and then that will be complemented by a more descriptive account.

The vorticity equation, Equation 9.6, is rendered analytically tractable by breaking the flow into a steady 'background' flow and a small wavelike 'perturbation'. Thus,

$$\mathbf{v} = U\mathbf{i} + \mathbf{v}'$$
$$\psi = -Uy + \psi' \tag{9.15}$$
$$\xi = 0 + \xi'$$

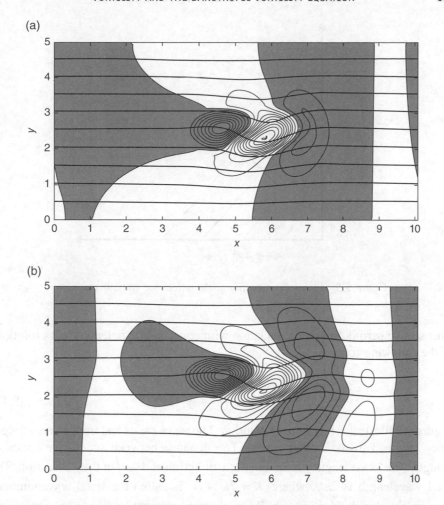

Figure 9.8 Streamfunction (thick contours) and vorticity (thin contours) plots from an integration of the barotropic vorticity equation, with $Be=5.0$, $Hi=3.0$ and $\tau_E=3.0$. (a) $t=1.0$, (b) $t=6.0$

The effect of the Be term in Equation 9.6 is isolated by assuming that Hi is small and that τ_E is extremely long. The perturbation flow is supposed to be weak, in the sense that $|\mathbf{v}'| \ll U$. The form of solution sought, Equation 9.15, is substituted into Equation 9.6, and any products of primed quantities are neglected on the assumption that they are very much smaller than the linear terms:

$$\frac{\partial \xi'}{\partial t} + U\frac{\partial \xi'}{\partial x} + Be v' = 0$$

or in terms of streamfunction:

$$\frac{\partial \xi'}{\partial t} + U\frac{\partial \xi'}{\partial x} + Be\frac{\partial \psi'}{\partial x} = 0. \tag{9.16}$$

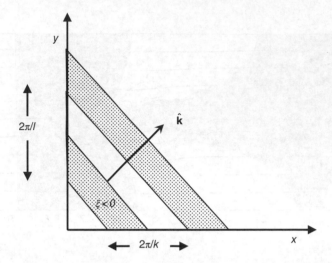

Figure 9.9   Plane-wave solution to the barotropic vorticity equation

This simple partial differential equation with constant coefficients admits solutions of the following form:

$$\xi' = Z \exp(kx + ly - \omega t)$$
$$\psi' = \Psi \exp(kx + ly - \omega t) \tag{9.17}$$

Figure 9.9 illustrates this form of solution. The wave crests and troughs are straight lines, inclined to the x- and y-axes. The distance between successive crests or troughs (the 'wavelength') is $2\pi/k$ in the x-direction and $2\pi/l$ in the y-direction. The total wavelength is $2\pi/K$ where $K = \sqrt{k^2 + l^2}$ is called the 'total wavenumber'. The unit vector $\hat{\mathbf{k}} = (k\mathbf{i}/K + l\mathbf{j}/K)$ is directed at right angles to the crests and troughs of the wave. The crests and troughs move along the x- and y-axes with phase speeds ω/k and ω/l respectively.[3] Substituting the supposed solutions, Equation 9.17, into Equation 9.16 yields a condition for the validity of these solutions, in the form of an algebraic relationship between the frequency ω and the wavenumbers k and l:

$$\omega = Uk - \frac{Bek}{\left(k^2 + l^2\right)} \tag{9.18}$$

This important equation is called the 'dispersion relationship'. Dividing by k, it may be rewritten in terms of the phase speed along the x-axis:

$$c_{px} = U - \frac{Be}{\left(k^2 + l^2\right)} = U - \frac{Be}{K^2} \tag{9.19}$$

Alternatively, Equation 9.19 can be written in terms of dimensional variables:

$$c_{px} = U - \frac{\beta}{K^2}$$

Two implications are immediately apparent from these forms of the dispersion relationship. First, relative to the background flow, Rossby waves can only travel in one direction along the x-axis, from east to west. This is in sharp contrast to most waves of everyday experience, such as sound and surface water waves, which propagate equally happily in any direction. Secondly, when the total wavenumber is small compared to Be, the phase speed depends strongly on the wavenumber. Such waves are said to be dispersive; an arbitrary disturbance, which may be Fourier-analysed into a sum of many different wavenumber components, quickly breaks up or 'disperses' as its different wavenumber components separate.

Equation 9.18 is in fact a complete description of the behaviour of small-amplitude Rossby waves in a barotropic fluid. However, the arguments which led to that equation are essentially algebraic and may not convey much physical insight into the processes responsible for the Rossby wave properties. As an alternative, we show how many of the Rossby wave properties implicit in the dispersion relationship may be deduced using vortex interaction arguments.

To simplify, consider a wave with $l = 0$, so that fluid elements are displaced only in the y-direction. Initially, consider a motionless fluid with zero relative vorticity. The fluid is displaced meridionally in a sinusoidal pattern, as illustrated in Figure 9.10. Write the displacement as

$$\Delta y = \Delta y_0 \sin(kx)$$

According to Equation 9.1, the displaced parcels must conserve their absolute vorticity f_0. So a parcel displaced a distance Δy from its initial latitude will acquire relative vorticity:

$$\Delta \xi = -\text{Be}\Delta y$$

The sign is important. Poleward displaced fluid acquires anticyclonic (negative) relative vorticity, and equatorward displaced fluid acquires cyclonic (positive) relative vorticity. So we have strips of fluid, perpendicular to the x-axis, with alternating positive and negative relative vorticities. Therefore,

$$\frac{\partial v}{\partial x} = -\text{Be}\Delta y_0 \sin kx.$$

Integrating this shows that each strip has a sheared meridional flow varying between $-\text{Be}\Delta y_0/k$ and $+\text{Be}\Delta y_0/k$. Consider the vicinity of $x = 0$. The line of fluid elements which initially lay along the line $y = 0$ now makes an angle α with the x-axis, where

$$\alpha = \Delta y_0 k$$

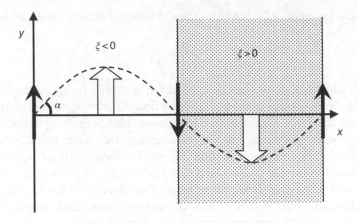

Figure 9.10 A Rossby wave, pictured as a set of parallel sheets of fluid displaced north and south of their initial latitudes. The bold arrows indicate the flow associated with the perturbation vorticity

where we are assuming that $\Delta y_0 \ll k^{-1}$. Considering the movement of the tangent at the intersection of the line of fluid elements with the x-axis, as fluid is advected by the meridional flow, so this intersection will migrate along the x-axis at the following speed:

$$c = -\frac{v}{\alpha}$$

This is the phase speed of the wave along the x-axis. We have just seen that $v = Be\Delta y_0/k$ and so

$$c = -\frac{Be}{k^2} \tag{9.20}$$

In fact, the whole sinusoidal pattern of displaced fluid elements will migrate along the x-axis with this speed. It is, of course, the Rossby phase speed predicted by Equation 9.19 for the case $l = 0$ and $U = 0$.

By way of an alternative, an interpretation of Rossby wave activity in terms of the Euler equations can also be developed. The question is: what provides the restoring force associated with the Rossby wave motion? The answer is that the pressure gradients provide the necessary restoring force, and we can deduce the pressure field from the vorticity field by considering geostrophic balance.

Take the same configuration as shown in Figure 9.10. The diagram is redrawn as Figure 9.11. The meridional velocity, shown by the dashed curve, is independent of y and is in geostrophic balance with a pressure field. That pressure field has a minimum where the cyclonic vorticity is largest and a maximum where the anticyclonic vorticity is largest. But the relationship between wind and pressure gradient involves the Coriolis parameter f. In the north of the domain, where f is larger, the pressure

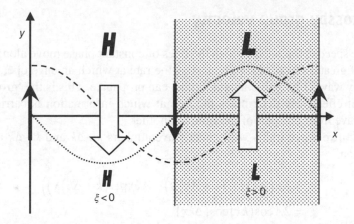

Figure 9.11 As Figure 9.10, but showing the meridional wind and the pressure field associated with a Rossby wave. The block arrows indicate the unbalanced north–south pressure gradient force

wave will have a larger amplitude than in the south, where f is smaller. Therefore, at the crests and troughs of the pressure wave, there will be an unbalanced north–south pressure gradient. This unbalanced pressure gradient force will accelerate or decelerate the meridional winds. The dotted line indicates this acceleration. It is in quadrature with the meridional wind wave and therefore will cause the entire pattern of v to migrate from east to west.

Rossby waves are ubiquitous in situations where the wavelength of disturbances is, in some sense, 'long'. What is meant by 'long' is that

$$K \sim \mathrm{Be}^{1/2}$$

or less. If $K \gg \mathrm{Be}^{1/2}$, the dispersion relationship reduces to $c_{px} = U$, that is, disturbances are simply advected by the flow.

An interesting isomorphism arises if Be is 0, but the bottom topography is simply

$$h(x,y) = Gy$$

Then, the barotropic vorticity equation, Equation 9.6, can be written as follows:

$$\frac{D\xi}{Dt} + \mathrm{Be}_{\mathrm{eff}} v = 0 \quad \text{where } \mathrm{Be}_{\mathrm{eff}} = \frac{1}{\mathrm{Ro}}\left(\frac{L}{D}\right)G \qquad (9.21)$$

This is exactly of the same form as the barotropic vorticity equation with an effective β proportional to the bottom slope. So an isolated hill, for example, can support cylindrical Rossby waves propagating around its edge. The flows over shallow topography described in Section 9.3 can be interpreted in terms of such trapped cylindrical Rossby waves.

9.6   Rossby group velocity

The phase speed is the rate at which the lines of constant phase move along the x- or y-axes. Of greater physical significance is the rate at which an envelope, enclosing the Rossby waves of significant amplitude, can propagate. This is the 'group velocity', and it effectively determines the rate at which information is carried by the Rossby wave from one part of the fluid to another.

Suppose two wave trains with zonal wavenumbers $k+\Delta k$ and $k-\Delta k$ are superposed. Then,

$$\psi = \mathrm{Re}\left(A\exp\left(i\left(k-\Delta k\right)x\right)+A\exp\left(i\left(k+\Delta k\right)x\right)\right)$$
$$= 2A\cos\left(kx\right)\cos\left(\Delta kx\right) \tag{9.22}$$

This represents a short-wave wavelength $2\pi/k$ modulated by a long-wave envelope of wavelength $2\pi/\Delta k$. While the individual crests and troughs move with the phase speed c_{px} along the x-axis, the speed of the envelope will be different. Rewrite Equation 9.17 but now include the time dependence of the two superposed waves:

$$\psi' = \mathrm{Re}\left(A\exp\left(i\left(\left(k+\Delta k\right)x-\left(\omega+\Delta\omega\right)t\right)\right)+A\exp\left(i\left(\left(k-\Delta k\right)x-\left(\omega-\Delta\omega\right)t\right)\right)\right)$$
$$= 2\cos\left(\Delta kx-\Delta\omega t\right)\mathrm{Re}\left(A\exp\left(i\left(kx-\omega t\right)\right)\right) \tag{9.23}$$

The envelope moves along the x-axis at a speed $\Delta\omega/\Delta k$. In the limit $\Delta k\to 0$, the envelope moves along the x-axis at speed

$$c_{gx} = \frac{\partial\omega}{\partial k} \tag{9.24}$$

Certain types of wave motion are non-dispersive, which means that the phase speed is independent of wavenumber k. In that case, and in that case only, the phase speed and the group velocity are identical. Many everyday wave motions belong to this category, among them sound waves and waves on a stretched string. However, long Rossby waves with $k\sim\mathrm{Be}^{1/2}$ or less are strongly dispersive.

For the moment, we consider the dimensional case, with dispersion relation (Figure 9.12)

$$\omega = Uk - \frac{\beta k}{\left(k^2+l^2\right)}.$$

The phase speed and group velocity in the x-direction are

$$c_{px} = U - \frac{\beta}{K^2},$$

$$c_{gx} = U + \frac{\beta\left(k^2-l^2\right)}{K^4}$$

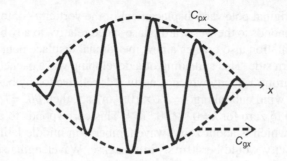

Figure 9.12 Schematic illustration of a packet of waves propagating along the x-axis. The individual crests and troughs move with the phase speed c_{px}, while the wave envelope moves with the x-component of the group velocity, c_{gx}

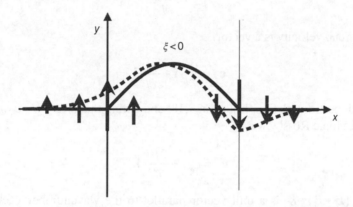

Figure 9.13 The development in time of an isolated meridional displacement of a vorticity contour (thick line) on a β-plane. Associated with the meridional motion (arrows), the vorticity contour changes to look like that given by the dashed line: the initial disturbance propagates to the west relative to any basic westerly flow, while an envelope of new disturbances spreads to the east relative to the basic flow. The former corresponds to the phase speed and the latter to the group velocity

Therefore, although the phase speed is always less eastwards than the flow, meridionally extended waves have an eastward group velocity that is greater than the flow speed – information is carried eastwards faster than the flow moves!

Figure 9.13 shows a simple example in which an isolated disturbance is introduced into a motionless fluid on a β-plane by displacing fluid northwards in a limited region. The heavy solid line marks the line of fluid elements which lay along the x-axis before the displacement. The northward displacement results in a region of negative relative vorticity. The meridional flow induced by this vorticity perturbation is indicated in the diagram. Consistent with the discussion earlier in this chapter, this meridional motion spreads beyond the region of the vorticity perturbation. It acts to move the vorticity contour poleward on the western side of the original perturbation and equatorward on the eastern side, as indicated by the dashed line.

Therefore, the original poleward displacement of the vorticity contour moves to the west. This corresponds to the westward phase speed relative to any basic zonal flow. The meridional motion also creates a new meridional displacement of the opposite sign on the eastern side. This spreading of a developing wave packet to the east corresponds to the group velocity which is eastwards relative to any basic zonal flow.

For a $20\,m\,s^{-1}$ wind and taking $\beta = 1.6 \times 10^{-11}\,m^{-1}\,s^{-1}$, then $\beta / k^2 = U$, and the phase speed is reduced to zero for $k = 0.9 \times 10^{-6}\,m^{-1}$. This corresponds to a wavelength of about $7000\,km$, which is about zonal wavenumber 4 in middle latitudes. The eastward group velocity would be $40\,m\,s^{-1}$ in this case. Wavelengths shorter than this wavelength would move eastwards, and those longer would move westwards.

A similar analysis for the propagation of wave packets in the y-direction yields

$$c_{gy} = \frac{\partial \omega}{\partial l}.$$

The total group velocity is a vector:

$$\mathbf{c}_g = \frac{\partial \omega}{\partial k}\mathbf{i} + \frac{\partial \omega}{\partial l}\mathbf{j} \tag{9.25}$$

From the dispersion relation, Equation 9.18, and noting that $\beta / K^2 = U - c_{px}$, it follows that for barotropic Rossby waves,

$$\mathbf{c}_g = c_{px}\mathbf{i} + \frac{2\beta k}{K^3}\hat{\mathbf{k}} \tag{9.26}$$

Here, $\hat{\mathbf{k}} = (k\mathbf{i} + l\mathbf{j})/K$ is a unit vector parallel to the wavenumber vector $k\mathbf{i} + l\mathbf{j}$. This expression implies that the zonal component of group velocity is always larger than the phase speed in the zonal direction. The meridional component of group velocity is poleward for $l > 0$, that is, when troughs and ridges of the wave are orientated north-west to south-east, and equatorwards for $l < 0$, that is, when the troughs and ridges are orientated south-west to north-east. Equation 9.26 can be manipulated further to give a form for $\mathbf{c}_g$ which can be illustrated by a simple graphical construction. If the angle between the x-axis and the vector $\hat{\mathbf{k}}$ is denoted by α, then $\cos(\alpha) = k/K$ and $\sin(\alpha) = l/K$. Substituting $U - c_{px}$ for β / K^2 and using some standard trigonometric identities, Equation 9.26 may be rewritten as follows:

$$\mathbf{c}_g = \left(U + \left(U - c_{px}\right)\cos\left(2\alpha\right)\right)\mathbf{i} + \left(\left(U - c_{px}\right)\sin\left(2\alpha\right)\right)\mathbf{j} \tag{9.27}$$

The construction, shown in Figure 9.14, works as follows. On axes representing the u and v components of velocity, draw a circle, radius $U - c_{px}$ centred on the point $(U, 0)$. A line from point $(c_{px}, 0)$ at an angle α denotes the direction of the unit vector $\hat{\mathbf{k}}$. Then, a line from the origin to the point where the wave vector intersects the circle gives the magnitude and direction of the group velocity. Notice that for sufficiently small K, it is possible for c_{px} to be negative.

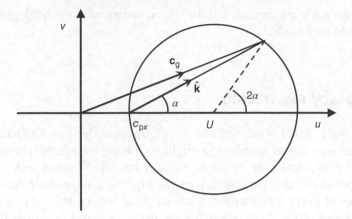

Figure 9.14   Graphical construction for the group velocity of barotropic Rossby waves

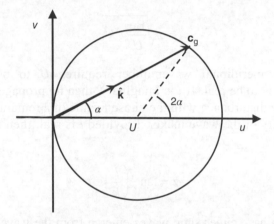

Figure 9.15   As Figure 9.14 but for the special case of steady Rossby waves with $c_{px} = 0$

A special case of particular importance is that of steady Rossby waves, that is, Rossby waves with zero phase speed. We have considered this case earlier in this section, where we found that they may typically have zonal wavenumber 4. As the atmosphere flows over fixed regions of vorticity forcing, such as mountain ranges and continent/ocean transitions, one might expect Rossby wave activity to be triggered. Over a period of time, the transient activity will average to close to 0, and the mean response will be a wave pattern with zero phase speed relative to the forcing. However, the fact that the phase speed is zero does not mean that the group velocity is 0. Such steady wave patterns can provide an important basis for teleconnection patterns, for the remote influence of localized forcing.

Figure 9.15 is the same as Figure 9.14, but for the case $c_{px} = 0$. The group velocity vector is now parallel to the wave vector. The maximum group velocity occurs when $\mathbf{c}_g$ is directed zonally; it then has a magnitude $2U$. The group velocity becomes smaller as the angle α increases, and there is no situation in which the group velocity is directed westwards.

9.7   Rossby ray tracing

Suppose a wave maker is inserted into a uniform westerly flow. The nature of the wave maker need not be specified; it might be forcing by a mountain as the flow passes over it, or some other localized vorticity forcing. We shall seek the remote steady response to this wave maker. We shall suppose that the wave maker excites disturbances of every wavenumber. Some of these will be able to propagate and some not. Trains of Rossby waves will leave the wave maker in various directions, depending upon their wave number. Choose a particular zonal wave number k. Then, from the dispersion relationship 9.18 with $\omega = 0$,

$$l = \pm \sqrt{\frac{\text{B}e}{U} - k^2} \tag{9.28}$$

Notice that the meridional wavenumber requires U to be positive and $k < (\text{B}e/U)^{1/2}$ if it is to be real. If l is imaginary, then no propagating solution is possible, and any disturbance with the chosen value of k remains trapped in the immediate vicinity of the wave maker. Provided l is real, then the angle of the wave vector is

$$\alpha = \tan^{-1}\left(\frac{l}{k}\right).$$

Since a real l can have either sign, waves emerge from the wave maker in pairs: one wave train, with $l > 0$, towards the pole and the other, with $l < 0$, towards the equator. The magnitude of the group velocity is given by the construction shown in Figure 9.16. After a certain time Δt, the envelope of the steady waves will have advanced into a circular region radius $U\Delta t$ extending east, north and south of the wave maker. Figure 9.16 illustrates the advance of a wavefront of $\omega = 0$ Rossby waves from a wave maker.

This argument will only work for a fairly short time from the initiation of the waves. Generally, Rossby waves will propagate into regions where the flow is different from that at their point of origin. A theory to deal with this, called 'ray tracing', is borrowed from optics. It considers how a wave propagates through a slowly varying medium. 'Slowly varying' means that the lengthscale over which the medium changes substantially is long compared to the wavelength of the waves themselves. This assumption is often only marginally satisfied for long Rossby waves, but nevertheless, the theory seems to give a satisfactory qualitative account of Rossby wave propagation over global scales.

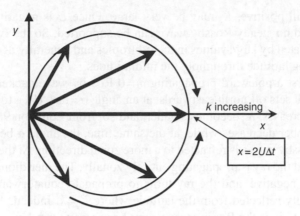

Figure 9.16   The envelope of steady Rossby waves emanating from a wave maker at the origin after time Δt

Suppose that the zonal flow is given by some $U = U(y)$. We must also recognize that the parameter Be also varies on a scale comparable to a, the Earth radius. Then, the linearized barotropic vorticity equation, Equation 9.16, becomes

$$\frac{\partial \xi'}{\partial t} + U(y)\frac{\partial \xi'}{\partial x} + \left(\mathrm{Be}(y) - \frac{\partial^2 U}{\partial y^2}\right)\frac{\partial \psi'}{\partial x} = 0 \qquad (9.29)$$

Now if U, Be and so on are slowly varying functions of y, we shall treat this equation as if it had constant coefficients, albeit different constant coefficients, at each latitude. As before, the dispersion relationship can be arranged into a diagnostic for the meridional wave number:

$$l = \pm\sqrt{\frac{\mathrm{Be} - U_{yy}}{U} - k^2} \qquad (9.30)$$

Now if a packet of waves propagates through a varying medium, its properties will evolve as it propagates. If the medium does not vary in the x-direction, then the wave will conserve its zonal wavenumber k. If the medium does not change in time, then the packet will conserve its frequency ω. However, the meridional wavenumber will change. In this simple case, it is sufficient simply to diagnose the value of l from Equation 9.30, using the local values of Be and U. The critical parameter which controls the meridional wavelength is

$$K_s^2 = \frac{\mathrm{Be} - U_{yy}}{U} \qquad (9.31)$$

The parameter K_s is called the 'total steady wavenumber'. Taking the upper tropospheric zonal wind as an example, U typically has a maximum in the subtropics and becomes weakly easterly near the equator. Near the tropical zero wind line, where U is still positive, K_s may be very large. Once U is negative, K_s becomes imaginary, and no steady Rossby wave can be supported. So the typical variation of K_s is dominated by large values in the subtropics and generally decreasing values with increasing latitude throughout the midlatitudes.

Consider first a poleward propagating $\omega = 0$ Rossby wave packet excited in the midlatitudes. It sets off north eastwards, at an angle $\alpha = \tan^{-1}(l/k)$, towards the pole. As latitude increases, K_s becomes smaller, and so, from Equation 9.30, the meridional wavenumber decreases while, at the same time, the angle α becomes smaller. The ray of Rossby waves is refracted to a more zonal direction. At the latitude where $K_s = k$, $l = 0$, and the ray is propagating entirely zonally. The meridional wavenumber then becomes negative, and the ray turns to propagate equatorwards. It has been totally internally reflected from the latitude where $K_s = k$. Indeed, K_s behaves as a refractive index since the Rossby wave: rays bend away from low values of K_s and towards high values.

Now consider the fortunes of a ray with negative l, whose group velocity has an equatorward component. Generally, as the wave packet propagates to lower latitudes, K_s will increase, and equatorwards of the subtropical jet, U will decrease. As K_s increases, l will become more negative, as will α. The packet propagates in an increasingly meridional direction. As is clear from Figure 9.14, the magnitude of the group velocity becomes smaller compared to U. Once equatorwards of the jet maximum, the magnitude of U itself becomes smaller, so the equatorward propagation of the wave packet becomes slower and slower. Indeed, the packet can never actually reach the zero wind line; the nearer it gets, its north–south scale collapses, and α tends towards $-90°$. The zero wind line is called a 'critical latitude' for steady Rossby waves, and according to linear theory, the critical latitude acts as an impervious barrier to propagating Rossby waves. The decreasing meridional scale of phase speed as a wave approaches a critical line suggests that they are likely to be absorbed there. However, nonlinear ideas also suggest that reflection can be possible.

We shall return to ray tracing and its application to large-scale atmospheric flow in Chapter 19.

9.8 Inflexion point instability

The ideas developed in this chapter can be combined to describe a generic form of fluid instability which operates in many situations where two sets of waves can have similar phase speeds and so can interact with each other. In particular, this arises when a basic undisturbed flow has a maximum of absolute vorticity. In other words, it arises in situations where the gradient of absolute vorticity changes sign. Later chapters will develop specific examples which include complications, such as rotation or stratification. In this section, rather than a formal stability analysis, qualitative arguments will be used to set out the basic physical mechanism responsible for instability. The argument will combine the two elements central to the arguments of

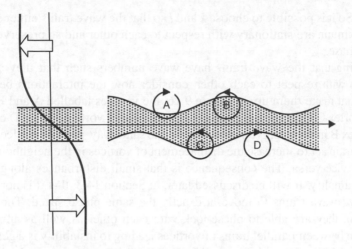

Figure 9.17 A schematic illustration of the mechanism for inflexion point instability in a sheared flow

the last section: propagation of Rossby-like waves and the interaction of discrete vortices.

For the present, we shall work in an inertial frame of reference and so consider the absolute vorticity. Consider an absolute flow $U(y)$ initially parallel to the x-axis and locally sheared, as illustrated in Figure 9.17. The shear has been drawn so as to imply a maximum of absolute vorticity at $y = 0$. Shading indicates the region of large vorticity in the diagram. A frame of reference is selected such that it moves with the flow at $y = 0$. Then, the flow for $y < 0$ is in the positive x-direction, while for $y > 0$, it is in the negative x-direction. The flow velocity parallel to the x-axis has an inflexion point at $y = 0$, a feature which gives the name to the instability.

Now suppose the flow is perturbed into a wave-like pattern, so that fluid is displaced alternately in the $y > 0$ and $y < 0$ directions. Displacement towards the centre of the shear zone brings smaller vorticity to a particular location, while displacement away from the centre brings larger values of vorticity there. Thus, the initial state is modified by a series of positive and negative vorticity anomalies. The dimensions of these anomalies can be characterized by a wavenumber k in the x-direction and a wavenumber l in the y-direction. Since they are located within a gradient of absolute vorticity, such anomalies will propagate as Rossby waves in the x-direction. Their phase speed relative to the local flow will be

$$c - U(y) = -\frac{\partial \zeta / \partial y}{\left(k^2 + l^2\right)} \tag{9.32}$$

Now for $y > 0$, $U(y)$ is negative and $\partial \zeta / \partial y$ is also negative. For an appropriate choice of k and l, the train of waves will be stationary relative to the midpoint of the shear zone. Exactly the same argument, but with all the signs reversed, applies for $y < 0$. So it is possible to choose k and l so that the wave trains either side of the shear maximum are stationary with respect to each other and so preserve the same relative phase.

Assuming that the wave trains have wave numbers such that they are indeed stationary with respect to each other, consider now the interactions between the vortices that make them up. In Figure 9.16, the vortices labelled A and B combine to push vortex C in the negative y-direction. Similarly, vortices C and D combine to push vortex B in the positive y-direction. In the same way, pairs of vortices all along the wave train act to increase the displacement of vortices in the neighbouring wave train, and vice versa. The consequence is that small disturbances along the shear zone will amplify. It will be discussed later, in Section 14.3, that it is not necessary for the two wave trains to move at exactly the same phase speed. Through their interaction, they are able to phase-lock with each other as well as amplify. This interaction between parallel trains of vortices leading to instability is a generic form of fluid instability, found in many different situations and contexts. In this kind of situation, it is variously termed inflexion point instability or shear instability. The basic criterion for the instability to exist is that the gradient of absolute vorticity should change sign, that is,

$$\frac{\partial \zeta}{\partial y} = 0 \text{ for some value of } y. \tag{9.33}$$

An alternative form of this criterion, in terms of the flow $u(y)$ parallel to the x-axis is that the second derivative $-\partial^2 u/\partial y^2$ should change sign for some value of y, u here being defined in an inertial frame of reference.

Equation 9.33 is easily modified for a rotating fluid. In terms of the relative vorticity, the criterion for instability is

$$\beta + \frac{\partial \xi}{\partial y} = \beta - \frac{\partial^2 u}{\partial y^2} = 0 \text{ for some } y \tag{9.34}$$

where $\beta = \partial f_0/\partial y$ is the poleward gradient of planetary vorticity. Since β is always positive, the effect of rotation is to inhibit instability. In fact, for typical midlatitude synoptic scale flows, it is quite unusual for a jet to be strong enough and sharp enough for $\partial^2 u/\partial y^2$ to exceed β, although the absolute vorticity gradient is often close to zero on the flanks of a strong jet. When the criterion 9.34 is met, the resulting instability in an atmospheric or ocean context is often called 'barotropic instability'.

A further generalization is needed to take account of stratification as well as rotation. Similar arguments to those which led to Equation 9.33 apply, except that now the gradient of potential vorticity, rather than absolute vorticity, must change sign. Much more will be said about potential vorticity in subsequent chapters. But one example serves to illustrate the effect at this point. In certain conditions where a quasi-geostrophic scaling is appropriate (see Chapter 12), the criterion for instability becomes

$$\beta - \frac{\partial^2 u}{\partial y^2} - \frac{f_0^2}{N^2} \frac{\partial^2 u}{\partial z^2} = 0 \text{ for some zonal line through the } (y, z) \text{ plane} \quad (9.35)$$

Here, the Brunt–Väisälä frequency N is taken to be constant. In the midlatitude atmosphere, this criterion is more readily met as a result of variations of vertical shear in the atmosphere than as a result of horizontal shears. That is, it is more likely that the third term $\left(f_0^2 / N^2 \right) \partial^2 u / \partial z^2$ should exceed β than the second term $\partial^2 u / \partial y^2$. In such a case, the instability arises from the interaction between vertically stacked trains of vortices, rather than the horizontally arranged vortices of barotropic instability. The instability which is generated when vertically stacked trains of vortices interact is called 'baroclinic instability'. Baroclinic instability is ubiquitous in the midlatitude atmosphere and will be the subject of much of the later chapters of this book.

Notes

1. 'To intervolve' is a rarely used verb coined by the poet John Milton; it means 'to wind together' and is a particularly apt description of the vortex merging process in geophysical flows.
2. A 'right cylinder' is a cylinder whose axis is perpendicular to its ends.
3. Note that the x- and y-phase speeds do not transform as components of a vector when the co-ordinate axes are rotated. We may not, therefore, speak of a phase velocity, only a phase speed in a specified direction.

10

Potential vorticity

10.1 Potential vorticity

This chapter introduces one of the most important unifying concepts in rotating fluid dynamics, the concept of potential vorticity. The principle is closely related to the Kelvin circulation theorem, mentioned in Section 8.9. However, in this section, the principle of potential vorticity conservation is approached rather intuitively. Subsequent sections make the discussion more rigorous and relate the properties of potential vorticity and circulation. Then, we shall explore some of the consequences of potential vorticity conservation.

Consider a parcel of dry air. On timescales short compared to a radiative relaxation time (typically around 30 days for the troposphere) and above the atmospheric boundary so that friction is small, such a parcel will be subject to a number of conservation principles. These are as follows:

1. It will conserve potential temperature since its motion is adiabatic.

2. It will conserve its mass.

3. The circulation around it will not change if the integral of the pressure gradient force around it is zero (see Section 8.8).

Figure 10.1 illustrates such a parcel, a cylinder confined between two nearby surfaces of constant potential temperature θ and at right angles to them. The first conservation principle says that the parcel must remain confined between the same two θ-surfaces. The length δh of the cylinder will vary according to the details of the flow. But it must always be bounded by the same θ surfaces.

The cross-sectional area of the fluid parcel projected onto the θ-surfaces is δA. As the flow evolves, δz (or ρ) may change. But δA must also change in step with such fluctuations if the mass is to remain constant. The mass δm of the parcel is

$$\delta m = \rho \, \delta A \delta h.$$

Fluid Dynamics of the Midlatitude Atmosphere, First Edition. Brian J. Hoskins & Ian N. James.
© 2014 John Wiley & Sons, Ltd. Published 2014 by John Wiley & Sons, Ltd.

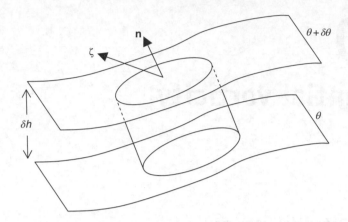

Figure 10.1 A parcel of fluid moves between two nearby surfaces of constant potential temperature, conserving its mass and the circulation around it

So the second conservation principle may be written as follows:

$$\delta m = \rho \delta A \delta h = \text{constant}. \tag{10.1}$$

Considering the third conservation principle, a circuit around the cylinder lies in a θ-surface, and so the circulation around it is constant in time, as shown in Section 8.8. The circulation around a circuit is equal to the integral across the area of the circuit of the component of absolute vorticity perpendicular to the θ-surfaces, $\zeta_n = \zeta \cdot \mathbf{n}$, where $\mathbf{n}$ is a unit vector perpendicular to the θ-surfaces. Therefore, for a small cylinder, the conservation of the circulation around it can be written as follows:

$$\zeta \cdot \mathbf{n} \delta A = \text{constant} \tag{10.2}$$

Equations 10.1 and 10.2 summarize the behaviour, and both include the area δA that is not itself of interest. Dividing one expression by the other gives

$$\frac{\zeta \cdot \mathbf{n}}{\rho \delta h} = \text{constant}. \tag{10.3}$$

The conservation expressed in Equation 10.3 is an amazingly simple summary of a combination of vorticity dynamics and thermodynamics. Referring to Figure 10.2a, if the θ-surfaces move apart, then as the cylinder is stretched, δh increases, and so the component of vorticity normal to the surfaces ζ_n increases (Figure 10.2b). If the θ-surfaces tilt, then the magnitude of the component of the vorticity normal to them is unchanged, but it now refers to the tilted direction (Figure 10.2c). Finally, if ρ increases due to compression, then the vorticity normal to the isentropes increases. This is like a conservation of angular momentum.

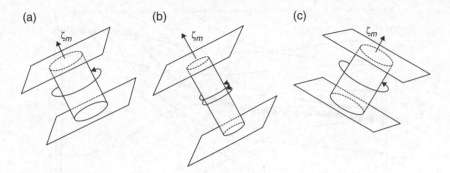

Figure 10.2 Illustration of the conservation following the motion of the expression in Equation 10.3. (a) A cylinder between two neighbouring isentropic surfaces. (b) The isentropes move apart, the cylinder is stretched and the normal component of vorticity increases. (c) The isentropes and the cylinder tilt, and the normal component of vorticity is the same but refers to the tilted direction

The conservation relation expressed by Equation 10.3 is of more general analytical use if it can be turned into a calculus expression applicable to an infinitesimally small cylinder. This can be done by multiplying Equation 10.3 by the constant $\delta\theta$ and noting that in the limit of a small cylinder,

$$|\nabla\theta| = \frac{\delta\theta}{\delta h}.$$

Also, the normal to the θ-surface can be written as

$$\mathbf{n} = \frac{\nabla\theta}{|\nabla\theta|}$$

This then gives the conservation of a quantity P following the motion:

$$\frac{DP}{Dt} = 0 \quad \text{where } P = \frac{1}{\rho}\zeta.\nabla\theta. \tag{10.4}$$

The quantity P is perhaps correctly called 'Ertel–Rossby potential vorticity' after the two who were first involved in its derivation. Rossby gave a discussion like that in this section, whereas Ertel gave a mathematical derivation like that in the next section. The name is often shortened to potential vorticity or just potential vorticity. It does not have the units of vorticity, but it is proportional to the vorticity the fluid element would have if $\nabla\theta$ and ρ were to take some standard values. Rossby first discussed it using a form of Equation 10.3 and, considering the value ζ_n would have if ρ and δh were given standard values, referred to this or, more precisely, the relative vorticity at a standard latitude, as the potential vorticity. However, the name was subsequently attached to P, and so now atmospheric scientists have to put up with a rather confusing piece of nomenclature.

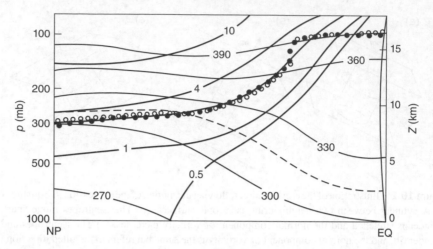

Figure 10.3   Potential vorticity (P) and potential temperature (θ) contours in a latitude–pressure section for the N hemisphere winter zonal average. The contours of P are drawn at 0, 0.5, 1, 2, 4, and 10 PVU, and those of θ every 30 K from 270 to 390 K. The approximate position of the tropopause is shown by a dotted line

In fact, the units of P are $\mathrm{K\,m^2\,kg^{-1}\,s^{-1}}$. To estimate a typical upper troposphere value for P, note that the tropospheric Brunt–Väisälä frequency squared $N^2 = (g/\theta)\partial\theta/\partial z$ is around $10^{-4}\,\mathrm{s^{-2}}$, which is consistent with $\partial\theta/\partial z$ about $3 \times 10^{-3}\,\mathrm{K\,m^{-1}}$. In the upper troposphere, ρ will be around $0.3\,\mathrm{kg\,m^{-3}}$, and the absolute vorticity may be taken as comparable to $f = 10^{-4}\,\mathrm{s^{-1}}$. Thus, a typical midlatitude upper tropospheric value of P will be around $10^{-6}\,\mathrm{K\,m^2\,kg^{-1}\,s^{-1}}$. This quantity has been adopted as a 'standard' value of potential vorticity, so that values are quoted in 'potential vorticity units' or 'PVU'. Figure 10.3 shows P and θ contours in a latitude–pressure section for a climatological zonal average northern hemisphere winter. In the troposphere, P increases from zero near the equator to higher values at high latitudes. In the vertical, P generally increases from the surface up to the tropopause. Even in this very smoothed section, a dramatic change of P occurs at the tropopause, where the stratification abruptly increases with height. Since in the lower stratosphere, N^2 is around $4 \times 10^{-4}\,\mathrm{s^{-2}}$, a typical midlatitude value of potential vorticity immediately above the tropopause will be around 4 PVU. The abrupt jump of P from around 1 PVU to 4 or more PVU is sometimes taken as a dynamical definition of the tropopause, with the actual critical value often taken to be 2 PVU.

10.2   Alternative derivations of Ertel's theorem

In this section, conservation of potential vorticity will be derived rigorously from the equations of motion as was first done in a more general context by Ertel. One aspect of the generalization is to consider an atmosphere in which there may be heating and friction. The general vorticity equation in height coordinates, Equation 8.10, can be written in terms of the vorticity divided by the density so as to absorb the divergence term:

$$\frac{D}{Dt}\left(\frac{1}{\rho}\zeta\right)=\left(\frac{1}{\rho}\zeta\cdot\nabla\right)\mathbf{u}-\frac{1}{\rho^3}\nabla\rho\times\nabla p+\frac{1}{\rho}\nabla\times\dot{\mathbf{u}} \qquad (10.5)$$

Now for any vector $\mathbf{V}$ and any scalar s, the following identity holds:

$$\mathbf{V}\cdot\nabla\left(\frac{Ds}{Dt}\right)=\mathbf{V}\cdot\left[\frac{D}{Dt}(\nabla s)\right]+\nabla s\left[(\mathbf{V}\cdot\nabla)\mathbf{u}\right]$$

So if $\mathbf{V}$ is the absolute vorticity ζ, this identity enables the vorticity equation, multiplied by ∇s, to be written in the following form:

$$\frac{D}{Dt}\left(\frac{1}{\rho}\zeta\cdot\nabla s\right)=\frac{1}{\rho^3}\left[(\nabla\rho\times\nabla p)\cdot\nabla s\right]+\frac{1}{\rho}(\nabla\times\dot{\mathbf{u}})\cdot\nabla s+\frac{1}{\rho}\zeta\cdot\nabla\left(\frac{Ds}{Dt}\right)$$

Now if s is a function of the state variables ρ and p only, the first term on the right-hand side is 0. In this case, in the absence of a frictional torque,

$$\frac{D}{Dt}\left(\frac{1}{\rho}\zeta\cdot\nabla s\right)=\frac{1}{\rho}\zeta\cdot\nabla\left(\frac{Ds}{Dt}\right).$$

This result that the material time derivative can be taken through the expression to apply to s only is a very surprising one due to Ertel. Returning to the frictional case and taking s to be the potential temperature θ, which is indeed a function of the state variables, then

$$\frac{D}{Dt}\left(\frac{1}{\rho}\zeta\cdot\nabla\theta\right)\equiv\frac{DP}{Dt}=\frac{1}{\rho}(\nabla\times\dot{\mathbf{u}})\cdot\nabla\theta+\frac{1}{\rho}\zeta\cdot\nabla\dot{\theta}. \qquad (10.6)$$

Here, $\dot{\theta}$ denotes the heating $D\theta/Dt$. If heating and friction are both 0, then the conservation of Ertel–Rossby potential vorticity, P, on fluid parcels, Equation 10.4, is recovered. Friction is mostly concentrated at the lower boundary in the atmosphere. Strong heating is associated with latent heat release and is often localized in fronts or relatively small regions of intense precipitation. Consequently, P is nearly conserved for much of the flow in the mid- and upper troposphere.

Since $\nabla\cdot\zeta\equiv0$, because ζ is a pure curl, a rearrangement of the expression for P in Equation 10.4 or 10.6 is

$$\rho P=\nabla\cdot(\zeta\theta). \qquad (10.7)$$

Integrate this equation over a volume V, which is bounded by a closed surface S, and use the Gauss divergence theorem:

$$\int_V \rho P\,dV=\oint_S (\theta\zeta)\cdot\mathbf{n}\,dS \qquad (10.8)$$

Also the right-hand side of Equation 10.6 multiplied by ρ can be written as a pure divergence since

$$\nabla \cdot ([\nabla \times \dot{\mathbf{u}}]\, \theta + \zeta \dot{\theta}) \equiv (\nabla \times \mathbf{u}) \cdot \nabla \theta + \zeta \cdot \nabla \dot{\theta}.$$

It follows from Equation 10.6 that

$$\frac{\mathrm{d}}{\mathrm{d}t}\left(\int_V \rho P \mathrm{d}V \right) = \oint_S [\theta\ (\nabla \times \dot{\mathbf{u}}) + \dot{\theta}\zeta\,] \cdot \mathbf{n}\, \mathrm{d}S. \tag{10.9}$$

This equation reveals that the mass-weighted potential vorticity in a volume V only changes in response to heating and friction on the bounding surface S. If friction or heating operate within V but are zero on the boundary S, the distribution of P may be altered, but the mass-weighted P integrated over the volume remains constant. For example, if there is an isolated region of heating, P will be decreased above the heating, but there will be a compensating increase of P below the heating maximum. The effects of friction and heating on potential vorticity will be discussed further in Section 17.6.

10.3   The principle of invertibility

The conservation of potential vorticity is a very strong constraint on atmospheric motion. It is also valuable in practice in the quality control of atmospheric analyses. Any non-conservation of analysed potential vorticity on short timescales almost certainly points to problems with data or with the data analysis, so that successive analyses are not dynamically consistent.

However, the importance of potential vorticity goes beyond its conservation properties. If the distribution of potential vorticity is known, then, with certain conditions, the distribution of wind and potential temperature can be deduced. This is called the 'principle of invertibility', and it adds greatly to the dynamical significance of potential vorticity. The principle is an extension of results we have already seen for the barotropic vorticity equation. There, the distribution of relative vorticity was related to the stream function field by a Poisson equation:

$$\nabla^2 \psi = \xi.$$

Given the vorticity field at a given time, the Poisson equation can be 'inverted' to give the stream function field at that time. From the stream function, the velocity components can be calculated, and so the advective processes affecting the vorticity deduced. Within the limitations of the barotropic vorticity equation, a complete description of the dynamics of the flow is achieved.

There are some provisos to the principle of invertibility. They are as follows:

1. Inverting the potential vorticity to obtain the wind and temperature fields involves solving a Poisson-like elliptical partial differential equation. This can only be done if complete and consistent boundary conditions are specified around the edge of the flow domain.

2. A balance condition, linking the vorticity and potential temperature fields, must be specified. In the simplest case, this might be straightforward geostrophic balance, where

$$\frac{\partial \xi}{\partial z} = \frac{g}{f\theta_0} \nabla_H^2 \theta$$

More complicated and generally nonlinear balance conditions, such as gradient wind balance, may be adopted.

3. The total mass distribution must be known as a function of θ.

The mathematical details of the inversion process depend upon the balance condition used and the level of approximation in the definition of potential vorticity. For the present, we shall illustrate inversion by reference to a steady circular vortex, using gradient wind balance.

For such a circular vortex, the wind and geopotential are related by

$$\left(f + \frac{v}{r} \right) v = \frac{\partial \phi}{\partial r} \tag{10.10}$$

The details of the analysis are summarized by Thorpe (1985). Defining a potential function $\Phi = \phi + v^2/2$, and using a transformed co-ordinate R for the distance from the centre of the vortex, where

$$r = \frac{R}{\left(1 + \frac{2}{f^2 R} \frac{\partial \Phi}{\partial R} \right)^{1/2}},$$

the relationship between gravitational potential Φ and potential vorticity can be written as follows:

$$\frac{1}{f^2} \frac{\partial^2 \Phi}{\partial R^2} + \frac{\left(1 + \frac{2}{f^2 R} \frac{\partial \Phi}{\partial R} \right)^2}{\rho P} \frac{\partial^2 \Phi}{\partial z^2} - \frac{3}{f^2 R} \frac{\partial \Phi}{\partial R} = 1 \tag{10.11}$$

Although it is nonlinear, this is an elliptic equation provided $P > 0$. It is also complicated by a singularity at $R = 0$. However, qualitatively, its solutions are not unlike those of the Poisson equation.

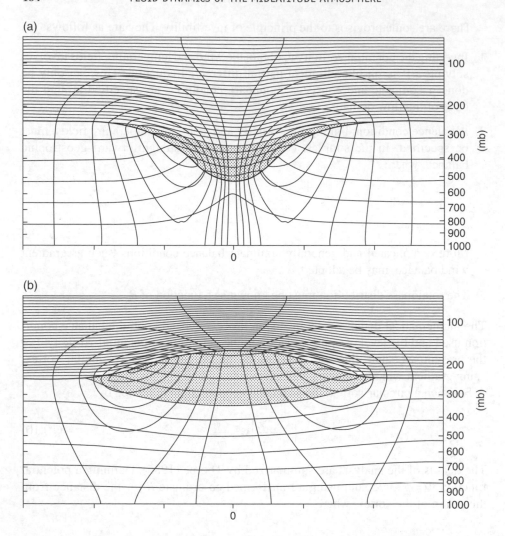

Figure 10.4 Inversion of a potential vorticity structure to give the wind and potential temperature field. The inversion is for circularly symmetric (a) depression and (b) elevation of the tropopause, using gradient wind balance. From Hoskins *et al.* (1985)

Figure 10.4 illustrates two solutions. Both imagine an initial motionless state, with high potential vorticity above the tropopause and low potential vorticity below. In this motionless state, contours of potential temperature and potential vorticity are parallel. Now suppose that there is a minimum in potential temperature on the tropopause centred at $r=0$, implying high potential vorticity there on isentropes that cross the tropopause. Figure 10.4 shows the resulting tangential winds and potential temperature assuming that Φ and its gradient tend to 0 as $r \to 0$. The region of anomalous potential vorticity is stippled. The tropopause is depressed in the region of the potential vorticity anomaly, and cyclonic winds circulate around it, with maximum values near the tropopause. Above and below the anomaly, the winds are reduced,

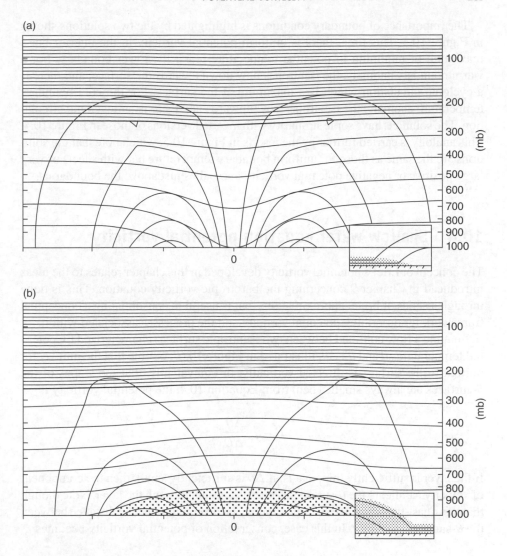

Figure 10.5 As Figure 10.4, but showing the effect of (a) a warm and (b) a cold anomaly on the lower boundary

as they are at large r. All this is a three-dimensional extension to the circulation around an isolated barotropic vortex, discussed in Section 8.2. At the same time, the potential temperature contours bow up towards the potential vorticity anomaly in the troposphere and down towards the anomaly in the stratosphere. The result is that $\partial v/\partial z$ and $\partial \theta/\partial R$ are in a state of balance, which may be derived from Equation 10.4.

In the case of Figure 10.4b, there is a potential temperature maximum on the tropopause and, therefore, negative potential vorticity anomalies on isentropes that cross the tropopause. The tropopause is elevated, and anticyclonic winds circulate around it. In thermal wind balance with them, isentropes bow upwards away from the potential vorticity anomaly in the stratosphere and downwards in the troposphere.

The importance of boundary conditions is highlighted by the two solutions shown in Figure 10.5. This time, there is uniform potential vorticity in the interior but a maximum or minimum in potential temperature along the lower boundary. For a warm boundary anomaly, the isentropes bow down towards the surface, and there is a cyclonic circulation which is a maximum at the surface. For the cold boundary temperature anomaly, the isentropes bow upwards and there is anticyclonic circulation. The solutions have some similarity with inverted versions of those in Figure 10.4. This analogy is carried further by the inserts in Figure 10.5, which note that the solutions are the same as those for uniform boundary temperature but with sheets of very large positive or negative potential vorticity anomalies just above the boundaries.

10.4　Shallow water equation potential vorticity

The concept of Ertel's potential vorticity developed in this chapter relates to the ideas introduced in Chapter 9 concerning the barotropic vorticity equation. This is done through a series of brutal approximations, but is useful in consolidating a conceptual framework for describing potential vorticity and the processes associated with it.

Imagine a flow confined between two isentropic surfaces characterized by potential temperatures θ and $\theta + \Delta\theta$, illustrated in Figure 10.6. Assume that the unit vector normal to the surface is directed vertically, or equivalently, that the slopes of the θ-surfaces are always small. Then, from Equation 10.4, the potential vorticity is

$$P = \zeta_n \frac{\Delta\theta}{\rho D}$$

If the layer is sufficiently thin, that is, if $\Delta\theta$ is sufficiently small, then the component of velocity parallel to the θ-surfaces is nearly constant through the layer. Note though that this quasi-two-dimensional velocity may be divergent since the distance between the θ-surfaces may vary. In this case, conservation of potential vorticity becomes

$$\left(\frac{\partial}{\partial t} + \mathbf{v} \cdot \nabla\right)\left(\frac{\zeta_n}{\rho D}\right) = 0. \tag{10.12}$$

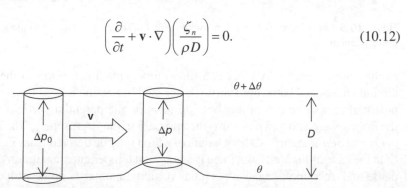

Figure 10.6　Potential vorticity in a shallow fluid layer

As an example, suppose the upper isentrope is flat, but that the lower isentrope rises and falls over some orography, height h, at the surface, providing a crude simulation of mountains on the lower boundary. Then Equation 10.12 becomes

$$\left(\frac{\partial}{\partial t} + \mathbf{v} \cdot \nabla\right)\left(\frac{\zeta_n}{\rho(D - h)}\right) = 0 \tag{10.13}$$

Finally, assuming the density ρ is constant, the role of potential vorticity in this situation is played by the conserved quantity:

$$P = \frac{f + \xi}{D - h} \tag{10.14}$$

If $|h| \ll D$, the potential vorticity equation, Equation 10.13, reduces to the barotropic vorticity equation, Equation 9.3.

The discussions of Chapter 9 reduce to conservation of potential vorticity in the form given in Equation 10.14. Making the β-plane approximation, and assuming the relative vorticity is 0 at the reference latitude and away from any orography, potential vorticity conservation means that

$$\xi = -\beta y - \frac{f_0 h}{D} \tag{10.15}$$

From this relationship, much of the discussion of flow over shallow orography and of Rossby wave propagation given in Sections 9.3 and 9.5 follows.

11
Turbulence and atmospheric flow

11.1 The Reynolds number

A dimensionless measure of the role of viscosity in a particular flow is the Reynolds number:

$$\mathrm{Re} = \frac{UL}{\nu} \tag{11.1}$$

where U is the magnitude of a characteristic velocity fluctuation, L is the length scale of that fluctuation and ν is the kinematic coefficient of viscosity. At first sight, the Reynolds number would appear to be the typical ratio of the acceleration terms, $\mathbf{u} \cdot \nabla \mathbf{u} \sim U^2/L$ in the Navier–Stokes equations, Equation 2.43, to the viscous term $\nu \nabla^2 \mathbf{u} \sim \nu U/L^2$. However, we shall show later that this is an over-simplification. Nevertheless, the Euler equations might be thought of as the limit as $\mathrm{Re} \to \infty$, and we might expect that as Re becomes large, solutions of the Navier–Stokes equation should tend towards solutions of the Euler equations. It turns out that this expectation too is misleading.

Figure 11.1 shows a schematic summary of a classical sequence of fluid experiments. Each frame illustrates the flow around a circular cylinder at different Reynolds numbers. In the definition of the Reynolds number for these flows, U is the uniform upstream flow speed, and L is the diameter of the cylinder. A beautiful and classic series of photographs of such flows is given in van Dyke (1982). Regardless of the size of the cylinder, the flow speed and the viscosity of the fluid, the character of the flow depends simply upon the Reynolds number.

At the smallest Reynolds number, $\mathrm{Re} \sim 1$, the flow is very nearly symmetric up- and downstream of the cylinder. Streamlines close to the axis of the system part, with a stagnation point on the leading face of the cylinder, and pass either side. Those close to the surface of the cylinder remain close to the surface until they reach the vicinity of the trailing face, where they break away into the fluid interior.

Fluid Dynamics of the Midlatitude Atmosphere, First Edition. Brian J. Hoskins & Ian N. James.
© 2014 John Wiley & Sons, Ltd. Published 2014 by John Wiley & Sons, Ltd.

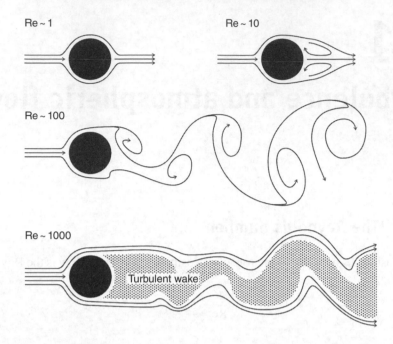

Figure 11.1 Schematic illustrations of flow past a circular cylinder at different Reynolds numbers

Paradoxically, the whole flow is qualitatively rather similar to 'potential flow', an idealized two-dimensional solution to the Euler equations which neglects viscosity entirely. Such very low Reynolds number flow is called 'creeping flow'.

As Re increases, the flow around the upstream half of the cylinder remains rather similar, with an upstream stagnation point and streamlines passing either side of the cylinder. The differences come on the trailing side of the cylinder and in the wake behind the cylinder. At Re around 4 or 5, streamlines on the trailing side of the cylinder break away from its surface, and enclose two small counter-rotating eddies, to produce a flow like that shown for Re ~ 10. As the Reynolds number increases further, the separation point moves towards the equator of the cylinder, and the length of the trailing vortices increases, roughly as $Re^{1/2}$. Above Reynolds numbers of 80 or so, the flow becomes unsteady; the flow shown for Re = 100 typifies this regime. The trailing vortices alternately break away from the rear of the cylinder and are swept away downstream. The result is a periodic 'Karman vortex street'. The regular periodicity of the vortex street breaks down as the Reynolds number approaches 1000, and the wake then becomes turbulent, with irregular fluctuations in both space and time. Even at such a very large Reynolds number, when it might be thought that the effect of viscosity was becoming negligible, the flow still varies with Re. For Re greater than 10 000, patches of turbulence break away quasi-periodically from the cylinder in a fashion reminiscent of the vortex street. But now it is circulating patches of turbulence rather than laminar eddies which are being shed. In all the cases illustrated, the upstream flow remains rather similar whatever the Reynolds number.

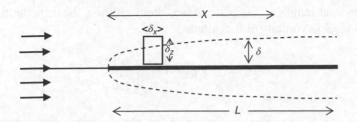

Figure 11.2 The Blasius boundary layer on a thin plate

The crucial point in all these flows, with the exception of creeping flow, is that the flow upstream of the cylinder is characterized by zero vorticity, while downstream, the flow has a complex vorticity structure which depends upon Re up to very large values of Re. Where has this vorticity come from?

Consider an even simpler flow, the uniform steady flow of fluid past a thin flat plate which is aligned parallel to the flow, illustrated in Figure 11.2. Upstream of the plate, the flow has uniform speed U parallel to the x-axis; it therefore has zero vorticity since it has neither shear nor curvature. At the surface of the plate, a no-slip boundary condition applies: the velocity components both parallel and perpendicular to the plate fall to zero at the surface of the plate. Now consider a streamline which at a distance X from the leading edge of the plate is close to the plate surface, say, a small distance δ from its surface, but which, immediately upstream of the plate, has flow speed U. Then, the average acceleration term must be

$$\mathbf{u} \cdot \nabla \mathbf{u} \sim \frac{U^2}{X}$$

In a steady state, this acceleration must be balanced, and the only process which can accomplish this is viscous stress. Assuming the flow changes from U to 0 over a distance δ, then

$$\nu \nabla^2 \mathbf{u} \sim \frac{\nu U}{\delta^2}$$

and so balancing these terms, the boundary layer thickness must be

$$\delta \sim \left(\frac{\nu X}{U} \right)^{1/2} \tag{11.2}$$

The boundary thickness increases with distance along the plate being zero at the leading edge and reaching a maximum thickness at the trailing edge of

$$\delta \sim \left(\frac{\nu L}{U} \right)^{1/2} = L\,\mathrm{Re}^{-1/2} \tag{11.3}$$

if L is the total length of the plate. Associated with the shear of the flow in the boundary layer is vorticity of magnitude

$$\xi \sim \mp \frac{U}{\delta} = \mathrm{Re}^{1/2}\left(\frac{U}{L}\right)\left(\frac{X}{L}\right)^{-1/2} \tag{11.4}$$

(where the minus sign refers to the top of the plate and the plus sign to its lower surface). Notice that this vorticity is infinite at the leading edge of the plate and drops to $(U/L)\mathrm{Re}^{1/2}$ at the trailing edge. Thin sheets of vorticity of this magnitude are shed into the wake behind the plate.

This simple, idealized flow is called a 'Blasius boundary layer', and it is full of implications. In fact, the boundary layers on a cylinder in the flows shown in Figure 11.1 would be very similar in their scale characteristics, though with the additional complication of curvature. The first thing it reveals is that there are at least two distinct space scales in the problem. There is the external scale L, imposed by the length of the plate, or, in the cylinder problem, by the diameter of the cylinder. But there is also an intrinsic length scale, characterizing the depth of the boundary layer. Because the viscous stress term contains higher derivatives than the remaining terms, the viscous stress can always balance the other forces acting, provided the vertical length scale is sufficiently small. In the Blasius problem, the viscous stresses are small throughout most of the fluid, but become large enough to balance the total forces acting within a small distance δ of the plate surface. So the Reynolds number, naïvely thought of as the typical ratio of viscous stresses to total acceleration, is better thought of as related to the ratio of the imposed and intrinsic scales in the flow:

$$\mathrm{Re} = \left(\frac{L}{\delta}\right)^2 \tag{11.5}$$

Secondly, the example gives us a clue as to the origin of vorticity in an initially irrotational flow. It is generated in boundary layers and may be shed into the fluid interior when separation takes place. In the cylinder flows of Figure 11.1, very little vorticity is generated for small Re, and it is confined to the region immediately downstream of the cylinder. As Re increases, the magnitude of the vorticity increases, and the separation point moves further up the cylinder, towards its equator. In the vortex street regime, the shed sheets of vorticity quickly wrap up to form separate eddies of alternate sign. Their scale is comparable to L, rather than the small thickness δ of the initial vortex sheet, and consequently, the vortices can survive for long distances downstream. As the Reynolds number increases, so the proportion of the downstream wake, which becomes full of vorticity, increases.

Having said this, it is important to avoid the error of supposing that viscous stresses generate vorticity. They do not. Rather, the effect of viscosity is to diffuse vorticity, to weaken the vorticity extrema and to spread them out spatially. To understand this, consider a tiny rectangular circuit, shown in Figure 11.2, of length δx

and height δz where $\delta z > \delta$. The circulation around this circuit is constant, with magnitude $U\delta x$, no matter where on the plate it is calculated. The vorticity, the circulation per unit area, is of order U/δ and so is unbounded sufficiently close to the leading edge of the plate, falling to around $(U/L)\text{Re}^{1/2}$ at the trailing edge. The total circulation has not changed; what has happened is that it has been spread over a greater region of the fluid, and the associated vorticity correspondingly weakened. The difference between the circuit immediately upstream of the leading edge, where the vorticity is 0, and immediately downstream, where it is unbounded sufficiently close to the leading edge, is not the action of viscous stress. It is the discontinuous application of the no-slip boundary condition as soon as the plate is encountered, which is responsible for the first appearance of vorticity in the flow near the plate surface. Viscous stresses then spread this vorticity to adjacent layers, attenuating its magnitude but preserving the total circulation.

In terms of mathematics, we note that since

$$\nabla^2 \mathbf{u} = \nabla\left(\nabla \cdot \mathbf{u}\right) - \nabla \times \xi$$

taking the curl of the momentum equation to obtain the vorticity equation introduces a viscous term $\nu\nabla^2\xi$ to the vorticity equation. This new term represents the diffusion of vorticity.

Traditional fluid dynamics serves to give a satisfactory description of the Blasius boundary layer upstream of the plate and for points well downstream of its leading edge. However, there are real difficulties very close to the leading edge where the vorticity tends to infinity and the boundary layer thickness tends to 0. Such a singularity means of course that the continuum hypothesis itself becomes inadequate immediately downstream of the point where the flow encounters the plate. There is a region, a few times the mean molecular spacing in width, where the establishment of the boundary layer should be described in terms of the ballistic behaviour of individual molecules. The Blasius layer is a salutary example of how we need to remain alert to the basic assumptions that are being made, even in apparently everyday situations.

In the flows considered in this chapter, whenever the Reynolds number exceeds a value of order 10^3, laminar flow breaks down and becomes turbulent. Large irregular fluctuations of the flow variables develop in both time and space. Parcels of fluid become shredded and inextricably muddled with shreds of other fluid parcels within a short time. Not unrelated is the behaviour of vortex tubes. They too become highly contorted and tangled together so that their identities become vague. A precise deterministic description of the flow quickly becomes impossible, and turbulent flows are generally described in terms of statistical concepts, in terms of means, of variances and co-variances of flow quantities.

Consider the Earth's boundary layer, which has a typical depth of around 10^3 m and across which the wind might increase from small values near the ground to $10\,\text{m s}^{-1}$ or more at the top of the boundary layer. The Reynolds number estimated from these figures is around 10^9. The atmospheric boundary layer is therefore highly turbulent. Wind in the boundary layer does not blow steadily. It fluctuates around some mean value, with irregular gusts and periods of relative calm. Similar fluctuations characterize conserved quantities such as potential temperature, humidity or tracer concentration. As these properties are advected by the turbulent wind field, so their values fluctuate erratically on a range of time and space scales.

11.2   Three-dimensional flow at large Reynolds number

Consider incompressible homogeneous flow, which is stirred into motion by some large-scale forcing $\mathbf{F}$. Ignore for the moment the effects of rotation and form an equation for total kinetic energy $\hat{E}$ by taking the scalar product of the Navier–Stokes equation with the velocity vector and integrating over the volume of the fluid, so that

$$\frac{\mathrm{d}\hat{E}}{\mathrm{d}t} = \int_V \left(\mathbf{F}\cdot\mathbf{u} - \nu\mathbf{u}\cdot\nabla\times\xi\right)\mathrm{d}V \qquad (11.6)$$

Other terms, such as those arising from the advection term and the pressure gradient term, integrate to 0 over the entire volume of the fluid by virtue of the boundary conditions. Now in a steady state, the two terms within the integrand must balance. But imagine separating the different scales of motion present by Fourier analysis of the velocity field and other flow variables. The term $\mathbf{F}\cdot\mathbf{u}$ will have a typical magnitude FU and will be large on the forcing scale. The second term has a typical magnitude $\nu U^2/L^2$ and will generally be negligible on the scale of $\mathbf{F}$, because the coefficient of viscosity for air is so small. Only for small values of L, that is, for small scales of motion, will this term become important. These scales might be millimetres for atmospheric flow. We deduce that for scales intermediate between the forcing scale and the viscous dissipation scale, energy is neither created nor destroyed but simply passed from one scale to another. On average, energy generated at large scales is passed by the turbulence to smaller scales until viscosity can destroy it. This process is called a 'turbulent cascade' of energy and is illustrated in Figure 11.3. The scales between the forcing scale with wavenumber K_f and the viscous dissipation scale with wavenumber K_ν are called the 'inertial subrange'. Neither the forcing nor the dissipation terms are important in the inertial subrange. Central to the cascade in the inertial subrange is vortex stretching, a fundamental process which reduces the scale of energetic vortices.

The distribution of energy according to scale is described by the spectral density of energy, $E(k)$, defined so that

$$\delta E = E(K)\delta K \qquad (11.7)$$

is the energy per unit mass between wavenumber k and $k + \delta k$. Note that if we assume the turbulence is 'isotropic', with properties independent of direction, then E is a

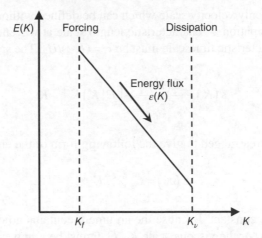

Figure 11.3 Schematic illustration of the turbulent cascade of energy

function only of the total wavenumber $K = |\mathbf{k}|$ and not of the individual components of the wavenumber. The rate at which energy is passing from large to small scales is called the energy flux and denoted by $\varepsilon(K)$.

Any attempt to predict $E(K)$ by direct integration of the Navier–Stokes equations would be a formidable problem and would involve numerical integration of the equations at a resolution high enough to span values of K ranging from K_f to K_ν. However, it turns out that considerable progress can be made using dimensional analysis. Suppose that $E(K)$ is some function of the energy flux ε, the wavenumber K and the forcing and dissipation wavenumbers K_f and K_ν:

$$E(K) = f\left(\varepsilon, K, K_f, K_\nu\right)$$

This relationship can be simplified further if the 'localization hypothesis' is invoked. The localization hypothesis suggests that in the 'inertial subrange' $K_f < K < K_\nu$, the spectral energy density does not depend directly on the forcing or the dissipation. If that is so, then $E(K)$ does not depend on either K_f or K_ν. Then,

$$E(K) = f(\varepsilon, K) \tag{11.8}$$

Now ε, the spectral energy flux, has dimensions of energy per unit mass per unit time, or velocity squared per unit time. It must be of order a characteristic velocity squared divided by a characteristic timescale. From Equation 11.7, a characteristic velocity for wavenumber K is

$$U_k \sim \left[E(K)K\right]^{1/2}$$

Indeed, this is the only velocity scale which can be defined without reference to the forcing or the dissipation. A characteristic length scale at wavenumber k is simply K^{-1}, and so a characteristic timescale must be $\tau_K = (K^{-1})/U_K$. The spectral energy flux is therefore

$$\varepsilon(K) \sim \frac{E(K)K}{\tau_K} \sim \left[E(K)K\right]^{3/2} K \qquad (11.9)$$

This expression is rearranged to give the following form of the energy spectrum:

$$E(K) = C_{3D}\varepsilon^{2/3}K^{-5/3} \qquad (11.10)$$

where C_{3D} is some constant. Because the argument contains no direct reference to the forcing scale K_f or the viscous scale K_n, C_{3D} must be a universal constant. This form of the turbulent energy spectrum is called the 'Kolmogorov spectrum'. Its characteristic $K^{-5/3}$ power law is observed in a variety of turbulent contexts. Such power law spectra imply 'self-similarity', that is, that similar structures exist in the flow at any scale in the inertial subrange. However, the assumptions made in deriving the Kolmogorov spectrum are quite severe. In geophysical situations, the turbulence intensity is likely to vary from place to place, violating the homogeneous assumption, and the three space dimensions are likely to behave differently so that the isotropic assumption is violated.

11.3   Two-dimensional flow at large Reynolds number

The Kolmogorov result for homogeneous, isotropic three-dimensional turbulence is extremely straightforward. Virtually no other result in turbulence theory is so straightforward. But relaxing any of the highly idealized assumptions in the Kolmogorov dimensional analysis is difficult, and progress is very limited.

For atmospheric flow on the synoptic or larger scales, the three-dimensional, isotropic assumptions are particularly suspect. We have already seen that both rotation and stable stratifications suppress vertical motion. As a result, there is a great asymmetry between the vertical and horizontal directions. In this section, we shall consider an idealized case of 'turbulent' motions of a purely two-dimensional flow. In such a case, the vertical component of the vorticity equation reduces to

$$\frac{\partial \zeta}{\partial t} + \mathbf{v} \cdot \nabla \zeta = 0 \qquad (11.11)$$

which is a simple, if nonlinear, conservation relationship. The stretching and tilting terms are 0 for a simple two-dimensional configuration. We have ignored the effects of friction although we shall see that some friction is needed on a sufficiently small scale for the system to have stationary statistics. The absence of vortex stretching is crucial. That is an essential process in the energy cascade of the Kolmogorov model: stretching of vortices leads to the collapse of their scale in that case. In two dimensions, vorticity is conserved. An individual vortex might be deformed, but it cannot be strengthened or weakened by the flow. As far as the integral properties of the flow are concerned, not only must the energy of the flow be conserved, so must its enstrophy $\zeta.\zeta$. This additional constraint changes the character of the turbulence greatly.

So the archetypal process in two-dimensional flow is vortex deformation. This can proceed to extraordinary lengths, with initially compact vortices ending up completely shredded into long, thin, sinuous streamers which fill the flow domain. Figure 11.4 illustrates a single vortex subject to constant deformation. An initially square vortex, width π/k is deformed, with the dilation axis in the y-direction. Initially, the total wavenumber $K = \sqrt{k^2 + l^2} = k\sqrt{2}$. Since vorticity is conserved, the area of the vortex must remain constant, which means that as its y-dimension expands, so its x-dimension must contract. Thus, when the deformation is α, the width of the vortex is $\alpha\pi/k$ in the x-direction and $\pi/\alpha k$ in the y-direction. The total wavenumber is therefore

$$K(\alpha) = k\sqrt{\alpha^2 + \frac{1}{\alpha^2}} \qquad (11.12)$$

Note that α^{-2} is the aspect ratio of the deforming eddy. Since vorticity is conserved by the vortex, its enstrophy remains constant. But the streamfunction will change; from Poisson's equation,

$$\psi = -\frac{\zeta}{K(\alpha)^2}$$

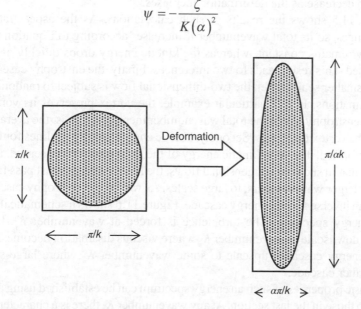

Figure 11.4   Schematic illustration of a vortex undergoing deformation in two-dimensional flow

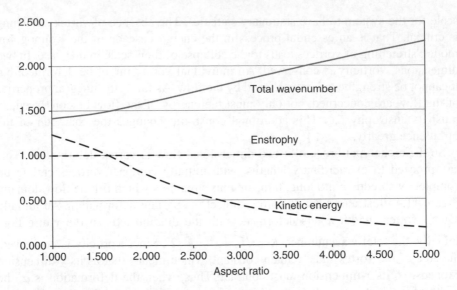

Figure 11.5   Variation of total wavenumber (solid), kinetic energy (dashed) and enstrophy (dotted) with the aspect ratio of a deforming vortex

Consequently, the kinetic energy of the vortex

$$E = \iint_A |\nabla \psi|^2 \, \mathrm{d}A$$

must also decrease as the deformation increases.

Figure 11.5 shows the results of such calculations. As the aspect ratio of the eddy changes, so its total wavenumber K increases according to Equation 11.2. The enstrophy remains constant, whereas the kinetic energy drops quickly as the eddy is deformed. This result leads to two inferences. Firstly, the enstrophy cascades from larger to smaller scales when the two-dimensional flow is subject to random fluctuating deformations. In this particular example, the vortex preserves its vorticity and hence its enstrophy, while the total wavenumber increases, that is, the average spatial scale of the vortex decreases. Secondly, kinetic energy rapidly declines for the vortex as it is deformed. The total kinetic energy of the fluid must be conserved, and so we conclude that in such two-dimensional flows, the kinetic energy must pass from small scales, or larger wavenumbers, to large scales. As well as the enstrophy cascade, there must be an inverse kinetic energy cascade. Figure 11.6 shows a schematic illustration of the energy spectrum. The turbulence is forced at wavenumber K_f. Enstrophy cascades downscale to wavenumber K_ν where viscous dissipation becomes effective. Kinetic energy cascades upscale to some wavenumber K_u where large-scale drag limits further cascade.

The basic properties of such an energy spectrum can be established using arguments parallel to those in the last section. At any wavenumber K, there is a characteristic eddy velocity given by

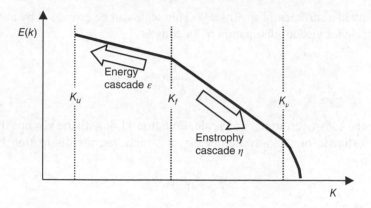

Figure 11.6   Schematic view of two-dimensional turbulence. The turbulence is forced at wave-number K_f; enstrophy cascades to wavenumber K_ν while energy cascades to wavenumber K_u

$$U_K \sim \left[E(K)K \right]^{1/2} \qquad (11.13)$$

The eddy has the characteristic spatial scale K^{-1}, and so it has a characteristic over-turning timescale given by

$$\tau_K \sim \frac{K^{-1}}{U_K} \sim \left[E(K)K^3 \right]^{-1/2} \qquad (11.14)$$

Now the enstrophy spectrum is related to the kinetic energy spectrum by

$$Z = \int Z(K)\mathrm{d}K = \int E(K)K^2 \mathrm{d}K$$

Assume that the enstrophy flux is related to the eddy overturning timescale, so that

$$\eta \sim \frac{Z(K)K}{\tau_K} \sim \frac{E(K)K^3}{\tau_K} \sim E(K)^{3/2}\,K^{9/2}$$

If η *is* independent of wavenumber, that is, if the cascade of enstrophy is purely inertial, then the energy spectrum is obtained by rearranging this last expression:

$$E(K) \sim C_{2D}\eta^{2/3}K^{-3} \qquad (11.15)$$

where C_{2D} is some universal constant. Comparing Equations 11.4 and 11.7, notice that for the enstrophy cascade inertial subrange, the eddy overturning timescale is independent of wavenumber. The cascading enstrophy will eventually be dissipated by viscosity at a sufficiently small scale. This scale can be estimated by noting that the timescale for viscous dissipation of an eddy is

$$\tau_v = \frac{1}{K^2 v}$$

Equating the eddy overturning timescale, Equation 11.4, with the viscous timescale gives an estimate of the wavenumber k_v at which viscous dissipation becomes important:

$$K_v \sim \eta^{1/6} v^{-1/2} \qquad (11.16)$$

The properties of the spectrum for $K < K_f$ are derived in a very similar way as for three-dimensional turbulence. The only difference is that ε is to be understood now as an upscale cascade of kinetic energy. The upscale flux of energy is

$$\varepsilon \sim \frac{E(K)K}{\tau_K} \sim E(K)^{2/3} K^{5/2}$$

And so, rearranging,

$$E(K) \sim C_{2D} \varepsilon^{2/3} K^{-5/3} \qquad (11.17)$$

The system can be closed if some linear drag law of the form $-\xi/\tau_D$, such as that suggested by Ekman pumping, be included in the vorticity equation. Although this term is scale invariant, it has the effect of preventing the cascade proceeding to very small wavenumber, since it becomes effective at dissipating eddies once the eddy overturning time becomes longer than the friction timescale τ_D. This occurs when $K < K_u$ where

$$K_u \sim \frac{1}{\varepsilon^{1/2} \tau_D^{3/2}} \qquad (11.18)$$

In an atmospheric context, this scale may not be relevant. As the scale of eddies increases, variations of the Coriolis parameter become more important, and for wavenumbers less than around $K_\beta \sim (\beta/U)^{1/2}$, the turbulent deformation of eddies is replaced by the dispersive radiation of Rossby waves (see Section 9.8).

The crucial difference between three- and two-dimensional turbulence lies in the flux of kinetic energy from scale to scale. For three-dimensional turbulence, the energy flux is systematically from large to small scales and ultimately to scales sufficiently small that viscosity dissipates the energy. Put another way, energy forced at some given scale only cascades to smaller scales, not to larger scales. Larger scales are therefore unaffected by events at very much smaller scales. This is not the case for two-dimensional turbulence. From some forcing scale, energy cascades upscale, eventually to arbitrarily large scales. Small details of the flow at some initial time will therefore eventually modify the large-scale flow.

The result of upscale energy cascade is that large-scale flow becomes inherently unpredictable. Consider any arbitrary flow, with energy on many scales. If this flow is observed, a minimum observable scale is defined by the spatial distribution of the sensors used. Flow features on smaller scales are unobservable and can be thought of as random small-scale forcing. If energy cascaded to smaller scales, these unobservable scales would not affect the larger observed scales. However, if energy cascades upscale, then eventually random and unpredictable small-scale forcings will become manifest on the larger observable scales. This is the origin of the picturesquely termed 'butterfly effect'. Even forcing as tiny as an insect flapping, its wings will eventually modify the large-scale flow and may, for example, be said to change the evolution of a developing circulation system in some distant part of the fluid.

These arguments have a strongly practical relevance to the business of weather forecasting. A modern weather prediction model is based on an equation set such as the primitive equations of Section 4.5, with the equations discretized in some way. Most straightforwardly, the flow variables are represented at nodes on some grid filling the domain under consideration, with the derivatives in the primitive equations replaced by finite-difference formulae. This means that events on scales smaller than the grid spacing are not represented in the model, even if they could be observed. They can be thought of as a random forcing of the flow at sub-gridscales. For short times, this forcing will have little effect on the large scales of interest. But eventually, the continual random forcing will have increasingly significant impacts on the large scale. In other, beyond a certain time, the forecast will become uncertain and eventually meaningless. The upscale cascade of energy implies a fundamental limit to the predictability of the flow. This is not a technical limitation, a result of poor models or inadequate measuring instruments. The inevitable uncertainties in any observation of the flow will eventually limit the value of a prediction even if the model could be made perfect.

11.4 Vertical mixing in a stratified fluid

Most of the Earth's atmosphere is stably stratified, which means that energy has to be supplied to an air parcel in order to displace it vertically. As a result, vertical mixing in the atmosphere is inhibited, despite the very large vertical shears which would suggest ubiquitous vertical mixing if the stratifications were weak. A necessary condition for shear instability to occur is that the kinetic energy associated with the shear is sufficient to supply the potential energy needed to overcome the stable stratification.

Consider a fluid parcel which is lifted a distance δz in a sheared, stratified flow. The buoyancy force acting upon the displaced parcel will be

$$b = -g\frac{\delta\theta}{\theta} = -\frac{g}{\theta}\left(\frac{\partial\theta}{\partial z}\right)\delta z = -N^2\delta z$$

The total work done per unit mass lifting the parcel to this position is $E_b = (N^2\delta z^2)/2$. The kinetic energy per unit mass associated with the shear over a layer δz deep is

$$E_k = \frac{1}{2}\left[\left(\frac{\partial u}{\partial z}\right)\delta z\right]^2$$

The shear can sustain continued lifting of the parcel provided $E_k > E_b$, that is, provided

$$\mathrm{Ri} = \frac{N^2}{\left(\partial u / \partial z\right)^2} < 1 \tag{11.19}$$

Ri is called the Richardson number, and it may be thought of a dimensionless measure of stratification. It was discussed before in Section 5.9 as a measure of vertical-to-horizontal thermal advection when Ro ~ 1, and in Section 7.4 as a dimensionless measure of stratification. A typical tropospheric value is obtained assuming N^2 of $10^{-4}\,\mathrm{s}^{-2}$ and a shear of $20\,\mathrm{m\,s}^{-1}$ over 10 km. This leads to Ri of 25, at least an order of magnitude larger than the maximum which would permit turbulence. The Richardson number is a most important parameter setting the large-scale circulation of a planet's atmosphere. In the boundary layer, the shear is greater and the stratification smaller. In the lowest kilometre of the midlatitude atmosphere, the Richardson number is often less than O(1), and there turbulence is ubiquitous.

However, the global value of Ri may be misleading, since the stability, but more especially the wind shear, can vary greatly from place to place. In Figure 11.7, the local Richardson number is plotted on the meridional plane. Even this involves a great of zonal averaging. The field is difficult to plot since Ri has very large values when the shear changes sign. Ri is large in the tropics and polar regions, where wind shear is small, and around the jet cores near 20 kPa where $\partial u/\partial z$ changes sign. It is small in the strongly sheared, weakly stratified parts of the upper midlatitude troposphere, beneath the main tropospheric jets. Here values as small as 1, but not as small as 0.25, are found in the winter hemisphere. Small values, less than 0.25, are found in the midlatitude boundary layer, and here at least shear-driven turbulence is expected to be ubiquitous.

The conclusion from Figure 11.7 is that the stratification of most of the troposphere and stratosphere outside the boundary layer is sufficient to stabilize the atmosphere to shear instability. However, the use of time and zonal mean fields undoubted exaggerates the stability of the flow. If three-dimensional synoptic fields are examined, then transient regions of small Ri in frontal regions will almost certainly be found. Theoretical examples will be given in Chapter 15. These patches of low Ri are regions where clear air turbulence is likely to occur, a matter of practical aviation forecasting and not merely of theoretical interest.

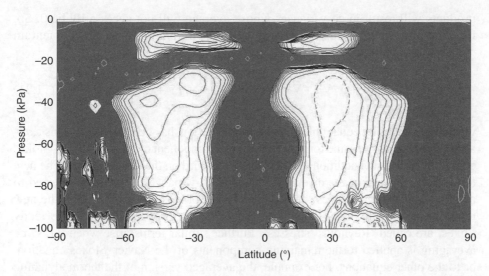

Figure 11.7   Richardson number based on the time and zonal mean fields of θ and u for the December–January–February season. Dashed contours for Ri of 5 and 10; solid contours every 10 from 20 until Ri = 100; shading indicates larger values

11.5   Reynolds stresses

As a result of the discussion in the earlier part of this chapter, it seems inevitable that atmospheric flow is always, to a degree, turbulent. Eddies exist on every scale, and energy and enstrophy are exchanged between the different scales. Yet for much of this book, we have proceeded as if the flow was smooth and laminar, as if parcels of fluid retained their identity for finite times, and we have pretended that fluid trajectories are smooth curves.

The approach will be to partition the flow into a mean part and a fluctuating part, so that

$$\mathbf{u} = \bar{\mathbf{u}} + \mathbf{u}' \qquad (11.20)$$

where the overbar denotes the average over of a range of space and timescales, and the prime denotes local deviations from this average. The choice of averaging scale has little theoretical basis and is generally made on the grounds of technical convenience. For example, in a numerical weather prediction or climate model, the overbar might denote averaging over a portion of space-time based on the model resolution and timestep. The prime denotes the local turbulent fluctuation around the mean state. Similar notation applies to all variables. Now by definition,

$$\overline{\mathbf{u}'} = 0$$

So if Equation 11.20 is substituted into the equations of fluid motion, and the equations averaged over our small space and time volume, the linear terms are simply replaced by an equivalent term in the mean quantities. However, any nonlinear products will result in co-variances whose average will not necessarily be 0. So, for example, applying this formalism just to the x-component of the momentum equation for incompressible flow,

$$\frac{\partial \overline{u}}{\partial t} + \overline{\mathbf{u}} \cdot \nabla \overline{u} + \nabla \cdot \left(\overline{u'u'} \right) = \frac{1}{\rho_0} \frac{\partial \overline{p}}{\partial x} + f\overline{v} + \nu \nabla^2 (\overline{\mathbf{u}}) \qquad (11.21)$$

where the continuity equation for incompressible flow has been assumed. This averaged equation has the same form as the original component of the full Navier–Stokes equation but with the addition of a new term, the third on the left-hand side. The new term takes the form of the divergence of eddy co-variances, sometimes referred to as the 'turbulent eddy fluxes', in this case fluxes of westerly momentum. The new terms in the Navier–Stokes equations, taking the form of the divergence of eddy fluxes, are called 'Reynolds' stresses'. Further similar terms arise when the same averaging is applied to the remaining components of the Navier–Stokes equation, or to the other equations. For example, the averaged version of the thermodynamic equation becomes

$$\frac{\partial \overline{\theta}}{\partial t} + \overline{\mathbf{u}} \cdot \nabla \overline{\theta} + \nabla \cdot \left(\overline{\mathbf{u}'\theta'} \right) = \overline{S} \qquad (11.22)$$

which includes the divergence of the eddy potential temperature fluxes.

Further deductive progress would require a detailed description of the turbulence, and this is in general extremely difficult. Ideally, one would like a relationship between the Reynolds' stresses and the mean fields of velocity, pressure and so on. Such a relationship is called a 'closure'. With its aid, the entire equation set could be written in terms of mean quantities, and we would have a complete, 'closed' set of equations. However, no rigorous way of deriving such relationships is known. In practice, a number of heuristic or empirical approaches of varying degrees of complexity have been proposed. These can yield adequate results over limited ranges of flow variables. In the remainder of this section, we merely sketch some approaches.

One attractive approach would be simply to ignore the sub-gridscale Reynolds' stresses, that is, to show that they are negligible compared with the accelerations represented by the resolved terms of the equation. This indeed is the approach taken by most teachers of atmospheric dynamics and in much analytic work. Can it be justified?

Suppose the flow is an example of two-dimensional turbulence, with its characteristic k^{-3} spectrum. It is forced at wavenumber k_f, and the enstrophy cascade is terminated by viscous dissipation at wavenumber k_d. Suppose furthermore that the flow is resolved down to some scale Δx. This means that the shortest wavelength resolved is $2\Delta x$, and the corresponding wavenumber is $k_r = \pi/\Delta x$. The kinetic energy associated with sub-gridscale motions is therefore

$$E_r = \int_{k_r}^{k_d} E(k)\, dk = K\eta^{2/3} \int_{k_r}^{k_d} k^{-3}\, dk = \frac{K\eta^{2/3}}{2} \left(\frac{1}{k_r^2} - \frac{1}{k_d^2} \right) \qquad (11.23)$$

Assuming $k_r \gg k_d$, the unresolved eddy kinetic energy can be written with adequate accuracy:

$$E_r = \frac{K\eta^{2/3}}{2}\frac{1}{k_r^2} \tag{11.24}$$

Now, the eddy fluxes depend upon the two components of eddy velocity, multiplied by an O(1) factor which depends upon the relative phase of the u and v waves. Therefore,

$$\overline{v'u'} \sim E_r$$

Assume that the eddy fluxes vary slowly, on a scale comparable to the scale of the system. Then, the typical acceleration due to sub-gridscale Reynolds' stresses will be

$$\dot{u}_r \sim \frac{\left|\overline{u'u'}\right|_r}{L} \sim k_f E_r$$

A similar argument will give an estimate for the acceleration due to resolved eddies for which $k_f < k < k_r$. The eddy kinetic energy associated with these scales is

$$E_f = \int_{k_f}^{k_r} E(k)\,\mathrm{d}k = \frac{K\eta^{2/3}}{2}\frac{1}{k_f^2}$$

which leads to an acceleration of

$$\dot{u}_f \sim \frac{\left|\overline{u'u'}\right|_f}{L} \sim k_f E_f$$

Comparing these two expressions for the acceleration affords a direct comparison of the typical accelerations due to Reynolds' stresses of the resolved motion with those due to the sub-gridscale motions:

$$\frac{\left|\dot{u}\right|_r}{\left|\dot{u}\right|_f} \sim \frac{k_f^2}{k_r^2} \sim \frac{\Delta x^2}{L^2} \tag{11.25}$$

Thus, provided that there is at least an order of magnitude difference in scale between L and Δx, the error made in ignoring the Reynolds' stresses due to sub-gridscale motion is probably not very important. However, many coarse resolution studies use low-resolution models, either in the interests of simplicity or to enable a large number of cases to be studied with limited computer resources. In many studies, the difference between L and Δx might only be a factor of two or three; in that case, substantial errors might result by neglecting the Reynolds' stresses due to sub-gridscale motions.

A very simple closure for the equations of atmospheric flow is an analogy with the kinetic theory of ideal gases. In a sheared flow, individual molecules carry their momentum for a short distance before colliding with other molecules and sharing their momentum with them. The mean distance between collisions is called the 'mean free path', and the average speed of the molecules is proportional to $T^{1/2}$, T being the temperature of the gas. In a sheared flow, molecules on average transport momentum across the shear, so that one layer of fluid exerts a stress on adjacent layers: this stress is quantified by Newton's law of viscosity. Kinetic theory gives a relationship between the dynamical coefficient of viscosity and the microscope properties of the molecules:

$$\nu = \frac{1}{2}\overline{\nu}l \qquad (11.26)$$

Here, l is the mean free path for the molecules, and $\overline{\nu}$ is their mean speed. This result predicts that the dynamic coefficient of viscosity for an ideal gas should increase with temperature, other factors being constant.

In a turbulent flow, something similar but on a much larger scale is going on. Blobs of fluid move more or less randomly through the fluid. They exert forces on neighbouring blobs as they go and are deformed. Eventually, a particular blob will lose its identity and mix with its environs, and any property it carries, such as potential temperature or tracer mixing ratio, or indeed momentum will be mixed with the fluid at this location. The average distance over which a blob preserves its identity is called the 'mixing length', denoted by λ, and the typical speed of the blob is given by the turbulent fluctuations of flow speed around the mean. So, by analogy with molecular viscosity, we may define an 'eddy viscosity' as

$$\nu_e = \frac{1}{2}\overline{\nu'^2}^{1/2}\lambda \qquad (11.27)$$

If ν_e is assumed constant, Equation 11.27 provides a crude closure. The Navier–Stokes equation for a turbulent flow then becomes

$$\frac{\partial \overline{u}}{\partial t} + \overline{\mathbf{u}} \cdot \nabla \overline{u} = \frac{1}{\rho_0}\frac{\partial \overline{p}}{\partial x} + f\overline{v} + \nu_e \nabla^2(\overline{\mathbf{u}}). \qquad (11.28)$$

It is exactly the same as the Navier–Stokes equation for laminar flow, but with an eddy viscosity coefficient in place of the molecular viscosity coefficient.

Ideally, theories of turbulence would provide estimates of λ and $\overline{v'^2}^{1/2}$ and hence of ν_e. Such approaches do not readily lead to simple expressions for the eddy viscosity. Instead, an estimate of a reasonable value for ν_e comes from the observed depth of the midlatitude boundary layer, which is of order 1 km. Equating this to the Ekman layer thickness discussed in Section 8.7, the effective eddy viscosity is

$v_e \sim H_E^2 f$, that is, around $100 \, \text{m}^2 \, \text{s}^{-1}$, around 8 orders of magnitude greater than the molecular viscosity.

Eddy viscosity in some form or other is frequently incorporated in numerical models of atmospheric flow, whether for short-term weather prediction or for longer-term global atmospheric circulation studies. Without it, energy and enstrophy would accumulate at the gridscale and eventually render all the fields noisy and unrealistic. By providing a means of removing such gridscale noise, eddy viscosity at least mimics turbulent cascades to small, unresolved scales of motion. The eddy viscosity term in Equation 11.28 is effectively a scale-selective filter, attenuating the small-scale features in the velocity field but leaving the larger, well-resolved scales more or less untouched. Various other filtering approaches are used in modern weather prediction and global circulation models, often with the aim of making the filtering more scale selective. But the aim is the same: to prevent variance building up on the smallest scales represented by the model, scales which will not be well handled by the discretization inherent in the model formulation.

It is clear that eddy viscosity is a gross over-simplification. The analogy with molecular viscosity and the kinetic theory of gases is weak. Parcels of air in a turbulent flow do not behave like molecules, conserving their properties until colliding with another parcel. Instead, air parcels in turbulent flow are only ever vaguely defined, and they steadily lose their identity in the surrounding fluid. In particular, air parcels do not conserve their momentum until they experience an elastic collision with another parcel. Instead, pressure forces between the parcel and the surrounding fluid continually change its momentum. The arguments for an eddy viscosity are at most order of magnitude estimates based on dimensional analysis. There are circumstances, some examples of which are discussed in Chapter 18, in which the concept of eddy viscosity can be very misleading.

Theme 3
Balance in atmospheric flow

12

Quasi-geostrophic flows

12.1   Wind and temperature in balanced flows

An exact balance between the horizontal pressure gradient term and the Coriolis term in the momentum defines a hypothetical velocity vector called the 'geostrophic wind'. Using the basic equation sets, Equations 7.35 and 7.36, for pressure coordinates, Equation 6.10, the geostrophic wind is

$$\mathbf{v}_g = -\frac{1}{f_0} \mathbf{k} \times \nabla_H \Phi' \qquad (12.1)$$

In Section 5.6, scale analysis suggested that synoptic scale flows in the midlatitudes should be close to geostrophic balance, that is, that the observed wind $\mathbf{v}$ should be approximately $\mathbf{v}_g$. More precisely, the magnitude of the ageostrophic wind $\mathbf{v}_a = \mathbf{v} - \mathbf{v}_g$ should be of order Rossby number times the magnitude of the geostrophic wind. Figure 12.1a is an example of upper air flow over the North Atlantic. It shows vectors of the horizontal component of the wind as well as contours of geopotential height. The wind is closely parallel to the height contours, with stronger winds in regions where the contours are closely spaced, in accordance with Equation 12.1. Figure 12.1b shows the ageostrophic wind deduced for the same case. First notice the different scale for the wind vectors. The magnitude of the ageostrophic wind is an order of magnitude smaller than the geostrophic wind. The direction of the ageostrophic wind indicates the direction in which fluid parcels are accelerating, which in turn has implications for the vertical motion field.

In the basic equation set, Equation 7.35, hydrostatic balance is

$$\frac{\partial \Phi'}{\partial z} = b \qquad (12.2)$$

where b is buoyancy. Eliminate the pressure between Equations 12.1 and 12.2 to give

Fluid Dynamics of the Midlatitude Atmosphere, First Edition. Brian J. Hoskins & Ian N. James.
© 2014 John Wiley & Sons, Ltd. Published 2014 by John Wiley & Sons, Ltd.

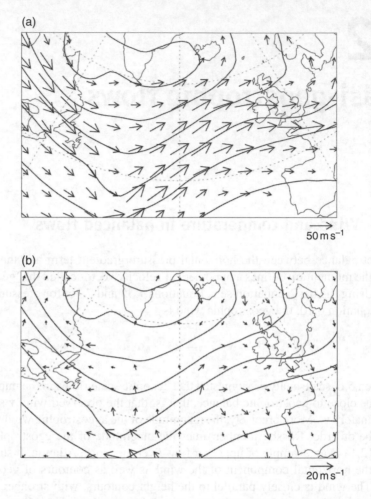

Figure 12.1 Showing contours of geopotential height and vectors of (a) geostrophic wind and (b) ageostrophic wind at 50 kPa at 00Z on 19 February 1997 during one of the FASTEX intensive observing periods

$$f_0 \frac{\partial \mathbf{v}_g}{\partial z} = -\mathbf{k} \times \nabla_H b \qquad (12.3)$$

An equivalent pressure co-ordinate version of these equations is

$$f_0 \frac{\partial \mathbf{v}_g}{\partial p} = \frac{R}{p} \mathbf{k} \times \nabla_H T \qquad (12.4)$$

or

$$f_0 \frac{\partial \mathbf{v}_g}{\partial \ln(p)} = -R\mathbf{k} \times \nabla_H T$$

According to the relationships (12.3) or (12.4), vertical variations of the geostrophic wind are proportional to the horizontal variations of the temperature field. The surface geostrophic wind is specified by the surface pressure field, from Equation 12.1, while the temperature and pressure fields are related by the various forms of the hydrostatic Equation 12.2. Therefore, the pressure field defines both the wind and the temperature fields. If geostrophic and hydrostatic balances are good approximations, these fields are not independent, and the number of variables in the problem is reduced.

Equation 12.3 or 12.4 represent 'thermal wind balance'. This balance can be thought of as a generalization of the Taylor–Proudman theorem, discussed in Section 3.6. For if $\nabla_H b = 0$, the thermal wind equation reduces to $\partial \mathbf{v}_g / \partial z = 0$. The implication then is that fluid moves in coherent vertical columns, with the same velocity at each level in the column.

If the equation of thermal wind balance is integrated in the vertical, the result is a relationship between the change of wind over the vertical interval and the mean temperature of the slab of air traversed. We follow the more usual convention by integrating the pressure co-ordinate version of the thermal wind equation from some low-level pressure p_2 to some upper-level pressure p_1. Then,

$$\Delta \mathbf{v}_g = \frac{R}{f_0} \mathbf{k} \times \nabla \left(\int_{p_1}^{p_2} T \mathrm{d}(\ln p) \right) \equiv \frac{R}{f_0} \mathbf{k} \times \nabla \overline{T} \qquad (12.5)$$

where $\overline{T}$ is a mean temperature through the slab of air. The quantity on the right-hand side of this equation is called the 'thermal wind' $\mathbf{v}_T$:

$$\mathbf{v}_T = \frac{R}{f_0} \mathbf{k} \times \nabla \overline{T}$$

The thermal wind is parallel to the contours of constant temperature (Figure 12.2). It is directed with cold air on the left in the northern hemisphere, and *vice versa* in the southern hemisphere. For geostrophic flow, the upper-level wind is the vector sum of the low-level wind and the thermal wind. If the low-level wind is parallel to the isotherms, then so must be the upper-level wind. In such a case, there is no advection of the isotherms throughout the slab of air. But if the low-level wind has a component parallel to the temperature gradient, so there is advection of temperature, then the upper-level wind will have the same component parallel to the temperature gradient. Throughout the slab, thermal advection will take place. Figure 12.3 illustrates two configurations. In Figure 12.3a, the wind has a component pointing from warm air towards cold air, leading to warm advection. In this case, the wind vector turns clockwise with height; it is said to 'veer'. In Figure 12.3b, cold advection takes place, and so the wind turns anticlockwise with height, or 'backs'.

The equations of thermal wind balance in their various forms can also be written in terms of the vertical component of the geostrophic vorticity

$$\xi_g = \mathbf{k} \cdot \left(\nabla \times \mathbf{v}_g \right) = \left(\frac{\partial v_g}{\partial x} - \frac{\partial u_g}{\partial y} \right)$$

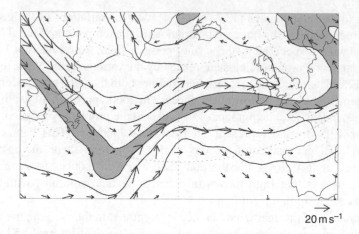

$\overrightarrow{}$
$20\,\mathrm{m\,s^{-1}}$

Figure 12.2 Showing the thickness of the 90 kPa to the 50 kPa layer, together with vectors of the wind shear across this layer. Other details as Figure 12.1

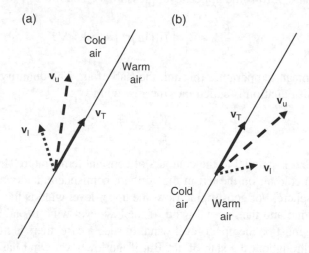

Figure 12.3 Thermal wind shear in (a) warm advection situations and (b) cold advection situations. The upper-level wind is the vector sum of the low-level wind and the thermal wind. In (a), the wind is said to veer, and in (b), it is said to back

For example, taking the curl of Equation 12.3 and using the identity that $\mathbf{k}\cdot\nabla\times\left(\mathbf{k}\times\nabla_H b\right)=\nabla_H^2 b$ give

$$f_0\frac{\partial\xi_g}{\partial z}=\nabla_H^2 b \tag{12.6}$$

As pointed out in Section 9.2, $\nabla_H^2 b$ generally has the opposite sign to b'. It follows that a system with a warm core generally has vorticity which becomes more anticyclonic with height, while the vorticity of a cold cored system becomes more cyclonic with height. For example, a cyclone with a cold core becomes more intense with height. In contrast, the intensely cold core Siberian anticyclone over northern Eurasia in winter is confined to the lower troposphere: in the middle and upper troposphere, the vorticity becomes cyclonic. A tropical cyclone has a warm core. Its cyclonic vorticity near the surface reverses and becomes anticyclonic near the tropopause.

12.2   The quasi-geostrophic approximation

Using the basic equation set, Equation 7.35, we first introduce the geostrophic streamfunction $\psi_g = \Phi'/f_0$. Then, Equations 12.1 and 12.2 become

$$u_g = -\frac{\partial \psi_g}{\partial y}, \quad v_g = \frac{\partial \psi_g}{\partial x}, \quad b = f\frac{\partial \psi_g}{\partial z}. \tag{12.7}$$

The geostrophic relative vorticity is $\xi_g = \nabla_H^2 \psi_g$, and the thermal wind equation

$$f\frac{\partial \xi_g}{\partial z} = \nabla_H^2 b \tag{12.8}$$

follows directly.

In Chapter 8, we showed how the vertical component of the vorticity was decoupled from the horizontal components of vorticity and that the generation of the vertical component of vorticity was a slow process. In this section, we continue this analysis of the vertical component of the vorticity equation. Following Equation 8.17, we write the vertical component of the vorticity equations:

$$\frac{\partial \xi}{\partial t} = -\mathbf{v} \cdot \nabla \zeta - w\frac{\partial \xi}{\partial z} + \frac{(f+\xi)}{\rho_R}\frac{\partial (\rho_R w)}{\partial z} \tag{12.9}$$

Here, the vertical component of the absolute vorticity is $\zeta = f_0 + \beta y + \xi$. This is the 'beta plane approximation' introduced in Section 4.6. If the meridional scale of motion is small so that $\beta L^2/U \ll 1$, then $|\beta y| \ll |\xi|$, and we have the f-plane approximation for which $\zeta = f_0 + \xi$. Firstly, note that in Equation 12.9, the horizontal velocity can be split into geostrophic and ageostrophic parts and that the typical magnitude of the ageostrophic wind is $O(\text{Ro})$ times the geostrophic wind. Secondly, this split implies that the vertical component of relative vorticity can be written in terms of the geostrophic and ageostrophic winds:

$$\xi_g = \frac{\partial v_g}{\partial x} - \frac{\partial u_g}{\partial y}, \quad \xi_a = \frac{\partial v_a}{\partial x} - \frac{\partial u_a}{\partial y}$$

Since $|\mathbf{v}_a| \sim \text{Ro}|\mathbf{v}_g|$, it follows that $|\xi_a| \sim \text{Ro}|\xi_g|$. Ignoring terms of $O(\text{Ro})$ and smaller means that the first two terms of Equation 12.9 can be rewritten:

$$\frac{\partial \xi_g}{\partial t} + \mathbf{v}_g \cdot \nabla_H \zeta_g + w \frac{\partial \xi_g}{\partial z} = \frac{(f + \xi)}{\rho_R} \frac{\partial (\rho_R w)}{\partial z} \tag{12.10}$$

We may think of the last term in Equation 12.10 as the sum of two terms, the stretching of planetary vorticity and the stretching of relative vorticity. Now the typical ratio of relative to planetary vorticity is

$$\frac{|\xi|}{f} \sim \frac{U}{fL} \sim \text{Ro}.$$

So, to order Rossby number, the stretching of planetary dominates over the stretching of relative vorticity. Finally, consider the ratio of the horizontal and vertical advection terms:

$$\frac{|w \partial \xi / \partial z|}{|\mathbf{v} \cdot \nabla \xi|} \sim \frac{W/D}{U/L}$$

But, as was shown in Equation 5.24, $W \sim \text{Ro}(D/L)U$ and so

$$\frac{|w \partial \xi / \partial z|}{|\mathbf{v} \cdot \nabla \xi|} \sim \text{Ro}$$

Putting these together, an approximate form of the equation for the vertical component of vorticity is

$$\frac{D_g \zeta_g}{Dt} = \frac{f}{\rho_R} \frac{\partial (\rho_R w)}{\partial z} \tag{12.11}$$

where $D_g/Dt \equiv \partial/\partial t + \mathbf{v}_g \cdot \nabla$ denotes a rate of change following the geostrophic wind. To $O(\text{Ro})$, Equation 12.11 is a consistently approximated form of the vorticity equation.

A similar scaling analysis can be applied to the thermodynamic equation, which may be written as follows:

$$\frac{\partial b}{\partial t} = -\mathbf{v} \cdot \nabla_H b - w \frac{\partial b}{\partial z} - w \frac{g}{\theta_R} \frac{d\theta_R}{dz} + S \tag{12.12}$$

Once more, split $\mathbf{v}$ into its geostrophic and ageostrophic parts and replace b by $g\theta'/\theta_0$ in the third term so that Equation 12.12 becomes

$$\frac{\partial b}{\partial t} = -\mathbf{v}_g \cdot \nabla_H b - w \frac{g}{\theta_0} \frac{\partial \theta'}{\partial z} - w \frac{g}{\theta_R} \frac{d\theta_R}{dz} + S$$

to order (Ro). The ratio of the two vertical advection terms is simply

$$\frac{\partial \theta' / \partial z}{\partial \theta_R / \partial z} \sim \frac{\Delta \theta_H}{\Delta \theta_V}$$

where $\Delta \theta_V$ is the typical change in potential temperature over a height scale H and $\Delta \theta_H$ is the typical fluctuation of potential temperature in the horizontal. Now $\Delta \theta_V$ is related to the stratification by

$$\Delta \theta_V = \frac{N^2 H \theta_0}{g}$$

while $\Delta \theta_H$ can be related to the typical horizontal wind using the thermal wind relationship:

$$\frac{\partial \theta'}{\partial x} \sim \frac{f \theta_0}{g} \frac{\partial u}{\partial z}$$

so that

$$\Delta \theta_H \sim \frac{fLU}{gH} \theta_0.$$

With a little manipulation,

$$\frac{\Delta \theta_H}{\Delta \theta_V} \sim \frac{1}{\text{RoRi}}$$

where $\text{Ri} = N^2 D^2 / U^2$ is the Richardson number. We shall make the assumption that this ratio is small, that is, the Richardson number $\text{Ri} \gg \text{Ro}^{-1}$, so that vertical advection of the reference profile dominates over vertical advection of potential temperature anomalies.

In fact, as given in Equation 5.45, the Richardson number, Rossby number and Burger number $\text{Bu} = N^2 D^2 / f^2 L^2$ are related as

$$\text{Ro}^2 \text{Ri} = \text{Bu}$$

So if there is the natural ratio of scales $D/L \sim f/N$ implying that $\text{Bu} \sim 1$, then $\text{Ri} \sim \text{Ro}^{-2}$ and $(\text{RiRo})^{-1} \sim \text{Ro}$. For the midlatitude atmosphere, Ro is typically 0.1, and Ri is typically around 100 and consistently $(\text{RoRi})^{-1} \sim 0.1$. With the assumption $\text{Ri} \gg \text{Ro}^{-1}$, we have an approximated thermodynamic equation in the compact and elegant form:

$$\frac{\partial \theta}{\partial t} + \mathbf{v}_g \cdot \nabla \theta' = -w \frac{N^2 \theta_0}{g} \qquad (12.13)$$

Note that in this equation, the stratification may vary with height but not with position in the horizontal. Equations 12.11 and 12.13 form the so-called quasi-geostrophic set.

They are actually a set of two equations in just two dependent variables since θ and ξ_g are related as they are both functions of ψ only. As $\mathbf{v}_g$ is also a function of ψ, there are just two unknowns, ψ and w.

Hence Equations 12.3 and 12.5 give a complete description of the flow to $O(\text{Ro})$. They are referred to as the 'quasi-geostrophic' equations. In the early days of numerical weather prediction, the quasi-geostrophic equations were actually used as the basis of forecasting models. They had the huge advantage of filtering out fast gravity wave motions, therefore permitting a longer timestep. As the power of computers increased, this advantage was outweighed by the approximations made in the equation set, and it became usual to base climate and weather prediction models on the full primitive equations. However, the quasi-geostrophic set remains of great value for teaching and theoretical discussions since they isolate low-frequency balanced motions, giving insight into those motions which dominate the spectrum of synoptic and global scales of motion in the atmosphere.

The scaling analysis of this section can equally be carried out using pressure as a vertical coordinate. The resulting vorticity equation is

$$\frac{\partial \xi_g}{\partial t} + \mathbf{v}_g \cdot \nabla_H \xi_g = f \frac{\partial \omega}{\partial p} \tag{12.14}$$

An advantage of this version is that the background variation of density with height no longer plays a role. The corresponding thermodynamic equation is

$$\frac{\partial \theta}{\partial t} + \mathbf{v}_g \cdot \nabla_H \theta = -\omega \frac{\partial \theta_R}{\partial p} \tag{12.15}$$

The static stability parameter $-\partial \theta_R / \partial p$ is a function of only p in the quasi-geostrophic approximation. It is always positive in a stably stratified atmosphere. However, as we noted in Section 6.2, $\partial \theta_R / \partial p$ varies sharply with height and, unlike $\partial \theta_R / \partial z$, cannot be treated as approximately constant.

The quasi-geostrophic equation set has been given in terms of the vorticity and thermodynamic equations. A consistent momentum equation can also be derived from Equation 5.25. Splitting the horizontal velocity into a geostrophic and $O(\text{Ro})$ ageostrophic parts and dropping all $O(\text{Ro})$ terms result in

$$\frac{D_g \mathbf{v}_g}{Dt} = -f \mathbf{k} \times \mathbf{v}_a. \tag{12.16}$$

The reader can show by direct differentiation that this leads back to the quasi-geostrophic form of the vorticity equation.

The physical nature of the quasi-geostrophic Equations 12.11 and 12.13, 12.14 and 12.15 is worth remarking upon. They have been written so that all the geostrophic terms are on the left-hand side, principally horizontal advection by the geostrophic wind. The right-hand sides, in contrast, depend upon the small ageostrophic component of the wind. This component tends to 0 as the Rossby number decreases. However, because the terms involving w are multiplied by the background rotation or stratification variables f or N^2, the total effect of these terms is to balance the horizontal advection; these terms are not in general negligible as Ro tends to 0, even though they involve the small ageostrophic flow. Thus, even in the limit of vanishingly small Rossby number, the development of the flow depends critically upon the vanishingly small ageostrophic flow. In the Earth's atmosphere, the geostrophic winds can be measured reasonably accurately, either by direct measurement of the total wind or by inference from the pressure field. The ageostrophic flow, on the other hand, is often smaller than the uncertainties in the wind measurements and so cannot be observed directly.

Two complementary lines of development from these quasi-geostrophic equations are possible. One is to use the thermal wind relationship to eliminate the time derivatives between Equations 12.11 and 12.13. The result is a diagnostic equation for the vertical velocity, epitomizing the ageostrophic flow, in terms of the geostrophic flow. This line of development will be studied in Chapter 13. The alternative is to eliminate the vertical motion between Equations 12.11 and 12.13. The result is a prognostic equation for the evolution of a quantity defined entirely in terms of geostrophic quantities which we shall call the 'quasi-geostrophic potential vorticity'. It is related to the Ertel potential vorticity discussed in Chapter 9. Applications of the quasi-geostrophic potential vorticity equation will be discussed in Chapter 14.

12.3   Quasi-geostrophic potential vorticity

Start with Equations 12.11 and 12.13, the latter in the form

$$w = -\frac{1}{N^2}\frac{D_g b}{Dt}.$$

Use this form of the thermodynamic equation to eliminate w from the vorticity equation:

$$\frac{D_g \zeta_g}{Dt} = \frac{f_0}{\rho_R}\frac{\partial}{\partial z}\left[\rho_R\left(-\frac{1}{N^2}\frac{D_g b}{Dt}\right)\right]$$

Now, in the quasi-geostrophic framework, N^2 and ρ_R are functions only of z. So these factors can be taken inside the D_g/Dt operator, which when expanded leads to

$$\frac{D_g\left(f+\xi_g\right)}{Dt} = -\frac{f_0}{\rho_R}\left\{\frac{D_g}{Dt}\left[\frac{\partial}{\partial z}\left(\frac{\rho_R b}{N^2}\right)\right] - \frac{\rho_R}{N^2}\frac{\partial u_g}{\partial z}\frac{\partial b}{\partial x} - \frac{\rho_R}{N^2}\frac{\partial v_g}{\partial z}\frac{\partial b}{\partial y}\right\}$$

But the thermal wind relationships, Equation 12.3, mean that the last two terms are equal in magnitude but opposite in sign, while the first term on the right can be combined with the left-hand side to give

$$\frac{D_g q}{Dt} = 0 \quad \text{where} \quad q = \zeta_g + \frac{f_0}{\rho_R}\frac{\partial}{\partial z}\left(\frac{\rho_R b}{N^2}\right) \tag{12.17}$$

That is, the quantity q is conserved following the geostrophic flow on pressure surfaces and is called the 'quasi-geostrophic potential vorticity'. In terms of the geostrophic streamfunction, Equation 12.17 becomes

$$q = f_0 + \beta y + \nabla_H^2 \psi_g + \frac{1}{\rho_R}\frac{\partial}{\partial z}\left(\rho_R \frac{f_0^2}{N^2}\frac{\partial \psi_g}{\partial z}\right) \tag{12.18}$$

If $D \ll H_\rho$, and if N^2 is constant, then this expression simplifies even further:

$$q = f_0 + \beta y + \nabla_H^2 \psi_g + \frac{f_0^2}{N^2}\frac{\partial^2 \psi_g}{\partial z^2} \tag{12.19}$$

With a rescaled, stretched vertical co-ordinate $z_g = Nz/f_0$, the last term would be $\partial^2 \psi_g/\partial z_g^2$, and comparison with the horizontal term $\partial^2 \psi/\partial x^2$ again highlights the natural ratio of vertical to horizontal scales discussed previously. For the troposphere, $z_g \sim 100z$. Equation 12.19 is a simple three-dimensional Poisson-like equation relating the geostrophic streamfunction to quasi-geostrophic potential vorticity.

Although they are often small compared with the advection terms, the friction and heating terms are often not negligible. Therefore, we write the quasi-geostrophic vorticity equation and thermodynamic equations in more general forms:

$$\frac{D_g \zeta_g}{Dt} = \frac{f_0}{\rho_R}\frac{\partial(\rho_R w)}{\partial z} + \dot{\zeta}; \quad \frac{D_g b}{Dt} = -N^2 w + \dot{b} \tag{12.20}$$

Here, $\dot{\zeta}$ represents the vorticity tendency due to friction. For the simple Ekman pumping discussed in Section 9.4, $\dot{\zeta}$ would simply be $-\xi_g/\tau_E$, while $\dot{b}$ represents the rate of change of buoyancy due to heating or cooling processes. Reworking the derivation of Equation 12.17 with these additional terms leads to

$$\frac{D_g q}{Dt} = \dot{\zeta} + \frac{f_0}{\rho_r}\frac{\partial}{\partial z}\left(\frac{\rho_r}{N^2}\dot{b}\right) \tag{12.21}$$

Mechanical friction, by reducing the relative vorticity contribution to q, will generally tend to dissipate potential vorticity anomalies on the friction timescale. The last term can usually be approximated by

$$\frac{f_0}{N^2}\frac{\partial \dot{b}}{\partial z},$$

showing that the effect of heating on q is proportional to the vertical gradient of heating. Consequently, an isolated, mid-tropospheric anomaly of heating will tend to generate a dipole of potential vorticity anomalies, with positive q tendency below the heating maximum and negative q tendency above.

In pressure co-ordinates, the thermodynamic equation is conveniently written as follows:

$$\frac{D_g \theta}{Dt} = -s^2 \omega \tag{12.22}$$

where s^2 is the (positive) static stability parameter $-\partial \theta_R / \partial p$. The potential vorticity equation can be derived just as earlier. The result is

$$\frac{D_g q}{Dt} = 0 \quad \text{with} \quad q = f_0 + \beta y + \nabla_H^2 \psi_g + \frac{\partial}{\partial p}\left(\frac{f_0^2}{s^2} \frac{\partial \psi_g}{\partial p} \right) \tag{12.23}$$

We recall that the static stability parameter s^2 varies rather sharply with pressure, especially in the upper troposphere, and so it may not be treated as a constant.

Within the quasi-geostrophic framework, q gives all the relevant information about the flow. That is, an invertibility principle applies to quasi-geostrophic potential vorticity in a similar way as it does to Ertel's potential vorticity. Equation 12.18 can be rewritten as

$$\nabla_H^2 \psi_g + \frac{1}{\rho_R} \frac{\partial}{\partial z}\left(\frac{\rho_R f_0^2}{N^2} \frac{\partial \psi_g}{\partial z} \right) = q(\mathbf{r}, t) - f_0 - \beta y$$

which is an elliptic diagnostic equation for ψ_g in terms of q and latitude y. For motions with a small vertical scale, and when N^2 is constant, Equation 12.18 is simply a Poisson equation. The problem is completed by specifying boundary conditions for ψ_g at the upper, lower and lateral boundaries of the domain. A direct consequence of the quasi-elliptic character of Equation 12.18 may be termed 'action at a distance'. A localized anomaly of q will affect the geostrophic flow at remote parts of the fluid, an influence which penetrates both in the horizontal and the vertical.

The boundary condition on horizontal surfaces may be obtained from the thermodynamic Equation 12.11, which with w set to 0 becomes simply horizontal advection of $\partial \psi_g / \partial z$.

12.4 Ertel and quasi-geostrophic potential vorticities

The quantity defined in Equation 12.18 or 12.23 has been called a 'potential vorticity'. The fundamental potential vorticity, namely, that due to Rossby and Ertel, was derived in Section 10.1. What is the relationship between Ertel's potential vorticity and the quasi-geostrophic potential vorticity derived in this section? Quasi-geostrophic potential vorticity is not a straightforward approximation of Ertel's potential vorticity, but as we shall show, they are related in certain parameter regimes.

Write Ertel's potential vorticity in the following form:

$$P = \frac{1}{\rho_R} \zeta \cdot \nabla \theta \qquad (12.24)$$

Already one approximation has been made: deviations from the reference density profile are taken to be negligible. We shall also make the hydrostatic approximation so that the absolute vorticity can be written as

$$\zeta = \left[-\frac{\partial v}{\partial z} \mathbf{i} + \frac{\partial u}{\partial z} \mathbf{j} + \left(f_0 + \beta y + \xi \right) \mathbf{k} \right]$$

As in Section 8.3, the horizontal component of vorticity do not include contributions from the horizontal gradients of vertical velocity. In fact as pointed out in Section 8.1, these terms are of order $Ro(D/L)^2$ compared to the vertical shear of the horizontal wind and would in any case be negligible. Substituting into Equation 12.24 and expanding, we get

$$P = \frac{1}{\rho_r} \left[\left(f_0 + \beta y + \xi \right) \left(\frac{\partial \theta_R}{\partial z} + \frac{\partial \theta'}{\partial z} \right) - \frac{\partial v}{\partial z} \frac{\partial \theta}{\partial x} + \frac{\partial u}{\partial z} \frac{\partial \theta}{\partial y} \right] \qquad (12.25)$$

Use the thermal wind equations in the form of Equation 12.3, to write the potential temperature gradients in the last two terms in terms of wind shears, so they become

$$-\frac{\partial v}{\partial z} \frac{\partial \theta}{\partial x} + \frac{\partial u}{\partial z} \frac{\partial \theta}{\partial y} = -\frac{f_0 \theta_0}{g_e} \left[\left(\frac{\partial v}{\partial z} \right)^2 + \left(\frac{\partial u}{\partial z} \right)^2 \right]$$

which scales as

$$-\frac{f_0 \theta_0}{\rho_R g_e} \left[\left(\frac{\partial v}{\partial z} \right)^2 + \left(\frac{\partial u}{\partial z} \right)^2 \right] \sim \frac{f_0 \theta_0}{\rho_R g} \frac{U^2}{H^2}$$

Compare this with the dominant term in Equation 12.25; this is

$$P \approx \frac{f_0}{\rho_R} \frac{\partial \theta_R}{\partial z} \sim \frac{f_0 \theta_0}{\rho_R g} N^2$$

Then, the ratio of the two terms is

$$\frac{-\left(f_0 \theta_0 / g \right) \left[\left(u_z \right)^2 + \left(v_z \right)^2 \right]}{f_0 \left(d\theta_R / dz \right)} \sim \frac{U^2}{N^2 H^2} = \mathrm{Ri}^{-1} \qquad (12.26)$$

So, if the Richardson number is large, the last two terms of Equation 12.25 are small, and Ertel's potential vorticity can be approximated as

$$P \approx \frac{1}{\rho_R} \left[\left(f_0 + \beta y + \xi \right) \left(\frac{d\theta_R}{dz} + \frac{\partial \theta'}{\partial z} \right) \right] \qquad (12.27)$$

Now each of the bracketed factors in Equation 12.27 contains two terms, one generally larger than the other. A first approximation to P retains only the two large terms and has already been invoked earlier. Denote it by P_R where

$$P_R(z) = \frac{f_0}{\rho_R}\frac{d\theta_R}{dz}. \tag{12.28}$$

A second approximation would also retain the large term from one factor multiplied by the small term from the other. In this way, a second approximation to P can be constructed:

$$P = P_R + P' \quad \text{where } P'(\mathbf{r},t) = \frac{1}{\rho_R}\left\{ f_0\frac{\partial\theta'}{\partial z} + (\beta y + \xi)\frac{d\theta_R}{dz}\right\}. \tag{12.29}$$

Since P_R is a function only of z, a consistent first approximation to the potential vorticity equation would combine vertical advection of the first approximation to P with horizontal advection of the second-order term, a procedure which is closely parallel to the quasi-geostrophic treatment of the thermodynamic equation. So write the potential vorticity equation as

$$\left(\frac{\partial}{\partial t} + \mathbf{v}\cdot\nabla\right)P' + w\frac{dP_R}{dz} = 0. \tag{12.30}$$

The vertical velocity is, from Equation 12.13,

$$w = -\left(\frac{d\theta_r}{dz}\right)^{-1}\left(\frac{\partial}{\partial t} + \mathbf{v}\cdot\nabla\right)\theta' \tag{12.31}$$

Substituting in Equation 12.30 for w, P_R and P leads to

$$\left(\frac{\partial}{\partial t} + \mathbf{v}\cdot\nabla\right)\left[\beta y + \xi + \frac{f_0}{d\theta_R/dz}\frac{\partial\theta'}{\partial z} + \frac{f_0\theta'}{\rho_R}\frac{\partial}{\partial z}\left(\frac{\rho_R}{d\theta_R/dz}\right)\right] = 0 \tag{12.32}$$

The last two terms in the curly brackets can be combined to give

$$\left(\frac{\partial}{\partial t} + \mathbf{v}\cdot\nabla\right)\left[\beta y + \xi + \frac{f_0}{\rho_R}\frac{\partial}{\partial z}\left(\frac{\rho_R\theta'}{d\theta_R/dz}\right)\right] = 0 \tag{12.33}$$

A constant f_0 can be added to the term in curly brackets without altering Equation 12.33, giving

$$\left(\frac{\partial}{\partial t} + \mathbf{v}\cdot\nabla\right)\left\{f_0 + \beta y + \xi + \frac{f_0}{\rho_R}\frac{\partial}{\partial z}\left(\frac{\rho_R b}{N^2}\right)\right\} = 0 \tag{12.34}$$

which is nearly the quasi-geostrophic potential vorticity equation. It differs from Equation 12.17 by including advection by the total horizontal wind rather than by the geostrophic wind; hence, Equation 12.34 will tend to Equation 12.17 in the limit of small Rossby number. However, the analysis shows that conservation of Ertel's potential vorticity implies conservation of quasi-geostrophic potential vorticity in the limit of small Ro, large Ri and when $Ri^{-1} \ll Ro$.

In fact, from this derivation, it is clear that

$$\left(\frac{\partial}{\partial t} + \mathbf{v} \cdot \nabla\right)\left\{\frac{f_0}{\rho_R}\frac{d\theta_R}{dz}q\right\} \simeq \left(\frac{\partial}{\partial t} + \mathbf{v} \cdot \nabla + w\frac{\partial}{\partial z}\right)P \equiv \left(\frac{\partial}{\partial t} + \mathbf{v} \cdot \nabla\right)_\theta P = 0.$$

This result shows that the quasi-geostrophic potential vorticity is not a formal approximation to the full potential vorticity. Rather, it is a quantity whose advection on horizontal surfaces mimics the advection of the full potential vorticity on isentropic surfaces.

13
The omega equation

13.1 Vorticity and thermal advection form

Vertical motion is arguably one of the most important, as well as one of the most elusive, of dynamical variables. At a practical level, vertical motion is tightly related to weather type. Upward motion leads to cooling by adiabatic expansion, and if an air parcel continues to rise, then ultimately it will become saturated. So cloudiness and rainfall are the products of either local or large-scale vertical motion. Downward motion is associated with adiabatic warming and sub-saturation of air parcels, and therefore with cloudlessness and warmth in the free atmosphere, although radiative cooling can lead to different conditions at the surface. At the same time, the dominance of the vortex stretching or squashing mechanism revealed by the scale analysis of the previous chapter shows that vertical motion is central to the evolution of the vorticity field and hence to the dynamical development of a flow.

The catch is that the vertical velocity is orders of magnitude too small to be measured directly save by very sophisticated and expensive instruments which cannot possibly form part of the routine observing network. Even the ageostrophic wind, from which the vertical velocity could be deduced, is comparable in magnitude to the uncertainty in observations of the horizontal components of the wind. What is needed is a way of inferring the vertical motion from quantities that can be measured with adequate accuracy.

The so-called omega equation uses the concept of balance to provide a simple way out of this impasse. Both the geostrophic vorticity and thermodynamic equations, Equation 12.11 and 12.13, can be written so they relate the rate of change of the geostrophic streamfunction to the vertical velocity. Eliminating the time derivatives between these equations leads to a diagnostic relationship between the vertical velocity and the geostrophic flow. The original analysis used the pressure coordinate version of the quasi-geostrophic equations, thereby giving a diagnostic relationship for the pressure vertical velocity ω. This is the origin of the name of the equation, the 'omega' equation.

Fluid Dynamics of the Midlatitude Atmosphere, First Edition. Brian J. Hoskins & Ian N. James.
© 2014 John Wiley & Sons, Ltd. Published 2014 by John Wiley & Sons, Ltd.

The key to eliminating the time derivatives is the thermal wind equation, Equation 12.6. The central assumption made is that throughout the evolution of the flow, thermal wind balance continues to hold; the vorticity and the buoyancy fields (or equivalently, the potential temperature field) must evolve in such a way that Equation 12.6 continues to hold. Thus,

$$f_0 \frac{\partial}{\partial z}\left(\frac{\partial \xi_g}{\partial t}\right) = \nabla_H^2\left(\frac{\partial b}{\partial t}\right)$$
(13.1)

This is clearly the case for geostrophic vorticity, but it is not so clear for the actual vorticity. On the face of it, the vorticity and buoyancy fields are subject to quite different and unrelated forcings, and so they would quickly be thrown out of thermal wind balance with each other. The result of such imbalances would be rather high-frequency motions typified by inertia-gravity waves, with frequencies between f and N. Empirically, rather little energy is observed to be associated with such high-frequency motions in most atmospheric situations. Of course, if real data is assimilated so there is effectively temporal smoothing on a timescale comparable to or greater than N^{-1}, balance could be an illusion of the data assimilation process rather than a real observation. But generally, the evidence is strong that such high-frequency motions as are excited in the large-scale atmosphere are quickly dissipated or dispersed, so that thermal wind balance prevails. So, according to this perspective, balance is a good working hypothesis and not an artefact of data processing.

From Equation 12.11, but including the effects of friction,

$$f_0 \frac{\partial}{\partial z}\left(\frac{\partial \xi_g}{\partial t}\right) = -f_0 \frac{\partial}{\partial z}\left(\mathbf{v}_g \cdot \nabla \xi_g\right) - f_0\beta \frac{\partial v_g}{\partial z} + f_0 \frac{\partial}{\partial z}\left(\frac{1}{\rho_R}\frac{\partial}{\partial z}\left(\rho_R w\right)\right) + f_0 \frac{\partial \dot{\xi}}{\partial z}$$
(13.2)

The poleward advection of planetary vorticity (the 'β-term') has now been included explicitly. F represents any frictional tendency of the vorticity. In the case of simple linear Ekman pumping (see Section 8.6), it would simply have the form $\dot{\xi} = -\xi_g/\tau_E$ (see Equations 8.40 and 8.41). Similarly, from the thermodynamic equation, Equation 12.13 leads to

$$\nabla_H^2\left(\frac{\partial b}{\partial t}\right) = -\nabla_H^2\left(\mathbf{v}_g \cdot \nabla_H b\right) - N^2\nabla_H^2 w + \nabla_H^2 \dot{b}$$
(13.3)

Then, the thermal wind equation in the form Equation 13.1 can be used to eliminate the time derivatives, leaving a relationship of the following form:

$$Lw = S\left(\psi_g\right)$$
(13.4)

where the operator L is

$$L = \nabla_H^2 + \frac{f_0^2}{N^2}\frac{\partial}{\partial z}\left[\frac{1}{\rho_R}\frac{\partial}{\partial z}\left(\rho_R \ldots\right)\right]$$

and the 'source function' S is

$$S\left(\psi_g\right) = \frac{f_0}{N^2}\frac{\partial}{\partial z}\left(\mathbf{v}_g \cdot \nabla_H \xi_g\right) - \frac{1}{N^2}\nabla_H^2\left(\mathbf{v}_g \cdot \nabla_H b\right) + \frac{f_0\beta}{N^2}\frac{\partial v_g}{\partial z} + \frac{f_0}{N^2}\frac{\partial \dot{\xi}}{\partial z} + \frac{1}{N^2}\nabla_H^2\dot{b}$$

Note that N^2 and ρ_R occur in different location with respect to the vertical derivatives than in the relationship between q and ψ in Equation 12.18. In the case of stable strati-fication, that is, for $N^2 > 0$, Equation 13.4 is an elliptic equation. On vertical scales, $D \ll H$ and with N^2 constant, Equation 13.4 reduces exactly to Poisson's equation with an appropriate (f_0/N) scaling of the vertical co-ordinate relative to the horizontal. But even in more general cases, Equation 13.4 remains very close to a Poisson equation, and the qualitative discussion of the nature of the solutions of Section 9.2 remains relevant. Thus, the vertical motion will in general be anti-correlated with the source function, with maxima in S corresponding to descent and minima to ascent. The wave-number dependence means that solving the ω-equation effectively filters the source term, emphasizing the large scales and attenuating the small scales. However, it would be incorrect to suppose that the vertical motion field will be particularly smooth. The source term involves first and second derivatives of terms that are already likely to be noisy in all but the simplest circumstances. The source term itself will be very noisy, and so there will generally be small-scale structure in the w field.

The operator L in Equation 13.4 is made up of a term involving horizontal deriva-tives and a term involving vertical derivatives. Their relative importance is deter-mined by the horizontal and vertical scales of the weather system. A scale analysis of the two terms yields

$$\frac{\left|\nabla_H^2 w\right|}{\left|\frac{f_0^2}{\rho_R N^2}\frac{\partial}{\partial z}\left(\rho_R\frac{\partial w}{\partial z}\right)\right|} \sim \frac{N^2 D^2}{f_0^2 L^2} \equiv \mathrm{Bu} \tag{13.5}$$

As discussed in Chapter 5, the usual midlatitude synoptic-scale values give $\mathrm{Bu} \sim 1$, so both terms are equally important for a typical synoptic-scale system. However, because of the quadratic dependence upon L and D, the situation changes rapidly for different scales. For example, a system such as a shallow cold-cored anticyclone would have large L but small D; in that case, the second term would dominate, and the omega equation would reduce to

$$\frac{1}{\rho_R}\frac{\partial}{\partial z}\left(\rho_R\frac{\partial w}{\partial z}\right) \approx \frac{N^2}{f_0^2}S$$

Notice that this is not a particularly clear way of writing the approximated omega equation. Each term in S contains N^2 in its denominator, so in this limit, the omega equation actually does not depend upon the stratification. On the other hand, a deep but horizontally confined system would be dominated by the first term so that the omega equation could then be approximated by

$$\nabla_H^2 w \approx S$$

However, these are extreme cases. For most purposes, the scale of synoptic weather systems is such that Bu ~ 1, that is, that the two terms on the left-hand side of the omega equation are comparable. Indeed, this natural ratio (N/f_0) of horizontal to vertical scale is a primary result of quasi-geostrophic theory.

Scale analysis of the right-hand side of the omega equation gives an indication of which terms make the most important contributions to the source function S. As a preliminary, scale analysis of the thermal wind equation relates a typical horizontal velocity fluctuation U to a typical horizontal buoyancy fluctuation Δb_H. From Equation 12.3,

$$\Delta b_H \sim \frac{f_0 L U}{D}. \tag{13.6}$$

Substituting typical midlatitude values gives Δb_H of around $0.1\,\mathrm{m\,s^{-2}}$. This corresponds to a typical horizontal potential temperature fluctuation of about $3\,\mathrm{K}$. Write the source term as

$$S = S_1 + S_2 + S_3 + S_4 + S_5. \tag{13.7}$$

Then, the first term, involving the differential vorticity advection, scales as

$$S_1 = \frac{f_0}{N^2} \frac{\partial}{\partial z} \left(\mathbf{v}_g \cdot \nabla \xi_g \right) \sim \frac{f_0 U^2}{N^2 D L^2}. \tag{13.8}$$

With typical midlatitude values, S_1 has a magnitude of around $10^{-14}\,\mathrm{m^{-1}\,s^{-1}}$. Similarly, the second term, involving thermal advection, scales as

$$S_2 = -\frac{1}{N^2} \nabla_H^2 \left(\mathbf{v}_g \cdot \nabla b \right) \sim \frac{U \Delta b_H}{N^2 L^3}$$

which, making use of the thermal wind scaling, Equation 13.6, can be rewritten as

$$S_2 \sim \frac{f_0 U^2}{N^2 D L^2}. \tag{13.9}$$

This is of the same order of magnitude as S_1.

Now consider the β-term

$$S_3 = \frac{f_0 \beta}{N^2} \frac{\partial v_g}{\partial z} \sim \frac{f_0 \beta U}{N^2 D}$$

Compare this result with S_1,

$$\frac{S_3}{S_1} \sim \frac{\beta L^2}{U} \tag{13.10}$$

For synoptic scales, this dimensionless ratio is ~0.4, not a lot smaller than unity. But for smaller-scale systems, it rapidly becomes small, and so in such circumstances,

S_3 apparently may be neglected compared to S_1. However, this might be a misleading conclusion for, in many circumstances, as we shall see shortly, there is a great deal of cancellation between terms S_1 and S_2. Even a small S_3 may be significant compared to the residual between S_1 and S_2.

The remaining terms, depending upon friction and heating, respectively, are dealt with in a similar fashion. In general, both $\dot{\xi}$ and $\dot{\theta}$ are complicated functions of the flow. For the sake of simplicity in the present discussion, they may be represented schematically by 'Rayleigh friction' and 'Newtonian cooling':

$$\dot{\xi} = -\frac{\xi}{\tau_D}; \quad \dot{\theta} = \frac{\theta_E - \theta}{\tau_E}$$

Here, τ_D and τ_E are representative 'spin-up' and 'radiative equilibrium' timescales, respectively, and $\theta_E(\mathbf{r}, t)$ is a potential temperature distribution in radiative equilibrium. Typical values of τ_D are around 5 days and of τ_E around 30 days. Comparing the typical magnitude of terms S_4 and S_5 with S_1 gives

$$\frac{S_4}{S_1} \sim \frac{\tau_A}{\tau_D}$$

$$\frac{S_5}{S_1} \sim \frac{\tau_A}{\tau_E}$$

(13.11)

where $\tau_A = L/U$ is the advective timescale. In both cases, the ratio is that of the advective timescale to the relevant process timescale. The advective timescale is of order 1 day, a good deal smaller than either the friction or the radiative timescales. Both terms are likely generally to be small compared with S_1, but again cancellation between S_1 and S_2 could change this. Also, significant latent heat release would lead to the heating term being of order 1 importance.

So, on the face of it, S_1 and S_2 make the dominant contributions to the source term in the omega equation. It is instructive to consider a simple example, illustrated in Figure 13.1. Suppose a basic zonal flow does not vary with height and that it advects some anomalies of temperature. The diagram represents a vertical section through this flow, parallel to the streamlines. By thermal wind balance, in regions where there is a warm patch of fluid so that the temperature anomaly is positive, the vorticity must decrease with height. There will be a tendency for negative vorticity to overlie low-level positive vorticity. Similarly, in the vicinity of a cold anomaly, vorticity will decrease with height. Now consider point A. Here, there is a maximum of warm advection, so that $-\mathbf{v} \cdot \nabla b$ is a maximum. It follows that S_2 is a minimum in the vicinity of A, and therefore it will tend to force ascent. The vorticity advection is negative above point A and positive below it. It follows that $\partial(-\mathbf{v} \cdot \nabla \xi)/\partial z$ is negative at point A, so that S_1 is positive, corresponding to descent. Thus, S_1 and S_2 have opposite signs. In fact, by using the thermal wind equation to relate buoyancy anomalies to vorticity anomalies, it is easy to show that in this situation, S_1 and S_2 are equal in magnitude and opposite in sign. Their contributions to the omega equation exactly cancel, and so there is no vertical motion.

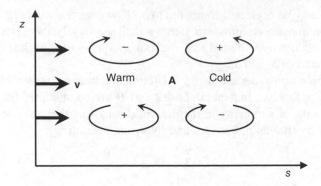

Figure 13.1   Advection of warm and cold anomalies. The co-ordinate s denotes distance along
a streamline

   This is not a surprising result. The flow illustrated in Figure 13.1 is in thermal
wind balance, and provided $\mathbf{v}$ is constant with height, the advection does not disturb thermal wind balance. The omega equation is essentially a balance condition;
it specifies a vertical velocity field which is required to maintain thermal wind
balance. If there is no tendency to disturb thermal wind balance, then there is no
need for any vertical motion, and so the total source function S must be zero. The
situation illustrated in Figure 13.1 is a rough approximation to many real situations in the troposphere. There is often substantial cancellation between terms S_1
and S_2.
   The tendency for S_1 and S_2 to cancel is clearly revealed by writing the relevant
terms. Expanding term S_1 gives

$$S_1 = \frac{f_0}{N^2} \frac{\partial \mathbf{v}_g}{\partial z} \cdot \nabla_H \xi_g + \frac{f_0}{N^2} \mathbf{v}_g \cdot \nabla \left(\frac{\partial \xi_g}{\partial z} \right) \qquad (13.12)$$

Similarly, the S_2 term can be expanded to give

$$S_2 = -\frac{1}{N^2} \left(\nabla_H^2 \mathbf{v}_g \right) \cdot \nabla_H b - \frac{1}{N^2} \mathbf{v}_g \cdot \nabla_H \left(\nabla_H^2 b \right)$$

After some manipulation, and making use of the thermal wind equation in both of
its two forms,

$$\nabla_H b = -f_0 \mathbf{k} \times \frac{\partial \mathbf{v}_g}{\partial z} \quad \text{and} \quad \nabla_H^2 b = f_0 \frac{\partial \xi_g}{\partial z}$$

S_2 can be rewritten as

$$S_2 = \frac{f_0}{N^2} \frac{\partial \mathbf{v}_g}{\partial z} \cdot \nabla_H \xi_g - \frac{f_0}{N^2} \mathbf{v}_g \cdot \nabla_H \left(\frac{\partial \xi_g}{\partial z} \right) + \text{D}. \qquad (13.13)$$

The term D depends upon the total deformation and the rotation of the deformation axis of dilation with height and can be written as

$$D = -8f_0\left(F^2\right)\frac{\partial\alpha}{\partial z}$$

(see Equations 2.15 and 2.16). The second terms in each of Equations 13.12 and 13.13 are identical but of opposite sign. In the simple example given earlier, these second terms were the only non-zero terms on the right-hand side of the omega equation, and so the cancellation between S_1 and S_2 was exact. But in many more complicated situations, this term is still large, and so there is a good deal of cancellation between S_1 and S_2.

13.2 Sutcliffe Form

There are various alternative ways of writing the right-hand side of the omega equation, some of which avoid or reduce the cancellation problem discussed earlier. This section and the next will consider some of these different versions of the omega equation.

In 1947, R.C. Sutcliffe of the UK Meteorological Office produced a theory of 'development', that is, of the conditions in which cyclonic or anticyclonic vorticity would intensify. Such changes of vorticity require vortex stretching or squashing, and so depend on the vertical variation of vertical velocity, $\partial w/\partial z$. In fact, Sutcliffe's development theory amounts to a derivation of the omega equation, albeit with some additional approximations to those made in the previous section. These approximations included the following:

1. A two-level representation of the atmosphere

2. Neglecting $\nabla_H^2 w$ compared to $(f^2/N^2)\partial^2 w/\partial z^2$ on the left-hand side of the omega equation, an approximation which is tantamount to assuming small horizontal length scales

3. Ignoring the β-effect

4. Neglecting terms involving the deformation

A form of the omega equation which is consistent with Sutcliffe's theory is derived in this section. The derivation retains the full vertical variation, but following assumption 4, neglects terms involving the deformation part of the flow. For clarity, the friction, heating and β terms will be neglected. Start from the full form of the omega equation, Equation 13.4, but with the advection terms rewritten according to Equations 13.12 and 13.13. Then, neglecting the deformation term gives as the right-hand side of the omega equation:

$$S_S = 2f_0\frac{\partial \mathbf{v}_g}{\partial z}\cdot\nabla_H\xi_g \tag{13.14}$$

Then, if s is a local co-ordinate which is parallel to contours of constant b and pointing eastwards, and using the thermal wind equation,

$$S_S = 2|\nabla_H b| \frac{\partial \xi_g}{\partial s} \qquad (13.15)$$

This 'Sutcliffe form' of the omega equation has a very simple interpretation. In regions of large baroclinicity, that is, of large $|\nabla_H b|$, ascent is forced ahead of a maximum of relative vorticity and descent behind it. Here, 'ahead' and 'behind' refer to locations along the b-contours. Equivalently, using the thermal wind relationship, we may conclude that ascent is forced where there is a strong positive advection of vorticity by the thermal wind and *vice versa*.

Clearly, the Sutcliffe form of the omega forcing, Equation 13.14, has no possibility of systematic cancellation. It requires only information about the buoyancy field (or potential temperature field) and the geostrophic wind at a single level. In fact, Sutcliffe himself used the 100–50 kPa thickness field to give thermodynamic information and 50 kPa height field to give information about the geostrophic wind and vorticity. It generally gives a good qualitative description of the vertical motion field for the larger scales. However, the neglect of the deformation part of the flow, the term D in Equation 13.13, means that the Sutcliffe form can be seriously in error for smaller-scale flows, such as those associated with frontal regions. This error is exemplified by the frontogenetic flow illustrated in Figure 13.2, where the thermal field is advected by a pure deformation flow. This basic flow may be written as follows:

$$u = -\alpha x, v = \alpha y$$

and it has zero relative vorticity. There is therefore no advection of vorticity parallel to the isotherms, and therefore the Sutcliffe form predicts no vertical motion in this

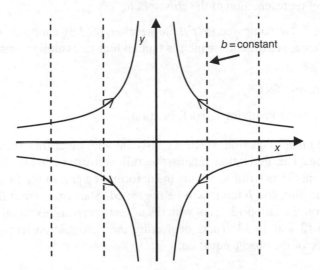

Figure 13.2   A frontogenetic flow

situation. Of course, there may be an additional contribution to the geostrophic wind field as a result of thermal wind balance with the thermal field. However, this contribution has only a component in the y-direction and does not vary in y. Therefore, there are no extrema of vorticity parallel to the isotherms and therefore no contribution to the vertical motion field. In this situation, the Sutcliffe form of the omega equation is seriously misleading. As we shall show in the next section, other forms of the omega equation, which do not neglect the contribution from the deformation part of the flow, predict strong vertical motions associated with a frontogenetic flow. Indeed, such vertical motion is why fronts are important elements of weather systems from the perspective of practical weather forecasting.

13.3 Q-vector form

An alternative derivation of the ω-equation starts from the quasi-geostrophic form of the momentum equations:

$$\frac{\partial \mathbf{v}_g}{\partial t} + \mathbf{v}_g \cdot \nabla_H \mathbf{v}_g + f_0 \mathbf{k} \times \mathbf{v} + \nabla \Phi = 0.$$

For present purposes, friction has been ignored. Combining the Coriolis and pressure gradient terms by introducing the ageostrophic wind as in Equation 12.16 gives the following compact form:

$$\frac{\partial \mathbf{v}_g}{\partial t} + \mathbf{v}_g \cdot \nabla_H \mathbf{v}_g + f_0 \mathbf{k} \times \mathbf{v}_a = 0. \tag{13.16}$$

If thermal wind balance is to be maintained, then

$$\frac{D_g}{Dt}(\nabla_H b) = -\frac{D_g}{Dt}\left(\mathbf{k} \times \frac{\partial \mathbf{v}_g}{\partial z}\right) \tag{13.17}$$

must hold as the flow evolves. Take the x-component of Equation 13.17. The x-derivative of the thermodynamic equation is

$$\frac{\partial}{\partial x}\left[\left(\frac{\partial}{\partial t} + u_g \frac{\partial}{\partial x} + v_g \frac{\partial}{\partial y}\right)b\right] = -N^2 \frac{\partial w}{\partial x}$$

Each of the horizontal advection terms on the left-hand side yields two terms. One is simply the advection of $\partial b/\partial x$. The other involves the x-derivatives of the advecting velocity. That is,

$$\frac{D_g}{Dt}\left(\frac{\partial b}{\partial x}\right) = Q_1 - N^2 \frac{\partial w}{\partial z} \tag{13.18}$$

where

$$Q_1 = -\frac{\partial u_g}{\partial x}\frac{\partial b}{\partial x} - \frac{\partial v_g}{\partial x}\frac{\partial b}{\partial y}. \tag{13.19}$$

Notice that Q_1 is defined entirely in terms of geostrophic quantities. It is the tendency of the geostrophic motion to change the buoyancy gradient in the x-direction. Now take $f_0 \partial/\partial z$ of the y-component of the momentum equation, Equation 13.16:

$$\frac{D_g}{Dt}\left(f_0\frac{\partial v_g}{\partial z}\right) = -f_0\frac{\partial u_g}{\partial z}\frac{\partial v_g}{\partial x} - f_0\frac{\partial v_g}{\partial z}\frac{\partial v_g}{\partial y} - f_0^2\frac{\partial u_a}{\partial z}. \tag{13.20}$$

Since the geostrophic wind is non-divergent, then from continuity,

$$-\frac{\partial v_g}{\partial y} = \frac{\partial u_g}{\partial x}$$

while from thermal wind balance,

$$-f_0\frac{\partial u_g}{\partial z} = \frac{\partial b}{\partial y}, \quad f_0\frac{\partial v_g}{\partial z} = \frac{\partial b}{\partial x}$$

Using these relationships, the two geostrophic terms on the right-hand side of Equation 13.20 can be written in terms of Q_1:

$$\frac{D_g}{Dt}\left(f_0\frac{\partial v_g}{\partial z}\right) = -Q_1 - f_0^2\frac{\partial u_a}{\partial z} \tag{13.21}$$

Comparing Equations 13.18 and 13.21, the geostrophic flow, summarized by Q_1, acts with opposite signs on the two sides of the thermal wind balance Equation 13.17. That is, the geostrophic flow acts continually to destroy thermal wind balance. In order to maintain that balance, agesotrophic circulations with their implication of vertical motion must develop.

An explicit expression for this vertical motion is given by subtracting Equation 13.21 from Equation 13.18 to get

$$N^2\frac{\partial w}{\partial x} - f_0^2\frac{\partial u_a}{\partial z} = 2Q_1 \tag{13.22}$$

A similar development from the y-component of Equation 13.16 leads to

$$N^2\frac{\partial w}{\partial y} - f_0^2\frac{\partial v_a}{\partial z} = 2Q_2 \quad \text{where} \quad Q_2 = -\frac{\partial u_g}{\partial y}\frac{\partial b}{\partial x} - \frac{\partial v_g}{\partial y}\frac{\partial b}{\partial y}. \tag{13.23}$$

Combining these two results,

$$N^2 \nabla w - f_0^2 \frac{\partial \mathbf{v}_a}{\partial z} = 2\mathbf{Q}, \quad \text{where} \quad \mathbf{Q} = Q_1 \mathbf{i} + Q_2 \mathbf{j} \qquad (13.24)$$

Take the divergence of this equation to obtain

$$N^2 \nabla_H^2 w - f_0^2 \frac{\partial}{\partial z}(\nabla_H \cdot \mathbf{v}_a) = 2\nabla_H \cdot \mathbf{Q} \qquad (13.25)$$

But mass conservation gives

$$\nabla \cdot \mathbf{v}_a + \frac{1}{\rho_R}\frac{\partial}{\partial z}(\rho_R w) = 0.$$

So finally, we obtain an alternative form of the omega equation as follows:

$$Lw = S_Q \qquad (13.26)$$

with L is as defined in Equation 13.4 and

$$S_Q = \frac{2}{N^2}\nabla_H \cdot \mathbf{Q} \qquad (13.27)$$

In fact, it can be shown, in terms of Equation 13.4, that $S_Q = S_1 + S_2$ exactly. So Equation 13.26 is an exact reformulation of the omega equation which avoids the cancellation problems implicit in Equation 13.4 and does not involve neglecting the deformation part of the flow, a neglect which is implicit in the Sutcliffe form. The β-effect and the friction and heating terms can be added directly to the Q-vector form of the omega equation.

From Equation 13.18 and its y-direction equivalent, $\mathbf{Q}$ is easily thought of as the tendency of the geostrophic motion to change the horizontal buoyancy gradient. If gridded data for the temperature and geostrophic wind are available, S_Q is readily computed directly, using simple finite-difference formulae to calculate the various terms making up the forcing of the omega equation. However, a qualitative estimate of $\mathbf{Q}$ gives more insight into the vertical motion field in many situations. To derive such an estimate, take local Cartesian co-ordinates $(\tilde{x}, \tilde{y})$ where the $\tilde{x}$-axis is parallel to the local b-contour and the $\tilde{y}$-axis is parallel to $-\nabla_H b$ (Figure 13.3). Then,

$$\mathbf{Q} = \left(-\frac{\partial \tilde{v}_g}{\partial \tilde{x}}\frac{\partial b}{\partial \tilde{y}}, -\frac{\partial \tilde{v}_g}{\partial \tilde{y}}\frac{\partial b}{\partial \tilde{y}} \right)$$

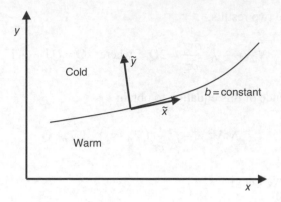

Figure 13.3   Local co-ordinates for estimating $\mathbf{Q}$

But $\partial b / \partial \tilde{y} = -|\nabla_H b|$ and, from continuity, $\partial \tilde{v}_g / \partial \tilde{y} = -\partial \tilde{u}_g / \partial \tilde{x}$ so that

$$\mathbf{Q} = -|\nabla_H b| \left(-\frac{\partial \tilde{v}_g}{\partial \tilde{x}}, \frac{\partial \tilde{u}_g}{\partial \tilde{y}} \right)$$

More compactly,

$$\mathbf{Q} = -|\nabla_H b| \mathbf{k} \times \frac{\partial \mathbf{v}_g}{\partial \tilde{x}} \equiv -|\nabla_H b| \mathbf{k} \times \frac{\partial \mathbf{v}_g}{\partial s} \tag{13.28}$$

where as usual $\mathbf{k}$ is a vertical unit vector. The latter form is valid at all points along the curved b-contour; s is sometimes called the 'thermal wind co-ordinate'. In words, one can estimate $\mathbf{Q}$ by taking the vector change of $\mathbf{v}_g$ along a contour of b (or equivalently, along a contour of T or θ) and rotating that vector anticyclonically (i.e., clockwise in the northern hemisphere) through 90°. A simple example is given below.

Once $\mathbf{Q}$ is known, then the omega equation suggests that there should be a negative correlation between $\nabla_H \cdot \mathbf{Q}$ and the vertical velocity. That is, we expect ascent in regions where there is convergence of $\mathbf{Q}$ and descent where there is divergence. So $\mathbf{Q}$ points towards regions of ascent and away from regions of descent. At the same time, mass continuity requires there to be horizontal components of the ageostrophic flow associated with these vertical motions. As shown by Equation 13.24, these will tend to be directed parallel to $\mathbf{Q}$ at levels below $\mathbf{Q}$ and antiparallel at levels above. Figure 13.4 illustrates. So each $\mathbf{Q}$-vector may be thought of as the centre of an associated overturning circulation.

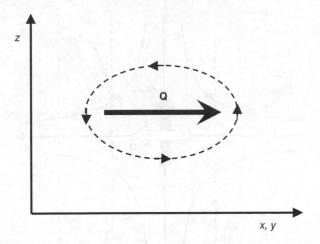

Figure 13.4 The ageostrophic meridional circulation associated with an isolated large **Q**-vector

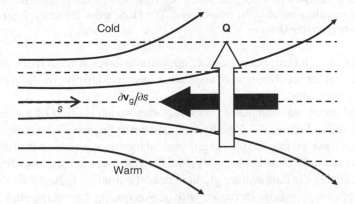

Figure 13.5 Schematic illustration of an upper jet exit in the northern hemisphere, showing the associated **Q**-vector. Solid contours represent the upper-level geopotential contours and the dashed line contours of constant b (which need not be straight lines)

Figure 13.5 shows a simple example of the use of the **Q**-vector to deduce vertical motion. It illustrates an upper-level jet exit in the northern hemisphere, in which the geostrophic wind becomes less strong moving from west to east along contours of constant b. Along the jet axis, $\partial \mathbf{v}_g / \partial s$, shown by the black arrow, is directed from east to west, parallel to the b-contours. The **Q**-vector, shown by the white arrow, is therefore directed across the jet, pointing from south to north. This implies that there will be ascent on the northern side of the jet exit and descent on the southern side. This same pattern is deduced from the Sutcliffe formulation. Because the upper-level ascent to the north of the jet exit implies vortex stretching in the lower troposphere and consequent intensification of cyclonic relative vorticity, the northern jet exit has been identified as a region of cyclonic development. In the same

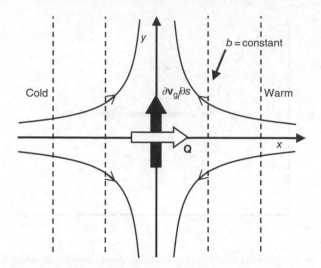

Figure 13.6 The **Q**-vector associated with a deformation flow in the northern hemisphere in which isotherms are parallel to the dilation axis. The black arrow shows $\partial \mathbf{v}_g / \partial s$ and the white arrow the corresponding **Q**-vector

way, the southern jet exit is a region of anticyclonic development. In the days before numerical weather prediction, such arguments were used to give qualitative guidance to weather forecasters.

A second important example of the **Q**-vector approach is given by the deformation flow illustrated in Figure 13.2. Recall that, as shown in Section 13.2, the Sutcliffe form of the omega equation failed to predict any vertical motion associated with a deformation flow. In this example, the dilation axis is parallel to the b-contours, and so the advection of b concentrates the temperature gradient near the dilation axis. Along the dilation axis, the geostrophic flow increases with y. According the recipe just given, the **Q**-vector therefore is at right angles to the dilation axis, pointing from cold air towards warm air. This is in fact obvious from the interpretation of **Q** as the geostrophic tendency to change the horizontal buoyancy gradient. An ageostrophic circulation around the dilation axis is implied, with rising motion in the warmer air and sinking motion in the colder air. Figure 13.6 illustrates this classic example of a flow which intensifies the temperature gradients and which is therefore called 'frontogenetic'. The quasi-geostrophic description of this chapter breaks down as the temperature gradients become stronger, and so extensions of this simple example will be developed in Chapter 15.

13.4   Ageostrophic flow and the maintenance of balance

Mathematically, the various forms of the omega equation determine the ageostrophic circulations needed to maintain thermal wind balance against any vorticity or temperature tendencies which act to disrupt the balance. In this section, the

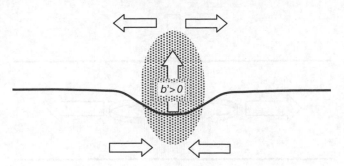

Figure 13.7 Schematic diagram showing the response of an initially motionless fluid to localized heating. The curve represents a θ-surface initially centred in the heating region

mathematical description of the previous sections will be complemented by a more intuitive approach, applied to the cases of buoyancy and vorticity sources.

For a first simple example, suppose some isolated heating acts on the atmosphere, via the thermodynamic equation, in the sense of generating a warm temperature anomaly. Since it does not appear in the vorticity or momentum equations, such heating on its own would throw the atmosphere out of thermal wind balance by creating a thermal anomaly without the balancing decrease in vorticity with height. According to the omega equation, ageostrophic circulations are required to maintain balance.

The situation is illustrated schematically by Figure 13.7. The warming effect of the heating can be reduced by adiabatic cooling associated with ascent in the region. This ascent must be associated with convergence and vortex stretching below the heating and divergence and vortex shrinking above. The stretching will act to create cyclonic vorticity below the heating, and the shrinking will create anticyclonic vorticity above the heating, as required for thermal wind balance. The ageostrophic circulation thus maintains balance both by reducing the thermal effect of the heating and by creating the balancing vorticity field. Whether balance is maintained principally by an adjustment of the vorticity field or the temperature field depends upon the vertical and horizontal scales of the anomaly and upon the values of f and N.

A similar argument applies to the case of an isolated cyclonic torque. Figure 13.8 illustrates such a situation. Thermal wind balance would require a warm anomaly above the developing vortex, where $\partial \xi_g / \partial z < 0$, and a cold anomaly beneath the vortex, where $\partial \xi_g / \partial z < 0$. However, the cyclonic tendency can be reduced by divergence and vortex shrinking associated with descent above and ascent below. Furthermore, this vertical motion will act in the sense of producing the balancing warm anomaly above and cooling below through adiabatic warming and cooling. Such an ageostrophic circulation can therefore maintain balance by both reducing the cyclonic tendency and producing the balancing temperature field.

Other similar examples can be constructed. In each case, the implied ageostrophic wind and vertical motion can be deduced from a consideration of the full omega equation in one or other of the forms given earlier. However, the qualitative arguments

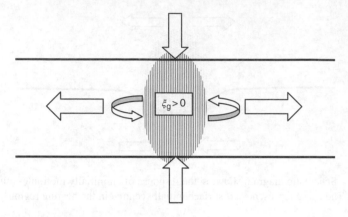

Figure 13.8   An isolated cyclonic vortex generated in an otherwise motionless fluid. The horizontal lines represent lines of constant potential temperature which initially does not vary in the horizontal

of this section give rise to a helpful generalization. Whenever some process, either dynamical or due to heating or dissipation, acts to throw the flow away from geostrophic balance, the ageostrophic circulations restore thermal wind balance in two ways. First, they act in such a way as to offset the original forcing. Second, they modify the directly unaffected field in such a way as to restore balance.

13.5   Balance and initialization

Central to the arguments of this, and most subsequent sections, is the concept of balance. That is, a near balance of the forces acting upon a fluid element means that its accelerations are, in some sense, small. So the atmosphere evolves slowly on synoptic timescales of days, rather than on the much faster timescales of minutes or less associated with gravity waves or sound waves. One such balance is hydrostatic balance. The near balance between the vertical components of the forces acting upon fluid elements relates the pressure field to the mass or temperature field. Hydrostatic balance is a good approximation for systems with vertical scales much smaller than their horizontal scale and has been assumed throughout this book. Imposing hydrostatic balance is equivalent to filtering sound waves, characterized by very short timescales, out of the dynamics.

Internal and external gravity waves, perhaps better termed buoyancy waves, are less easily removed. They are associated with imbalances between the horizontal components of the forces acting upon fluid elements and are characterized by timescales of the order of N^{-1}, typically around 10 minutes in the troposphere. The primitive equations permit such high-frequency motions, but such motions in the atmosphere account only for a very small part of the total energy. The velocity and pressure fields are such that these high-frequency motions are generally unimportant.

The simplest balance condition is geostrophic balance, in which the horizontal components of the pressure gradient force balance the Coriolis force:

$$\mathbf{v} = -\frac{1}{\rho f}\mathbf{k}\times\nabla p \equiv -\frac{1}{f}\mathbf{k}\times\nabla\Phi$$

An exact balance between the Coriolis force and the pressure gradient force implies that the acceleration of fluid elements is zero. That is, the flow cannot evolve. Furthermore, the wind implied by this relationship is non-divergent. Consequently, the implied vertical velocity is zero. So while geostrophic balance might hold approximately, the importance of the ageostrophic motion is clear. Atmospheric motions do evolve, and ageostrophic motions, while small and difficult to measure, are crucial to that evolution. This chapter has been about the ageostrophic motion and in particular its vertical component. The vertical component of the ageostrophic motion is implicit in quasi-geostrophic theory and underlies the evolution of the geostrophic motion.

Charney (1955) considered an alternative approach, seeking implications for the relationship between the pressure field and the horizontal components of the wind field if, instead of taking the divergence itself to be 0, the change of divergence following the motion is 0. This is a higher-order balance condition than geostrophic balance. Approximating the wind by its rotational component, described by a streamfunction ψ, this nonlinear balance condition can be written as follows:

$$\nabla_H^2\Phi = \nabla_H\cdot\left(f\nabla_H\psi\right) + \left[\frac{\partial^2\psi}{\partial x^2}\frac{\partial^2\psi}{\partial y^2} - \left(\frac{\partial^2\psi}{\partial x\partial y}\right)^2\right]$$

When the nonlinear terms are small, this expression reduces to simple geostrophic balance. When fluid parcels move in circular trajectories at constant speed, it reduces to gradient wind balance, discussed in Section 5.4. But the nonlinear balance condition can be applied in more general circumstances. For example, flow in a jet exit requires a component of flow perpendicular to the isobars to provide the necessary deceleration of fluid elements. This ageostrophic flow has a non-zero divergence and so predicts vertical motions in association with the jet exit. Of course, these features are also given by quasi-geostrophic theory. One interesting aspect of the Charney approach is that it raises the prospect of a sequence of higher-order balance assumptions, associated with setting higher-order material derivatives of the divergence to zero.

The Charney balance equation was used at one time to give fields that were suitable for initializing numerical weather forecast models. A later approach that is still of interest when considering the idea of balance in the atmosphere is called 'normal-mode initialization'. The state of the model atmosphere is described by the values of wind, temperature and so on at each gridpoint, a set of numbers which can be represented by a column vector $\mathbf{x}$ of K elements. Suppose the primitive equations are linearized about a state of rest. Then, the numerical prediction model can be written schematically as follows:

$$\frac{d\mathbf{x}}{dt} = i\mathbf{L}\mathbf{x} \qquad (13.29)$$

Here, $\mathbf{L}$ is a $K \times K$ symmetric matrix describing the linearized model. Its eigenvalues ω_k give the frequency of the various wave motions supported by the model, and the corresponding eigenfunctions describe the structure of the mode. These eigenfunctions are orthogonal, and the so-called normal modes are of several kinds. Some are low-frequency modes, the various Rossby modes supported by the system and should be retained in the forecast. They have frequencies smaller than or of order Ω, comparable to the rotation rate of the Earth. Others represent high-frequency gravity waves, with frequencies comparable to N, around two orders of magnitude larger than the Rossby frequencies. These gravity modes have little energy in the atmosphere as observed, and so they are modes whose amplitudes are to be reduced in the initial fields. Suppose there are J such high-frequency gravity modes. Then, the field can be partitioned into a slowly varying Rossby wave part $\mathbf{x}_R$ and a high-frequency gravity wave part $\mathbf{x}_G$:

$$\mathbf{x} = \mathbf{x}_R + \mathbf{x}_G = \mathbf{x}_R + \sum_{j=1}^{J} \alpha_j \mathbf{g}_j \qquad (13.30)$$

Here, the $\mathbf{g}_j$ are the high-frequency eigenfunctions of L, and the α_j are their amplitudes. Because the normal modes are orthogonal, it follows from Equation 13.30 that the amplitudes are given by

$$\alpha_j = \mathbf{g}_j \cdot \mathbf{x}$$

and the high-frequency gravity wave part of x, $\mathbf{x}_G$, can be calculated. Linear normal-mode initialization simply means setting $\mathbf{x}_G$ to zero, so that

$$\mathbf{x}_I = \mathbf{x} - \mathbf{x}_G \qquad (13.31)$$

The magnitude of the difference between $\mathbf{x}_I$ and $\mathbf{x}$ is small, generally smaller than the observational uncertainties in the various elements of $\mathbf{x}$. Notice that in the linearized system of Equation 13.29, any residual gravity wave modes evolve according to

$$\frac{d\alpha_j}{dt} = -i\omega_j \alpha_j$$

so that the magnitude of their amplitude does not change.

Linear normal-mode initialization certainly reduces the high-frequency gravity wave activity in a forecast, but it does not entirely eliminate it. The reason for this is that Equation 13.29 is an approximation to the actual model equations. Schematically, we might write the full equations as

$$\frac{d\mathbf{x}}{dt} = i\mathbf{L}\mathbf{x} + \mathbf{N}(\mathbf{x}) \tag{13.32}$$

Here, the column vector $\mathbf{N}$ depends upon the state $\mathbf{x}$ and describes the tendencies due to nonlinear terms which were neglected in linear normal-mode initialization. It may also contain forcing terms due to heating, friction and so on. Substituting the decomposition 13.30 into the nonlinear equations and projecting onto the J gravity modes which are to be initialized give

$$\frac{d\alpha_j}{dt} = i\omega_j\alpha_j + N_j\left(x_R,\alpha_1,\alpha_2,\ldots,\alpha_J\right) \tag{13.33}$$

where N_j is the projection of $\mathbf{N}$. Clearly, even if the $\alpha_j=0$ initially as a result of linear normal-mode initialization, gravity wave activity will quickly develop because the N_j are non-zero. Nonlinear initialization involves generating a non-zero set of α_j such that $d\alpha_j/dt$ is zero, or at least small, that is, choosing α_j so that

$$N_j\left(\mathbf{x}_R,\alpha_1,\alpha_2,\ldots,\alpha_J\right) + i\omega_j\alpha_j = 0 \tag{13.34}$$

Since this equation is nonlinear, possibly highly nonlinear, a direct solution of Equation 13.34 is not generally possible. Instead, an iterative procedure to reduce the gravity wave activity may be employed:

$$\alpha_j^{n+1} = -\frac{1}{i\omega_j}N_j\left(x_R,\alpha_1^n,\alpha_2^n,\ldots,\alpha_J^n\right) \tag{13.35}$$

In practice, only two or three such iterations are needed to improve the forecast greatly.

Such initialized fields minimize the high-frequency component of the solutions to the governing equations and so represent a slowly varying or nearly balanced state. This will not be a state simply of geostrophic or gradient wind balance, for other terms which can excite gravity wave motions are also taken into account. Consequently, the initialized winds, while close to geostrophic, will have a significant ageostrophic component, including vertical velocity.

Modern forecast systems no longer use such initialization systems explicitly. But the notion of balance is inherent in the procedures used.

14

Linear theories of baroclinic instability

14.1 Qualitative discussion

The theme of this chapter is the way in which small-amplitude disturbances on an initially uniform, vertically sheared flow can evolve. In some circumstances, initially small disturbances amplify: in other words, the flow may be unstable. At the level of quasi-geostrophic theory, the omega equation and the vorticity equation together provide a complete set of predictive equations. For a given geostrophic flow at some time t, the omega equation with suitable boundary conditions determines the vertical velocity w. Vertical motion leads to vortex stretching or shrinking. The vorticity equation then predicts $\xi_g = \nabla^2 \psi$ at some later time $t + \delta t$. With horizontal boundary conditions on ψ, the Laplacian operator can be inverted to obtain ψ and thus the geostrophic flow from the vorticity field at the new time. By repeating this sequence, the evolution of the flow can be predicted.

For the present qualitative discussion, assume that the vertical velocity w is zero at $z = 0$, H_T. Consider the configuration shown in Figure 14.1a. The basic state has significant baroclinicity, with a thermal wind orientated in the x-direction. Hence flow parallel to the x-axis strengthens with height. In the upper troposphere, a weak vorticity anomaly has positive v ahead of it and negative v behind it. Consequently, the change in v in the direction of the thermal wind is northwards. The Q-vector is therefore approximately in the x-direction, implying ascent ahead and descent behind, as indicated in the diagram. At the low levels, this vertical motion gives a positive vorticity tendency due to vortex stretching ahead of the upper vortex and negative vorticity tendency due to vortex shrinking behind the upper vortex, as also shown in Figure 14.1b. As low-level vorticity is generated by these tendencies, the low-level wind perturbation and Q-vectors must look much as shown in Figure 14.1c. The vertical motion forced from the lower-level vorticity anomalies intensifies the upper cyclone by stretching and also intensifies the upper anticyclone by shrinking. Thus, the vertical motion forced from the low levels acts to reinforce the upper-level vorticity anomalies. The cycle of feedbacks can generate stronger and stronger

Fluid Dynamics of the Midlatitude Atmosphere, First Edition. Brian J. Hoskins & Ian N. James.
© 2014 John Wiley & Sons, Ltd. Published 2014 by John Wiley & Sons, Ltd.

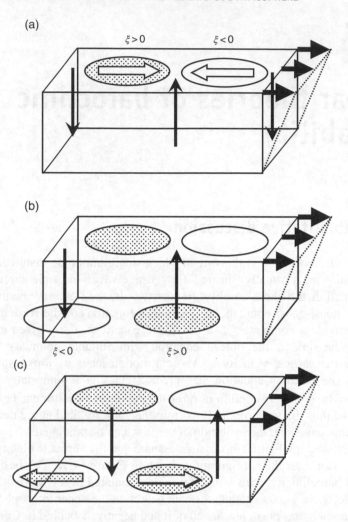

Figure 14.1 Three stages in the self-development of an unstable baroclinic wave. A basic westerly flow, increasing with height, has weak upper-level vorticity anomalies superimposed upon it. Shading denotes positive (cyclonic) vorticity. (a) The pattern of Q-vectors indicates ascent and descent beneath vorticity anomalies. (b) The stretching/shrinking associated with the vertical motion induces low-level vorticity anomalies. (c) The Q-vector pattern associated with the low-level anomalies indicates vertical motions whose stretching/shrinking signature amplifies the upper-level anomalies

vorticity anomalies until other processes come into play. This positive feedback process is sometimes referred to as 'self-development' and is one view of the essence of the baroclinic instability process.

Two contrasting aspects of the role of the basic baroclinicity are apparent:

1. Large baroclinicity is required to ensure that the vertical velocities are substantial.

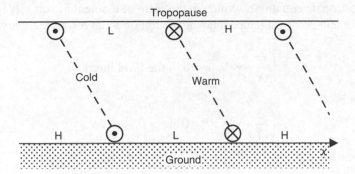

Figure 14.2 Schematic illustration of the longitude–height structure of an unstable baroclinic disturbance. The dashed lines indicate the centre of the troughs and ridges at each level

2. The consequent large differential advection of the upper-level pattern in the x-direction could mean that the low-level vorticity maximum is realized almost directly beneath the upper-level maximum. In such a case, the positive feedback shown in Figure 14.1c would be extremely small.

The qualitative discussion of this section indicates that a growing wave-like system may be possible. If it is, then its structure along the direction of the thermal wind must be qualitatively as shown in Figure 14.2. The pressure distribution follows from the vorticity distribution, via the geostrophic relationships. The westward tilt with height is that required for 'self-development'. The temperature structure in the mid-troposphere follows from the hydrostatic relationship which may be written in the following form:

$$\frac{\partial \phi'}{\partial z} = g \frac{\theta'}{\theta_0} = b' \qquad (14.1)$$

In the region of the northward flow at mid-levels, there is a relatively low-pressure anomaly beneath and a high-pressure anomaly above, so the air must be relatively warm. Furthermore, in this region, there is ascent forced from the upper- and lower-level cyclones. Therefore, relatively warm air is moving polewards and rising. Similarly, behind the cyclone region, relatively cool air is moving equatorwards and sinking. This net poleward and upward transport of internal energy is crucial to the growth of the system.

14.2 Stability analysis of a zonal flow

In this section, we investigate further the possibility of the growth of baroclinic disturbances on a steady zonal current. The qualitative discussion of the previous section will now be put on a more rigorous mathematical basis, starting from the quasi-geostrophic equations, written in the form of a potential vorticity equation in the interior and potential temperature advection on a flat lower boundary.

$$\left(\frac{\partial}{\partial t} + \mathbf{v} \cdot \nabla \right) q = 0 \text{ in the fluid interior} \tag{14.2}$$

$$\left(\frac{\partial}{\partial t} + \mathbf{v} \cdot \nabla \right) b = 0 \text{ at } z = 0 \tag{14.3}$$

where

$$\mathbf{v} = \mathbf{k} \times \nabla \psi$$

$$b = f_0 \frac{\partial \psi}{\partial z}$$

$$q = f_0 + \beta y + \frac{1}{\rho_R} \frac{\partial}{\partial z} \left(\rho_R \frac{f_0^2}{N^2} \frac{\partial \psi}{\partial z} \right)$$

A top boundary condition is also required. For the present, assume that Equation 14.3 applies at $z = H$ also. A final assumption is that the y-dependence of the problem is trivial. Either the flow is cyclic in y with period Y, or that $v = 0$ at $y = \pm Y/2$ so that ψ has no x-dependence on these surfaces.

The basic zonal flow with streamfunction $\overline{\psi}(y,z)$ is characterized by

$$\overline{\mathbf{v}} = \left(\overline{u}, \overline{v} \right) = \left(-\frac{\partial \overline{\psi}}{\partial y}, 0 \right) \tag{14.4a}$$

$$\overline{b} = f_0 \frac{\partial \overline{\psi}}{\partial z} \tag{14.4b}$$

$$\overline{q} = f_0 + \beta y + \frac{\partial^2 \overline{\psi}}{\partial y^2} + \frac{1}{\rho_R} \frac{\partial}{\partial z} \left(\rho_R \frac{f_0^2}{N^2} \frac{\partial \overline{\psi}}{\partial z} \right) \tag{14.4c}$$

This form of the basic flow clearly satisfies the Equations 14.2 and 14.3, there being no gradient in $\overline{q}$ and $\overline{b}$ in the direction of flow.

Now consider a perturbation to this flow:

$$\psi = \overline{\psi} + \psi'$$
$$\mathbf{v} = (\overline{u} + u', v')$$
$$b = \overline{b} + b' \tag{14.5}$$
$$q = \overline{q} + q'$$

where

$$u' = -\frac{\partial \psi'}{\partial y}, v' = \frac{\partial \psi'}{\partial x} \qquad (14.6a)$$

$$b' = f_0 \frac{\partial \psi'}{\partial z} \qquad (14.6b)$$

$$q' = \nabla_H^2 \psi' + \frac{1}{\rho_R} \frac{\partial}{\partial z}\left(\rho_R \frac{f_0^2}{N^2} \frac{\partial \psi'}{\partial z} \right) \qquad (14.6c)$$

All the perturbation quantities are in general functions of x, y, z and t. Insert these forms of solution into the basic Equation 14.2 and move all terms which are quadratic in the perturbations to the right-hand side. The result is

$$\left(\frac{\partial}{\partial t} + \bar{u} \frac{\partial}{\partial x} \right) q' + v' \frac{\partial \bar{q}}{\partial y} = -\mathbf{v}' \cdot \nabla q'. \qquad (14.7)$$

For sufficiently small perturbations, the quadratic term on the right-hand side will be negligible. Its neglect will linearize the perturbation equation:

$$\left(\frac{\partial}{\partial t} + \bar{u} \frac{\partial}{\partial x} \right) q' + v' \frac{\partial \bar{q}}{\partial y} = 0 \qquad (14.8)$$

Similarly, the boundary condition, Equation 14.3, when linearized becomes

$$\left(\frac{\partial}{\partial t} + \bar{u} \frac{\partial}{\partial x} \right) b' + v' \frac{\partial \bar{b}}{\partial y} = 0 \text{ on } z = 0, H \qquad (14.9)$$

On $y = \pm Y/2$, the boundary conditions are either cyclic or have the form $\psi' = 0$. The perturbation equation, Equation 14.8, together with the boundary conditions (14.9) is quite general. We now seek a particular form of solution, a so-called normal-mode solution, which has the following form:

$$\psi' = \text{Re}\left(\Psi (y,z) e^{ik(x-ct)} \right) \qquad (14.10)$$

Such a solution has a fixed structure in y and z and a sinusoidal structure in x with wavelength $2\pi/k$; the pattern moves parallel to the x-axis with phase speed c. In general, both the wave amplitude Ψ and the phase speed c are taken to be complex, while the wavenumber k is taken to be real and positive. Thus, setting $c = c_R + ic_I$,

$$ik(x-ct) = kc_i t + ik(x-c_R t)$$

and so the perturbation may be written

$$\psi' = \mathrm{Re}\left(\Psi(y,z)e^{\sigma t}e^{ik(x-c_R t)}\right)$$

This is interpreted by saying that the mode moves towards the east with speed c_R, the phase speed, and that it amplifies with growth rate $\sigma = kc_i$. For such a sinusoidal normal-mode solution, taking a derivative with respect to x is equivalent to multiplying by ik. Similarly, taking a time derivative is equivalent to multiplying by $-ikc$.

With these comments in mind, the normal-mode solution, Equation 14.10, is substituted into Equation 14.8. Every term in the result contains the factor $\mathrm{Re}(e^{ik(x-ct)})$, and so this has been omitted for clarity:

$$(-ikc + ik\bar{u})\left\{-k^2\Psi + \frac{\partial^2\Psi}{\partial y^2} + \frac{1}{\rho_R}\frac{\partial}{\partial z}\left(\rho_R \frac{f_0^2}{N^2}\frac{\partial\Psi}{\partial z}\right)\right\} + ik\Psi\frac{\partial\bar{q}}{\partial y} = 0$$

or

$$-k^2\Psi + \frac{\partial^2\Psi}{\partial y^2} + \frac{1}{\rho_R}\frac{\partial}{\partial z}\left(\rho_R \frac{f_0^2}{N^2}\frac{\partial\Psi}{\partial y}\right) + \frac{1}{\bar{u}-c}\frac{\partial\bar{q}}{\partial y}\Psi = 0 \qquad (14.11)$$

Similarly, substituting in the boundary condition, Equation 14.9, gives

$$\frac{\partial\Psi}{\partial z} - \frac{1}{\bar{u}-c}\frac{\partial\bar{u}}{\partial z}\Psi = 0 \text{ on } z = 0, H \qquad (14.12)$$

On the boundaries $y = \pm Y/2$, either cyclic boundary conditions apply, or $\Psi = 0$. We denote any solution to the linearized perturbation equations as (Ψ, c). Taking the complex conjugate (denoted by a *) of Equations 14.11 and 14.12, and using the fact that $\Psi/(\bar{u}-c) = \Psi^*/(\bar{u}-c^*)$, we see that if (Ψ, c) is a solution, then so is (Ψ^*, c^*). Its phase speed is the same, but its growth rate is minus that of the original mode. Thus, modes are either neutral, with σ or equivalently c_i zero, or they occur in pairs, one growing and the other decaying. This symmetry breaks down if further terms, representing dissipative processes, are retained in the perturbation equations. However, if the dissipation is weak, an approximate symmetry holds.

To make further progress in determining Ψ and c, specific forms for $\bar{u}$ and $\bar{q}_y$ need to be chosen. However, some general deductions can be made without deriving specific solutions by generating an energy equation from Equation 14.11.

The technique used is a common one in such stability problems. Multiply the interior equation, Equation 14.11, by $\rho_R \Psi^*$ and integrate over the y, z domain. To simplify the calculation, assume that there is a rigid upper boundary at $z = H$, so that Equation 14.12 also applies at that level. Note first that $\Psi\Psi^* = |\Psi|^2$ and that integration by parts gives

$$\int_{-Y/2}^{+Y/2} \frac{\partial^2 \Psi}{\partial y^2} \Psi^* dy = \left[\frac{\partial \Psi}{\partial y} \Psi^* \right]_{-Y/2}^{+Y/2} - \int_{-Y/2}^{+Y/2} \frac{\partial \Psi}{\partial y} \frac{\partial \Psi^*}{\partial y} dy = - \int_{-Y/2}^{+Y/2} \left| \frac{\partial \Psi}{\partial y} \right|^2 dy$$

Also

$$\int_0^H \frac{1}{\rho_R} \frac{\partial}{\partial z} \left(\rho_R \frac{f_0^2}{N^2} \frac{\partial \Psi}{\partial z} \right) \rho_R \Psi^* dz = \left[\rho_R \frac{f_0^2}{N^2} \frac{\partial \Psi}{\partial z} \Psi^* \right]_0^H - \int_0^H \rho_R \frac{f_0^2}{N^2} \frac{\partial \Psi}{\partial z} \frac{\partial \Psi^*}{\partial z} dz$$

Using Equation 14.12 to substitute for $\partial\Psi/\partial z$ at $z = 0$, H in the first term on the right-hand side then gives

$$\int_0^H \frac{1}{\rho_R} \frac{\partial}{\partial z} \left(\rho_R \frac{f_0^2}{N^2} \frac{\partial \Psi}{\partial z} \right) \rho_R \Psi^* dz = \left[\rho_R \frac{f_0^2}{N^2} \frac{1}{\bar{u} - c} \frac{\partial \bar{u}}{\partial z} |\Psi|^2 \right]_0^H - \int_0^H \rho_R \frac{f_0^2}{N^2} \left| \frac{\partial \Psi}{\partial z} \right|^2 dz$$

Using these last two results, the integral of Equation 14.11 multiplied by $\rho_R \Psi^*$ is

$$-\int_0^H \int_{-Y/2}^{Y/2} \left\{ \rho_R \left[k^2 |\Psi|^2 + \left| \frac{\partial \Psi}{\partial y} \right|^2 + \frac{f_0^2}{N^2} \left| \frac{\partial \Psi}{\partial z} \right|^2 \right] \right\} dydz +$$

$$\int_{-Y/2}^{Y/2} \left[\rho_R \frac{f_0^2}{N^2} \frac{1}{\bar{u} - c} \frac{\partial \bar{u}}{\partial z} |\Psi|^2 \right]_0^H dy + \int_0^H \int_{-Y/2}^{Y/2} \left\{ \rho_R \frac{1}{\bar{u} - c} \frac{\partial \bar{q}}{\partial y} |\Psi|^2 \right\} dydz = 0 \tag{14.13}$$

The first complete integral is positive definite. In fact, it is equal to $4E$ where E is the total eddy energy averaged with respect to x. A positive definite quantity which we may call the wave activity may be defined in the following form:

$$A = \rho_R \frac{|\Psi|^2}{|\bar{u} - c|^2} \tag{14.14}$$

Then, using the identity $(\bar{u} - c)^{-1} = (\bar{u} - c^*) / |\bar{u} - c|^2$, Equation 14.13 can be written as follows:

$$-4E + \int_{-Y/2}^{Y/2} \left[\frac{f_0^2}{N^2} \frac{\partial \overline{u}}{\partial z} (\overline{u} - c^*) A \right]_0^H dy + \int_0^H \left\{ \int_{-Y/2}^{Y/2} \frac{\partial \overline{q}}{\partial y} (\overline{u} - c^*) A \ dy \right\} dz = 0 \quad (14.15)$$

If the wave exists, then E is real and positive. Therefore, the real part of Equation 14.15 is

$$\int_{-Y/2}^{Y/2} \left[\frac{f_0^2}{N^2} \frac{\partial \overline{u}}{\partial z} (\overline{u} - c_R) A \right]_0^H dy + \int_0^H \left\{ \int_{-Y/2}^{Y/2} \frac{\partial \overline{q}}{\partial y} (\overline{u} - c_R) A \ dy \right\} dz = 4E > 0 \quad (14.16)$$

and the imaginary part is

$$c_I \left\{ \int_{-Y/2}^{Y/2} \left[\frac{f_0^2}{N^2} \frac{\partial \overline{u}}{\partial z} A \right]_0^H dy + \int_0^H \left(\int_{-Y/2}^{Y/2} \frac{\partial \overline{q}}{\partial y} A \ dy \right) dz \right\} = 0 \quad (14.17)$$

For a neutral mode, c_I is zero and so Equation 14.17 is satisfied trivially, leaving only Equation 14.16 to be satisfied. If the wave is growing or decaying, Equation 14.17 gives

$$\int_{-Y/2}^{Y/2} \left[\frac{f_0^2}{N^2} \frac{\partial \overline{u}}{\partial z} A \right]_0^H dy + \int_0^H \left(\int_{-Y/2}^{Y/2} \frac{\partial \overline{q}}{\partial y} A \ dy \right) dz = 0 \quad (14.18)$$

Equation 14.18 is called the Charney–Stern necessary condition for instability.

The condition for the existence of instability divides into a number of statements, any one of which must be satisfied. If, for example, $\partial \overline{u} / \partial z$ is zero on the boundaries $z = 0$, H, then the first integral vanishes automatically, and the condition for instability is that the second integral be zero. Given that A is positive definite, this requires that the poleward gradient of potential vorticity $\partial \overline{q} / \partial y$ should change sign somewhere in the (y, z) plane. On the other hand, if $\partial \overline{q} / \partial y$ is zero throughout the (x, y) plane, then instability is still possible provided $-\partial \overline{u} / \partial z$ at $z = 0$ and $\partial \overline{u} / \partial z$ at $z = H$ have opposite signs. If $\partial \overline{q} / \partial y$ is non-zero, then a full statement of the Charney–Stern condition takes the following form:

For instability, $\partial \overline{q} / \partial y$ in the interior, $-\partial \overline{u} / \partial z$ at $z = 0$ and $\partial \overline{u} / \partial z$ at $z = H$ (if there is a boundary at $z = H$) must somewhere be both positive and negative.

We will discuss some examples of the application of this criterion in Sections 14.3–14.5. The condition is based on the possibility that eddies can extract energy from the zonal flow. However, it does not guarantee that coherent eddy structures will be found that can actually grow. The Charney–Stern criterion is therefore a necessary, but not a sufficient condition, for instability.

The discussion of this section has revealed the central role played by the poleward gradient of potential vorticity. It is worth examining this quantity more closely. Taking the y-derivative of Equation 14.4c and making use of the hydrostatic equation and the thermal wind equation, $\bar{q}_y$ can be written in terms of the zonal mean zonal wind $\bar{u}$:

$$\frac{\partial \bar{q}}{\partial y} = \beta - \frac{\partial^2 \bar{u}}{\partial y^2} - \frac{1}{\rho_R}\frac{\partial}{\partial z}\left(\rho_R \frac{f_0^2}{N^2}\frac{\partial \bar{u}}{\partial z}\right) \tag{14.19}$$

In the limit where the vertical extent of the region of interest is small compared to the scale heights H_ρ and H_θ, Equation 14.19 simplifies further to

$$\frac{\partial \bar{q}}{\partial y} = \beta - \frac{\partial^2 \bar{u}}{\partial y^2} - \frac{\partial^2 \bar{u}}{\partial z_g^2}$$

where $z_g = (N/f_0)z$ is the natural, quasi-geostrophic scaling of z. However, more generally, the vertical variations of ρ_R and N need to be taken into account. In that case, Equation 14.19 is written as the sum of four terms:

$$\frac{\partial \bar{q}}{\partial y} = \beta - \frac{\partial^2 \bar{u}}{\partial y^2} - \frac{f_0^2}{N^2}\frac{\partial^2 \bar{u}}{\partial z^2} - \frac{1}{\rho_R}\frac{\partial \bar{u}}{\partial z}\frac{\partial}{\partial z}\left(\frac{\rho_R f_0^2}{N^2}\right) \tag{14.20}$$

Figure 14.3 shows the total $\bar{q}_y$ and its component terms. The fields have not been plotted at levels below 80 kPa, where small values of N in the atmospheric boundary layer make the calculation of the third and fourth terms extremely ill-conditioned. In the troposphere, the smallest term is generally β, the horizontal curvature term $\bar{u}_{yy}$ is somewhat larger, but in most places, the third and fourth terms, involving the vertical variations of $\bar{u}$ and N, are dominant. In the region of the tropospheric jets, both the third and fourth terms are large, amounting to several times the local value of β. In the stratosphere, the balance is rather different. The factor $\left(f_0^2 / N^2\right)$ falls by a factor typically as large as four or more from the troposphere into the lower stratosphere, and consequently, the third term in Equation 14.20 is generally less important than the second at such levels.

A useful interpretation of the vertical derivative term, the third on the right-hand side of Equation 14.19, follows from the thermal wind relationship. Given that this third term is the dominant part of $\bar{q}_y$ in the midlatitude troposphere and tropopause, we consider the mass-weighted mean $\bar{q}_y$ between two levels, z_- and z_+:

$$\int_{z_-}^{z_+} \rho_R \bar{q}_y \, dz \approx -\rho_R \frac{f^2}{N^2}\frac{\partial \bar{u}}{\partial z}\bigg|_{z_-}^{z_+}$$

The thermal wind equation enables the vertical wind shear to be related to the horizontal potential temperature gradients, while N^2 is proportional to the vertical gradient of potential temperature. With these relationships substituted in, we have

$$\int_{z_-}^{z_+} \rho_R \bar{q}_y \, dz \approx \rho_R f \left. \frac{\partial \theta / \partial y}{\partial \theta / \partial z} \right|_{z_-}^{z_+} \tag{14.21}$$

But

$$-\frac{\partial \theta / \partial y}{\partial \theta / \partial z} = \left. \frac{\partial z}{\partial y} \right|_\theta$$

is the slope of an isentrope. So Equation 14.21 says that the average $\bar{q}_y$ over a layer of the atmosphere is proportional to the change of the slope of the isentropes across the layer. One example is seen in Figure 14.3c and d, where large values of $\bar{q}_y$ are seen around the tropopause, above which the isentropes are much flatter than below.

The northward potential vorticity gradient, $\partial \bar{q} / \partial y$, is mostly positive throughout the troposphere. It has large values near the jet core and lower, sometimes negative, values in the lower troposphere beneath the main jets. More importantly than these lower tropospheric values, $-\partial \bar{u} / \partial z$ is negative at the lower boundary, with temperatures decreasing towards the pole and westerlies increasing with height. This means that the necessary condition for instability is satisfied. From a different perspective, at the tropopause, which dynamically can act as an upper boundary with w becoming small if not zero, $\partial \bar{u} / \partial z$ is positive, reinforcing the conditions for instability.

The various criteria for instability can be unified by means of a mathematical artifice. Equation 14.18 includes boundary values of $\partial \bar{u} / \partial z$ (or equivalently, by the thermal wind relationships, of $\partial \bar{\theta} / \partial y$) at $z=0$ and $z=H$. Now suppose that this real boundary condition actually were to apply a short distance, δz, in from the boundary and that the hypothetical boundary condition was simply $\partial \bar{u} / \partial z = 0$ at $z = 0, H$. There, of course, the first integrals in Equation 14.18 would vanish. However, in the thin, δz-thickness layer between the physical and hypothetical boundaries, there would be an additional contribution to the potential vorticity gradient $\bar{q}_y$:

$$-\frac{f_0^2}{N^2} \frac{\bar{u}_z}{\delta z} \text{ at } z = 0$$

$$\frac{f_0^2}{N^2} \frac{\bar{u}_z}{\delta z} \text{ at } z = H$$

More elegantly, using the notation of Dirac's delta function, as $\delta z \to 0$, we may write

$$\frac{\partial \bar{q}}{\partial y} = \beta - \frac{\partial^2 \bar{u}}{\partial y^2} - \frac{1}{\rho_R} \frac{\partial}{\partial z} \left(\rho_R \frac{f_0^2}{N^2} \frac{\partial \bar{u}}{\partial z} \right) - \left(\delta(0) - \delta(H) \right) \frac{f_0^2}{N^2} \frac{\partial \bar{u}}{\partial z} \tag{14.22}$$

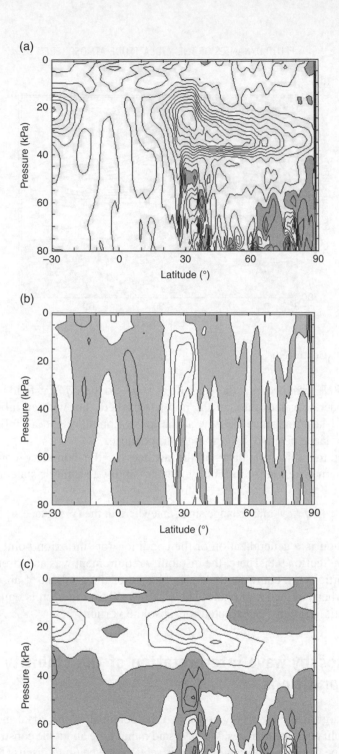

Figure 14.3 Latitude–pressure section of $\overline{q}_y$ for zonal mean Northern Hemisphere winter conditions. Contour interval is $0.2(\Omega/a)$; negative values shaded. (a) The total $\overline{q}_y$, (b) $-\overline{u}_{yy}$, (c) $-\left(f_0^2 / N^2\right)\overline{u}_{zz}$ and (d) $(1/\rho_R)\overline{u}_z\left(\rho_R f_0^2 / N^2\right)_z$

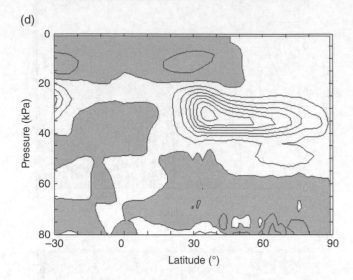

Figure 14.3 (Continued)

where the $\delta(H)$ term occurs if there is a lid at $z = H$. So the $\partial \overline{q} / \partial y$ field is characterized by moderate values through the mid-troposphere, but with an infinitesimally
thin layer of large negative $\partial \overline{q} / \partial y$ at the lower boundary and an infinitesimally
thin layer of large positive $\partial \overline{q} / \partial y$ at the tropopause.

So, accepting this formalism of the artifice for the boundary condition, the
Charney–Stern criterion for instability reduces simply to a single statement:

$$\partial \overline{q} / \partial y \text{ must change sign somewhere in the } (y, z) \text{ plane.}$$

This condition is a generalization of the condition for 'inflexion point instability'
discussed in Section 9.8. There, the instability requirement was an extremum in the
absolute vorticity or, equivalently, a sign change of the gradient of absolute vorticity, somewhere in the (y, z) plane. The Charney–Stern criterion is equivalent, but
with 'potential vorticity' substituted for 'absolute vorticity'.

14.3   Rossby wave interpretation of the stability conditions

So far, the argument has largely concerned the potential of the zonal mean flow to
support amplifying disturbances. We have said rather little about the constraints on the
scale or shape of those disturbances. In this section, we begin to discuss the implications of the zonal mean flow for the nature of growing, unstable disturbances.

Consider first the case of neutral modes in an atmosphere which has zero
temperature gradient on the boundaries, or, equivalently, zero vertical wind shear at

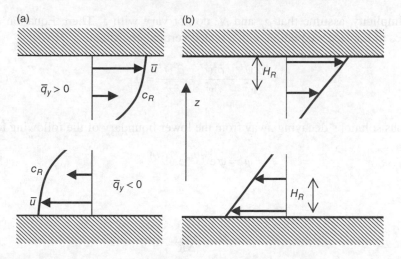

Figure 14.4 Illustrating the configuration of neutral modes (a) when $\partial \overline{u} / \partial z$ is zero at the boundaries; here, the open arrows depict the direction of the potential vorticity gradient and (b) when $\partial \overline{q} / \partial y$ is zero in the fluid interior

the boundaries, shown in Figure 14.4a. The first integral in Equation 14.16 must be zero. For the second integral to be equal to the positive definite quantity $4E$, then in the region where the amplitude is concentrated, as measured by the wave action A, $\partial \overline{q} / \partial y$ and $\overline{u} - c_R$ must have the same sign. For the simple barotropic Rossby waves described in Section 8.5, $\partial \overline{q} / \partial y = \beta$ and $c_r = \overline{u} - \beta / K^2$ so that $\overline{u} - c_R$ does indeed have the same sign as $\partial \overline{q} / \partial y$.

An alternative case is for a fluid with zero $\partial \overline{q} / \partial y$ but with boundary temperature gradients, shown in Figure 14.4b. In this case, the second integral in Equation 14.16 vanishes, but neutral modes can still exist. Their existence requires that $\partial \overline{u} / \partial z$ and $\overline{u} - c_R$ should have the same signs near an upper boundary but opposite signs near a lower boundary.

Rossby-like waves with this property are easily illustrated. Consider a zonal flow with zero $\partial \overline{q} / \partial y$ and constant $\partial \overline{u} / \partial z$. Since $\overline{q}$ is constant, the perturbation q is zero for any meridional displacement. Consequently, the perturbation potential vorticity equation reduces to

$$q' = 0$$

or from Equation 14.6c,

$$q' = \nabla^2_H \psi' + \frac{1}{\rho_R} \frac{\partial}{\partial z} \left(\rho_R \frac{f_0^2}{N^2} \frac{\partial \psi'}{\partial z} \right) = 0 \qquad (14.23)$$

For simplicity, assume that ρ_R and N^2 do not vary with z. Then, Equation 14.23 reduces to Laplace's equation with a scaled vertical coordinate $z_s = (N/f_0)z$:

$$\frac{\partial^2 \psi'}{\partial x^2} + \frac{\partial^2 \psi'}{\partial y^2} + \frac{\partial^2 \psi'}{\partial z_s^2} = 0 \qquad (14.24)$$

This has solutions decaying away from the lower boundary of the following form:

$$\psi' = \psi e^{-z/H_R} e^{i(kx+ly)} \qquad (14.25)$$

where

$$H_R = \frac{f_0}{NK}$$

The phase speed is found by substituting this form of solution into the boundary condition at the lower surface, Equation 14.9:

$$c_R = H_R \frac{\partial \bar{u}}{\partial z} \qquad (14.26)$$

The phase speed is simply equal to the flow speed a height H_R above the surface. This level is sometimes called the 'steering level'; it is as if the eddies were advected by the flow at that level. This neutral mode is a 'trapped mode', with an amplitude which is maximum at the boundary and falls off exponentially with height. For waves with no meridional structure, $K = k$ in the definition of H_R, and Equation 14.26 implies that the wave frequency is

$$\omega = \frac{f_0}{N} \frac{\partial \bar{u}}{\partial z}$$

Since this is independent of wavenumber, the group velocity in the x-direction is zero: wave activity moves with the speed of the flow at the boundary. Note that in this example, $\partial \bar{u} / \partial z$ is positive, while $\bar{u} - c_R$ is negative at the lower boundary, in agreement with the condition derived in the previous paragraph. A similar analysis applies in cases with an upper boundary, so that upper boundary-trapped Rossby wave activity moves with the basic flow speed at that boundary.

Consider for a moment a flow with zero $\partial \bar{u} / \partial z$ at the boundary but non-zero $\partial \bar{q} / \partial y$ throughout the fluid. Then, the Charney–Stern criterion for instability reduces to

$$\int_0^H \int_{-Y/2}^{Y/2} \left(\frac{\partial \bar{q}}{\partial y} A \right) dydz = 0$$

which is possible only if $\partial \bar{q} / \partial y$ takes both signs in the region of significant A. Such a sign reversal may be due primarily to the $\bar{u}_{yy}$ term in the potential vorticity gradient, in which case the instability is called 'barotropic instability'. A simple example is flow on an f-plane which is independent of height, so that $\bar{q}_y = -\bar{u}_{yy}$. Suppose, as shown in Figure 14.5, that $\bar{u}_{yy}$ has opposite signs for $y>0$ and $y<0$. Then, it is possible for Rossby waves existing in the region where $\bar{q}_y$ is negative to have a very similar phase speed to a separate train of Rossby waves centred in the region where $\bar{q}_y$ is positive. Indeed, if the x- and y-wavenumbers are chosen carefully, the phase speeds can be identical. The instability can then be viewed as a positive feedback between the two trains of Rossby waves, producing phase locking and growth.

A discussion of the mechanism for this was given in Section 9.8. Here we shall consider it in more detail. Figure 14.6 shows two trains of Rossby waves, stationary with respect to one another, either side of a maximum of potential vorticity. Eddies A and B belong to a train northward of the potential vorticity maximum, while eddies C and D are part of a train located to the south of the potential vorticity maximum. The crucial point was discussed extensively in Section 9.2. The circulations associated with eddies A and B, say, extend well beyond the confines of their respective vorticity distributions. In particular, there will be induced velocities near the centre of eddy D due to the eddies A and B in the northern region. The sense of that circulation is predominantly southwards, that is, it will advect higher potential vorticity air from nearer the maximum of the potential vorticity distribution and will therefore amplify the vorticity anomaly of vortex D. In the same way, the circulation associated with eddy D and its companions further east in the diagram will act to reinforce the vorticity anomaly associated with eddy B. Similar arguments predict an intensification of the negative vorticity anomalies of eddies A and C.

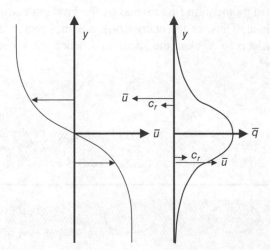

Figure 14.5 Sheared barotropic flow showing (left) the zonal wind profile and (right) the corresponding potential vorticity profile. For the baroclinic case, the y-axis must be relabelled as the z-axis

Thus, provided the two wave trains maintain their phase relative to one another, a positive feedback means that all the eddies will amplify. This is the basis of 'barotropic instability'. It depends crucially on the dispersive nature of Rossby waves, which allows us to choose wavenumbers such that the trains have similar phase speeds and therefore have the possibility of remaining in the same relative phase to one another. The diagram has been drawn with about 66° phase difference between the two wave trains. For such a phase difference that is less than 90°, the eddies A and B induce a maximum southward motion and therefore q tendency which is slightly to the west of D. This and the similar action near C act to reduce the eastward phase speed of this wave train. Similarly, the action of C and D is to increase the eastward phase speed of the wave train including A and B. Therefore, the interaction helps both waves move against the shear in the basic state and in the sense of increasing the phase difference between the two wave trains. If the phase difference was a little greater than 90°, a similar argument shows that the interaction leads to growth and movement in the sense of reducing the phase difference. If the phase difference were reduced to 0 or increased to 180°, then the feedback between them would not amplify the waves at all, but it would modify their phase speeds more strongly in the sense of respectively increasing or decreasing the phase. For a 90° phase difference, the interaction gives maximum growth but no effect on the phase speeds. Thus, the interaction between the two wave trains can promote phase locking between them, a necessary condition for growth, as well as amplification of the waves.

Figure 14.6 also shows dashed lines of constant phase. These have a characteristic slope from south-west to north-east, in the opposite sense from what one might have expected if the disturbance was simply sheared passively by the background flow. Such a 'counter shear' tilt is the characteristic of growing waves. It is associated with a poleward momentum flux carried by the growing eddies in such a sense as to reduce the shear. Thus, as the energy of the eddies increases, the northward flux of momentum acts to weaken the shear, and hence the kinetic energy, of the background flow.

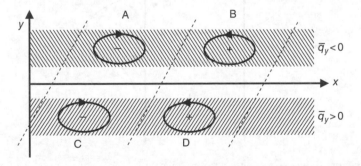

Figure 14.6   Sheared barotropic flow, showing the interactions between trains of vortices either side of the potential vorticity maximum

Finally, note that the arguments of this section are linear and really only apply to the initial stages of the evolution of an unstable flow. The vortices amplify because meridional displacements bring air with more extreme potential vorticity to a given latitude. As the total meridional displacement of fluid elements approaches the width of the potential vorticity anomaly, this supply of high or low potential vorticity becomes exhausted. Growth will cease, and more complex dynamics, associated either with vortex interactions or with Rossby wave radiation, will take over.

When the sign change of $\partial \bar{q} / \partial y$ is associated with the third, vertical curvature term in Equation 14.19 (and so represents a vertical stratification of $\partial \bar{q} / \partial y$; see Figure 14.5), the instability is termed 'baroclinic instability'. One interpretation of what is happening is precisely parallel to the earlier discussion of barotropic instability. Two Rossby wave trains, now at different levels rather than at different latitudes, interact with one another, inducing positive feedbacks which promote phase locking and amplification in appropriate circumstances. The basis of the argument is the quasi-elliptic relationship between streamfunction and potential vorticity anomaly, Equation 14.6c. If vertical variations of ρ_R and N^2 are negligible, Equation 14.6c is a Poisson equation. Even if ρ_R and N^2 vary substantially, the qualitative nature of the solutions to Equation 14.6c is still similar to those of the Poisson equation. For all the qualitative arguments in this chapter, the essential character of the relationship between potential vorticity anomaly and streamfunction is 'action at a distance', that is, a localized potential vorticity anomaly is associated with induced circulations at points remote from the anomaly. This influence affects both the horizontal, as discussed in Section 9.2, and also the vertical. The ratio of the depth of the vertical to the width of the horizontal response is (f_0/N).

Consider the flow shown in Figure 14.7. It is bounded by rigid surfaces at $z = \pm H/2$ and is sheared with height so that it is easterly for $z < 0$ and westerly for $z > 0$ with an inflexion point at $z = 0$, and zero shear at each boundary. There is assumed to be no variation in the y-direction. The potential vorticity in the upper layer increases in the y-direction by virtue of negative $\bar{u}_{zz}$ at these levels. The potential vorticity gradient is reversed in the lower layer. So in the upper layer, a train of Rossby waves will propagate to the west relative to the mean flow at that level. In the lower layer, Rossby waves will propagate to the east relative to the local zonal flow. It is possible to choose x- and y-scales for the Rossby waves such that they will be stationary with respect to one another. The situation is of course precisely analogous to the barotropic situation illustrated in Figure 14.5, but with the two trains of Rossby waves stacked on top of one another rather than side by side.

The interactions between the two wave trains are illustrated by means of a longitude–height section in Figure 14.8. The circulation around each extremum of potential vorticity extends to either side of that extremum and, by influencing neighbouring vortices, ensures that the train of vortices propagates relative to the mean flow. The argument for this is essentially that set out in Section 9.5. But the circulation also extends in the vertical and affects the vortices in the adjacent level. So, for example, vortices A and B in Figure 14.8 both induce northward motion near the centre of vortex C. Now in the lower layer, $\bar{q}_y < 0$, and so this northward motion

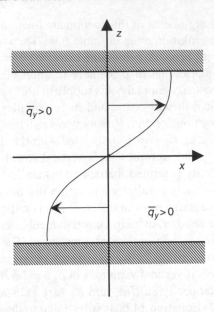

Figure 14.7   A vertically sheared flow on an f-plane. If the flow is uniform in the y-direction, the potential vorticity gradient will be positive in the upper half and negative in the lower half of the fluid

will advect air with larger potential vorticity from lower latitudes, thereby strengthening vortex C. In the same way, the combined effect of vortices C and D is to induce northward motion near the centre of vortex B, in the upper layer. This time, $\bar{q}_y < 0$. So the induced circulation advects low potential vorticity air into the centre of vortex B. Such advection strengthens the negative anomaly of q which already exists there. In this way, a consistent positive feedback exists between the upper and lower wave trains, increasing the amplitude of both. The amplification depends upon the characteristic phase tilt, from east to west with height, of the velocity and vorticity fields of the eddies. Just as in the barotropic case, the phase tilt is in the opposite sense from that which might have been expected on the basis of the shear of the basic flow. In the baroclinic case, the phase tilt is associated with poleward eddy temperature fluxes. Baroclinic waves systematically move cold air equatorwards and warm air polewards, reducing the temperature gradient or, equivalently, the vertical wind shear.

Figure 14.8 has been drawn with a 90° phase lag of the upper wave over the lower wave, and this leads to pure growth. If the phase lag differed from 90°, the induced flow would also act to modify the phase, as well as the amplitude, of the wave in the other layer. In other words, the feedback between the trains of waves would modify their phase speeds and promote phase locking. Again, such phase locking is analogous to equivalent phase locking in the barotropic case.

The arguments in this section have been largely qualitative. However, they can be used to give some order-of-magnitude estimates of the characteristics of unstable baroclinic waves. The phase speed of a Rossby wave is

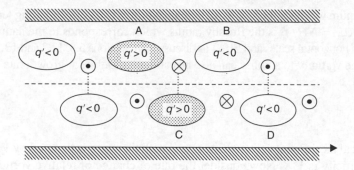

Figure 14.8 Longitude–height section of two Rossby wave trains in the flow illustrated in Figure 14.7

$$c_R = U - \frac{\overline{q}_y}{K^2}$$

(see Equation 8.19), where $K = \sqrt{k^2 + l^2}$ is the total horizontal wavenumber of the waves. First, note that if K is too large, that is, the wavelength is too small, the waves in each layer will have a phase speed close to the local flow speed. Furthermore, the modes will be shallow, and they will only affect each other weakly. Phase locking will be difficult, and so one might suppose that the baroclinic instability mechanism will not be very effective. The same is true for very long waves, that is, for waves with extremely small values of wavenumber. In that case, the wave phase speeds will be extremely large and in the opposite direction to the local flow speed. Again, it will be difficult to have effective phase locking, and baroclinic instability will not be large. In terms of the flow presented in this section, with the zonal wind changing from $-\Delta U$ to ΔU between the lower and upper boundaries, the optimum phase locking will be for waves whose phase speed is close to the mean flow speed, that is, for

$$K^2 \sim \frac{\overline{q}_y}{\Delta U}$$

Now for the basic flow envisaged in Figure 14.7, where the potential vorticity gradient is associated with the vertical curvature term in Equation 14.6c, the potential vorticity gradient is

$$\overline{q}_y \sim \frac{f_0^2}{N^2} \overline{u}_{zz} \sim \frac{f_0^2}{N^2 H^2} \Delta U$$

It follows that

$$K \sim \frac{f_0}{NH} \equiv K_R \qquad (14.27)$$

is the optimum wavenumber for baroclinic instability. The corresponding length scale $L \sim L_R = K_R^{-1} = NH/f_0$ is the Rossby radius which corresponds to the natural ratio of vertical and horizontal scales and Bu~1 as discussed in Sections 5.8, 12.2 and 13.1. A time-scale for the vorticity to amplify can also be estimated. Define a 'growth rate' as

$$\sigma \sim \frac{1}{\xi}\frac{\partial \xi}{\partial t}$$

If the typical meridional velocity associated with the wave is denoted by V', then the typical vorticity is $V'K$. Now consider the rate of change of relative vorticity at one level due to velocity induced by the wave at the upper level. For the most rapidly growing wave, this will be of the following order:

$$\frac{\partial \xi'}{\partial t} \sim \frac{\partial q'}{\partial t} \sim \varepsilon V'\frac{\partial \overline{q}}{\partial y} \sim \varepsilon V'\frac{f_0^2}{N^2}\frac{\Delta U}{H^2}$$

Here, ε where $0<\varepsilon<1$ is a number which measures the attenuation of the circulation induced by the upper-layer wave train at the level of the lower-layer train. Setting $\varepsilon=1$ will give an upper bound on the growth rate. Combining this last result with Equation 14.27 gives the following growth rate:

$$\sigma \sim \frac{1}{\xi'}\frac{\partial \xi'}{\partial t} \sim \frac{\varepsilon}{K}\frac{f_0^2}{N^2 H^2}\Delta U \sim \varepsilon \frac{f_0}{N}\overline{u}_z \qquad (14.28)$$

These arguments are quite general and give a characteristic space and timescale for baroclinic instability. For the Earth's midlatitudes, a typical length scale K^{-1} is around 500 km and a typical timescale σ^{-1} is around 1 day.

    The qualitative argument given in this section has much in common with the argument given in Section 14.1 and Figure 14.1. There, we used the omega equation and vortex stretching to illustrate the positive feedback possible between two verti-cally stacked trains of waves. The present arguments based on potential vorticity give a more direct approach to the vorticity tendencies. But more importantly, the potential vorticity argument has shown directly how the Rossby wave dispersion relationship predicts that the two trains of waves can become phase locked. In Figure 14.1 et seq, we merely presumed such phase locking.

14.4 The Eady model

The preceding sections have dealt in generalities. To proceed further, we need to consider the stability properties of specific zonal flows $\overline{u}(y,z)$ along with the den-sity and stratification profiles ρ_R and N. Our aim is to calculate the growth rates and structures of unstable modes which can grow on such a flow. This turns out to be difficult. Very few $\bar{u}(y,z)$ are amenable to an analytic stability calculation. Instead, one is usually driven to discretize the problem in some way and to use numerical techniques to determine eigenvalues and eigenvectors. However, the idealized problems which do admit of direct solution give considerable insight and are suggestive of more general ideas.

The simplest such model is that due to Eady (1949). He considered flow on an f-plane, confined between rigid surfaces a distance H apart. For simplicity, suppose that the variation with height of ρ_R can be neglected and N^2 is constant. It is also convenient to use a coordinate system fixed in the mid-level flow and with the origin of the z-axis at that level, so the flow is bounded by rigid boundaries at $z=\pm H/2$. The basic state has a uniform vertical shear and may be written as follows:

$$\bar{u}(z) = \Lambda z \qquad (14.29)$$

The meridional buoyancy gradient is constant, with

$$-\frac{\partial \bar{b}}{\partial y} = f_0 \frac{\partial \bar{u}}{\partial z} = f_0 \Lambda \qquad (14.30)$$

corresponding to a uniform poleward temperature gradient everywhere (Figure 14.9). For this basic state, the interior $\partial \bar{q} / \partial y$ is zero (see Equation 14.19).

The Eady model satisfies the Charney–Stern criterion for instability. The vertical shear $\partial \bar{u} / \partial z$ is positive at $z=+H/2$ while $-\partial \bar{u} / \partial z$ at $z=-H/2$ is negative. The interior potential vorticity gradient is 0. Therefore, the instability relies on the interactions between two Rossby wave trains trapped at the top and bottom boundaries.

The analysis is closely similar to that already carried out for a boundary Rossby wave in such a system (see Equations 14.23–14.26). Because the interior potential vorticity gradient is 0, it follows that the potential vorticity perturbation must also be 0:

$$q' = \nabla_H^2 \psi' + \frac{f_0^2}{N^2} \frac{\partial^2 \psi'}{\partial z^2} = 0 \qquad (14.31)$$

while the boundary conditions, Equation 14.9, are

$$\left(\frac{\partial}{\partial t} + \bar{u} \frac{\partial}{\partial x} \right) b' - f_0 \Lambda v' = 0 \quad \text{at } z = \pm \frac{H}{2} \qquad (14.32)$$

Note that

$$b' = f_0 \frac{\partial \psi'}{\partial z}, \quad v' = \frac{\partial \psi'}{\partial x}$$

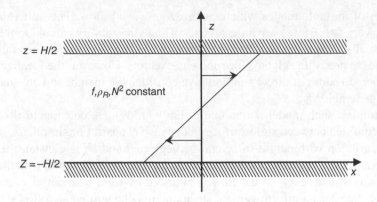

Figure 14.9   The configuration envisaged in the Eady model

The problem is made dimensionless, using H for the vertical length scale, $L_R = NH/f_0$, for the horizontal length scale and an arbitrary amplitude A for the perturbation streamfunction. The variables are transformed as

$$\tilde{z} = z / H, \tilde{x} = x / L_R, \tilde{y} = y / L_R, \tilde{t} = t / (L_R / \Delta U) = t (f_0 \Lambda / N), \tilde{\psi}' = \psi' / A \text{ and}$$
$$\tilde{q} = q' / (A / L_R^2).$$

With these transformations, the perturbation equations become

$$\left(\frac{\partial}{\partial \tilde{t}} + \tilde{z} \frac{\partial}{\partial \tilde{z}} \right) \tilde{q}' = 0 \quad \text{where } \tilde{q}' = \frac{\partial^2 \tilde{\psi}'}{\partial \tilde{x}'^2} + \frac{\partial^2 \tilde{\psi}'}{\partial \tilde{y}'^2} + \frac{\partial^2 \tilde{\psi}'}{\partial \tilde{z}'^2} \tag{14.33a}$$

$$\left(\frac{\partial}{\partial \tilde{t}} + \tilde{z} \frac{\partial}{\partial \tilde{z}} \right) \frac{\partial \tilde{\psi}'}{\partial \tilde{z}'} - \frac{\partial \tilde{\psi}'}{\partial \tilde{x}} = 0 \quad \text{at } \tilde{z} = \pm \frac{1}{2} \tag{14.33b}$$

Note that there are no parameters in the transformed equations. A single solution describes all possible cases.

Equation 14.33 can only be satisfied if $\tilde{q}' = 0$ everywhere. Assume solutions of the following form:

$$\tilde{\psi}' = \text{Re} \left\{ \Psi (\tilde{z}) \exp \left(i \left(\tilde{k} \tilde{x} + \tilde{l} \tilde{y} - \tilde{k} \tilde{c} \tilde{t} \right) \right) \right\} \tag{14.34}$$

It then follows that

$$\frac{d^2 \Psi}{d\tilde{z}^2} - \tilde{K}^2 \Psi = 0 \tag{14.35}$$

where $\tilde{K} = \left(\tilde{k}^2 + \tilde{l}^2\right)^{1/2}$ is the total dimensionless horizontal wavenumber. Instead of posing solutions with exponential growth and decay with z, as we did for boundary-trapped waves, here it is convenient to look for solutions of the following form:

$$\tilde{\Psi}(\tilde{z}) = a\cosh\left(\tilde{K}\tilde{z}\right) + b\sinh\left(\tilde{K}\tilde{z}\right)$$

where a and b are arbitrary constants, to be determined from the boundary conditions, Equation 14.3b. Substituting this solution into Equation 14.3b gives

$$\tilde{K}\tilde{c}\left(a\sinh\left(\tilde{K}\tilde{z}\right) + b\cosh\left(\tilde{K}\tilde{z}\right)\right) - \tilde{z}\tilde{K}\left(a\sinh\left(\tilde{K}\tilde{z}\right) + b\cosh\left(\tilde{K}\tilde{z}\right)\right) +$$

$$\left(a\cosh\left(\tilde{K}\tilde{z}\right) + b\sinh\left(\tilde{K}\tilde{z}\right)\right) = 0 \text{ on } \tilde{z} = \pm\frac{1}{2}$$

This expression contains terms such as $\cosh\left(\tilde{K}\tilde{z}\right)$, $\tilde{z}\sinh\left(\tilde{K}\tilde{z}\right)$, which are even functions of $\tilde{z}$, and others, such as $\tilde{z}\cosh\left(\tilde{K}\tilde{z}\right)$, $\sinh\left(\tilde{K}\tilde{z}\right)$, which are odd functions of $\tilde{z}$. If the boundary condition is to be satisfied at both the upper and lower boundaries, it follows that the sum of the odd terms and the sum of the even terms must separately be zero at $\tilde{z} = 1/2$:

$$\tilde{K}\tilde{c}\left(\frac{b}{a}\right) = \frac{\tilde{K}}{2}\tanh\left(\frac{\tilde{K}}{2}\right) - 1, \quad \tilde{K}\tilde{c}\left(\frac{a}{b}\right) = \frac{\tilde{K}}{2}\coth\left(\frac{\tilde{K}}{2}\right) - 1 \qquad (14.36)$$

Since the amplitude is arbitrary, only the ratio a/b is relevant for the structure of the wave. The imaginary part of the phase speed c determines whether or not the mode is unstable. Eliminating a/b between the two equations 14.36 gives, after a little manipulation,

$$\tilde{c}^2 = \frac{1}{\tilde{K}^2}\left(\frac{\tilde{K}}{2} - \coth\left(\frac{\tilde{K}}{2}\right)\right)\left(\frac{\tilde{K}}{2} - \tanh\left(\frac{\tilde{K}}{2}\right)\right) \qquad (14.37)$$

Now the first term in brackets is positive for $\tilde{K}$ greater than 2.399 and negative otherwise. The second bracketed term is negative for all $\tilde{K}$ and the first bracketed term has a zero at $\tilde{K} = 2.399$. Two cases are distinguished:

1. $\tilde{K} < 2.399$: these long waves have $\tilde{c}^2$ negative or $\tilde{c}$ imaginary. This corresponds to pairs of growing and decaying modes.
2. $\tilde{K} > 2.399$: these short waves have $\tilde{c}^2$ positive or $\tilde{c}$ real. This corresponds to pairs of neutral modes, propagating at phase speeds $\pm\tilde{c}$.

Figure 14.10 shows the growth rate on the wavenumber plane. The critical $\tilde{K}$, $\tilde{K}_c = 2.399$, is referred to as the 'shortwave cut-off', being the wavenumber beyond which baroclinic instability ceases. The most unstable wave has zonal wavenumber $k = 1.61 K_R$, $l = 0$ and a growth rate of $0.31(f_0 / N)\bar{u}_z$. These values are consistent with the scaling derived in the previous section, Equations 14.27 and 14.28). The reason for the shortwave cut-off can be understood in terms of the interaction of Rossby wave trains trapped on the two boundaries that underlies the instability. These waves have a ratio f/N for their vertical and horizontal scales. For short wavelengths, the mutual interaction of the waves over the depth H is too weak to give phase locking.

The structure of the waves is obtained by evaluating the ratio (b/a) from Equations 14.36. Divide the two equations to eliminate c and obtain

$$\left(\frac{b}{a}\right)^2 = \frac{\dfrac{\tilde{K}}{2}\tanh\left(\dfrac{\tilde{K}}{2}\right) - 1}{\dfrac{\tilde{K}}{2}\coth\left(\dfrac{\tilde{K}}{2}\right) - 1} \tag{14.38}$$

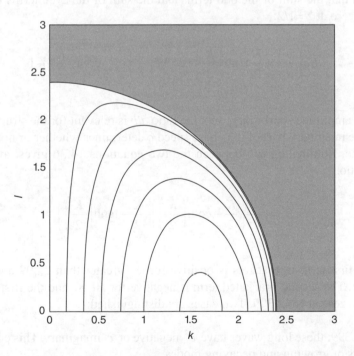

Figure 14.10 Growth rate for the Eady model as a function of zonal wavenumber k and meridional wavenumber l. Contour interval $0.05(f_0 / N)\bar{u}_z$. Shading indicates regions where there is no instability

For unstable waves with $\tilde{K} < 2.399$, the numerator on the right-hand side of Equation 14.38 is negative. Then,

$$\frac{b}{a} = \pm i \sqrt{\frac{1 - \frac{\tilde{K}}{2}\tanh\left(\frac{\tilde{K}}{2}\right)}{\frac{\tilde{K}}{2}\coth\left(\frac{\tilde{K}}{2}\right) - 1}} = \pm i\alpha \qquad (14.39)$$

The positive root corresponds to the growing mode. Arbitrarily fix the phase of the wave by assuming that a is real. Then, the normal-mode perturbation, Equation 14.34, will be

$$\tilde{\psi}'(x,y,z) = ae^{\pm\sigma t}\cos(ly)\left(\cosh(\tilde{K}\tilde{z})\cos(\tilde{k}\tilde{x}) \mp \alpha\sinh(\tilde{K}\tilde{z})\sin(\tilde{k}\tilde{x})\right) \qquad (4.40)$$

The signs imply that the $\tilde{\psi}'$ field tilts to the west with height for growing modes and to the east with height for decaying modes. Fields of $\tilde{v}'$ and $\tilde{b}'$ are obtained by differentiating Equation 14.40 with respect to $\tilde{x}$ or $\tilde{z}$ respectively. The vertical velocity for the Eady wave can be obtained from the omega equation with constant ρ_R and a source term which in dimensional terms reduces to the following form:

$$S = 2f\Lambda\frac{\partial^2 v}{\partial x^2} \qquad (14.41)$$

This equation, with boundary conditions $w=0$ on the two boundaries, can be solved analytically.

Figure 14.11 shows the structure of the most unstable Eady wave, with $k=1.61$ and $l=0$. The wave has maximum amplitude in streamfunction, v and b perturbations at the lower and upper boundaries. Their amplitudes are somewhat smaller at midlevels, but the vertical velocity is a maximum here. The poleward velocity wave, like the streamfunction perturbation, tilts to the west with height. For the fastest growing mode, the upper- and lower-level streamfunction waves are displaced relative to one another by exactly one quarter of a wavelength. The vertical velocity waves also tilt westwards with height but by a smaller amount. The temperature wave, on the other hand, tilts towards the east with height. To the east of the surface, low warm air is moving polewards and rising in agreement with the qualitative discussion in Section 14.1 and Figure 14.2. In fact, at the middle level, all three fields are exactly in phase for the most unstable wave. A consequence of the variation of both amplitude and phase with height is that for this wave, the co-variance of the

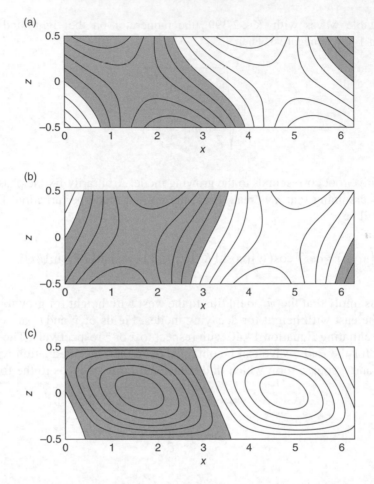

Figure 14.11 The structure of the fastest growing Eady wave. (a) Poleward velocity component, (b) buoyancy perturbation and (c) vertical velocity. In (a), the low and high in streamfunction and pressure are located on the zero contours of the poleward wind. In each case, the field has been arbitrarily normalized so that its maximum value is 1; contour interval 0.2; negative values shaded

poleward velocity and temperature perturbation, the so-called poleward temperature flux, is actually constant with height.

For wavelengths other than that of the fastest growing mode, the structure of the wave changes, so that the correlation between the velocity and temperature fields becomes smaller. For long waves, with $K \to 0$, the parameter $\alpha \to \infty$. The westward phase tilt of velocity wave increases to half a wavelength, while the eastward tilt of the temperature wave drops to zero. Towards the shortwave cut-off, as $K \to 2.399$, $\alpha \to 0$. In this limit, the phase tilt of the velocity wave is zero, while the eastward tilt of the temperature wave has increased to half a wavelength.

Despite the simplicity of its assumptions, the Eady model does involve significant algebraic complexity. Other idealized models of baroclinic instability will be
discussed in the next section. What seems remarkable about them is that they all
lead to similar values for the growth rate and wavelength of the most unstable
waves, despite great disparity in the mathematical details of extracting normal
modes. However, the basic discussion of necessary structure in Section 14.1 and the
scales for growth rates and horizontal length scales in Equations 14.27 and 14.28
make this less surprising.

14.5   The Charney and other quasi-geostrophic models

Charney's model of baroclinic instability was in fact the first to be published. It is,
like Eady's, based on extremely simple assumptions. The main differences are that
the density decays with height, that the upper boundary is removed to infinity and
that a non-zero beta effect is included. Otherwise, the assumptions are similar to
those of Eady, namely, a wind which has a constant vertical shear but no horizontal
shear and a stratification, as measured by the Brunt–Väisälä frequency N, which is
constant with height. The model develops baroclinic instability because of the
phase locking and interaction of an interior Rossby wave with a trapped wave at
the lower boundary. It is unstable according to the Charney–Stern criteria. Despite
its similarity to the Eady model, and the modest nature of the extensions it makes
to the underlying assumptions, the Charney model is considerably more complex
algebraically.

In the Charney configuration, the temperature or buoyancy gradient on the lower
boundary is

$$\frac{\partial \bar{b}}{\partial y} = f_0 \frac{\partial \bar{u}}{\partial z} \equiv f_0 \Lambda \tag{14.42}$$

Taking the density $\rho_R(z) = \rho_0 e^{-z/H}$, the poleward gradient of potential vorticity is

$$\frac{\partial \bar{q}}{\partial y} = \beta - \frac{1}{\rho_R}\frac{\partial}{\partial z}\left(\rho_R \frac{f_0^2}{N^2}\frac{\partial \bar{u}}{\partial z} \right) = \beta + \frac{f_0^2 \Lambda}{N^2 H} \tag{14.43}$$

For typical midlatitude values, with Λ taken as $10^{-3}\,\mathrm{s^{-1}}$, the term $f_0^2 \Lambda / N^2 H$ in
Equation 14.43 is comparable to β. The analysis starts from Equation 14.11 with 14.12
on $z = 0$. Since density decreases exponentially with height, and since the energy of
vertically propagating disturbances should be conserved, the perturbation amplitude
is written as

$$\tilde{\Psi} = \Psi e^{z/2H} \tag{14.44}$$

The resulting amplitude equation is clarified if it is written in dimensionless form, denoted by *, with vertical distance scaled by H, horizontal distance by the Rossby radius $L_R = NH/f$ and the phase speed c by ΛH. Then, the amplitude equation is

$$\frac{d^2\tilde{\Psi}}{dz^{*2}} - \left(\alpha^2 - \frac{1+\beta^*}{z^* - c^*} \right) \tilde{\Psi} = 0 \tag{14.45}$$

with the boundary conditions

$$\frac{d\tilde{\Psi}}{dz^*} + \left(\frac{1}{2} - \frac{1}{z^* - c^*} \right) \tilde{\Psi} = 0 \text{ on } z^* = 0 \tag{14.46}$$

$$\tilde{\Psi} \text{ finite as } z^* \to \infty$$

Here, $\alpha^2 = K^{*2} + 1/4$, and $\beta^* = \beta(N^2H)/(f_0^2\Lambda)$ is a dimensionless measure of β. Equations 14.45 and 14.46 may be solved analytically by means of hypergeometric functions. Today, with the advent of fast computers, they are more usually solved numerically. The amplitude equation, Equation 14.45, is in the form of an eigenvalue equation. The phase speed, which is generally complex, is calculated from the eigenvalues. The imaginary part of the phase speed is related to the growth rate σ of unstable modes:

$$\sigma = -k\text{Im}(c)$$

Note that this is non-dimensionalized by $f_0\Lambda/N$, just as in the Eady problem.

Consider first the case in which $\beta=0$, but in which the density decays with height and the upper boundary is removed to infinity. This is an interesting intermediate case between the Eady model and the usual form of the Charney model with non-zero β. The growth rate curve is shown in Figure 14.12. Although the β effect is absent, instability is still possible, for there remains a positive poleward gradient of potential vorticity in the fluid interior by virtue of the variation of density with height, term 2 of Equation 14.43. So according to the Charney–Stern theorem, instability is still possible. For longer waves, the growth rate is quite similar to that in the Eady model. However, the curves differ markedly for short waves. The short-wave cut-off of the Eady model is absent. Even the shortest wavelength perturbations, which according to Equation 14.45 will be increasingly shallow as K increases, will still feel this interior potential vorticity gradient, and so they will be weakly

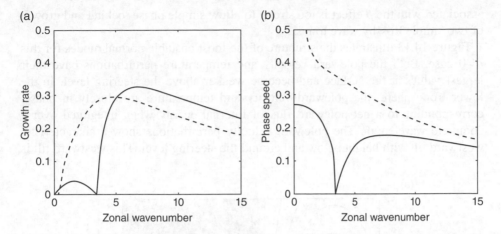

Figure 14.12   Showing (a) the growth rate and (b) the phase speed of the most unstable normal mode at different zonal wavenumbers, according to the Charney model. Dashed curves are for $\beta=0$, and solid curve is for $\beta=8.03\times10^{-12}\text{m}^{-1}\text{s}^{-1}$

unstable. In contrast, as discussed in the previous section, in the Eady problem, there would be insufficient interaction between the boundary Rossby waves to enable instability.

As the wavenumber increases, and the unstable modes become shallower, the phase speed decreases. This behaviour is in marked contrast to the unstable Eady modes, the phase speed of which is constant for all unstable wavenumbers, and equal to the mean westerly flow speed. For the Charney modes with $\beta=0$, the phase speed is equal to the westerly flow at some level in the fluid interior. It is as if the mode was being advected by the flow at this level. The level at which the flow speed is equal to the phase speed is therefore called the 'steering level'. The steering level also marks the characteristic depth of the unstable mode. So, as the zonal wavenumber increases, the modes become more shallow, the steering level becomes lower and so the phase speed becomes smaller. The long waves, in contrast, have deeper structures, with their steering level progressively higher as the wavelength increases. Consequently, their phase speed increases as the zonal wavenumber becomes smaller.

The solid curves in Figure 14.12 show a case comparable to that considered by Charney in his original paper, with a value for β typical of midlatitudes. For large zonal wavenumber k, the growth rate and phase speeds are similar to those of the $\beta=0$ modes. For such short wavelengths, the effect of β is small. Unstable modes exist for the largest wavenumbers; as in the $\beta=0$ case, there is no shortwave cut-off. Modes have their maximum growth rate for k corresponding to six or seven waves around a latitude circle. This time, the modes have a minimum wavenumber for instability, sometimes called a 'longwave cut-off'. In the case illustrated, this corresponds to wavenumber 3–4. At longer wavelengths, the westward propagation associated with the β effect is too strong to allow simple phase locking and growth of two simple Rossby wave trains.

Figure 14.13 illustrates the structure of the most unstable normal modes for this $\beta \neq 0$ case. Both the poleward velocity and temperature perturbations have their largest values at the surface and become weaker above the steering level. In the lower troposphere, the poleward velocity and temperature are closely in phase, corresponding to a net poleward flux of internal energy when integrated over a complete wavelength. The poleward velocity perturbations show a characteristic westward tilt with height below and around the steering level. This westward tilt is

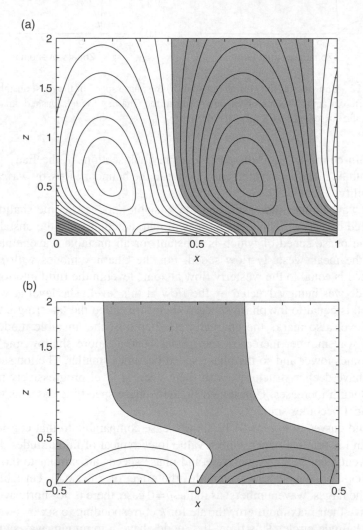

Figure 14.13 The structure of the most unstable Charney mode. (a) Poleward velocity and (b) temperature perturbation. In both cases, the fields have been arbitrarily normalized to have a maximum value of 1.0, contour interval 0.1; negative values shaded

consistent with the temperature perturbation being in phase with the velocity perturbation and so with the existence of a poleward flux of internal energy. A comparison with the fastest growing Eady mode, Figure 14.11, shows that the structure of the Eady mode in the lower half of the domain is qualitatively very similar to the structure of the most unstable Charney mode below the steering level, with largest amplitudes near the boundaries, and the same characteristic phase tilt with height.

At wavelengths longer than the longwave cut-off, weakly unstable modes exist. These longwave modes, sometimes called 'Green modes', have an entirely different structure to the most unstable modes. Instead of a maximum amplitude at the surface, their maximum amplitude is at very high levels. Their phase speed is large, considerably greater than that of the more unstable classic Charney modes. The structure of the lower part of the most unstable of these Green modes is shown in Figure 14.14. The modes are physically unrealistic: in order to reproduce their properties consistently, the vertical domain extended to 20 units in these calculations, corresponding to some 360 km; only the lowest quarter of the domain is shown in the diagram. Notice that at upper levels, the temperature and velocity waves have little phase tilt with height and are virtually in quadrature. This means that the net poleward internal energy flux is close to zero away from the lower boundary. In the lowest part of the domain, the amplitudes are relatively weak, but the characteristic westward tilt of the velocity wave and the eastward tilt of the temperature wave are still to be seen. The Green modes are perhaps a curiosity and are likely to be swamped by more unstable modes growing on more realistic basic states.

14.6 More realistic basic states

In this section, we return to the problem of the stability of a general zonal jet $\bar{u}(y,z)$, which was discussed in Section 14.2. In general, analytical solutions do not exist, and numerical methods have to be used to determine the form of unstable modes and their growth rates for particular jet profiles.

A simple example is given of how such a procedure can be formulated. The perturbation equations, Equations 14.8 and 14.9, can be approximated using finite differences on this grid. First, note that within a quasi-geostrophic framework, the problem can be reduced to a set of equations in only one dependent variable, the perturbation geostrophic streamfunction ψ', for from that, the perturbation potential vorticity and buoyancy are defined using Equation 14.6.

Let us return to the instability problem defined by Equation 14.8 with Equation 14.9 where $\bar{u}$ is a general function of y and z. Assume that the perturbation varies sinusoidally in x, so that the perturbations may be written as follows:

(a)

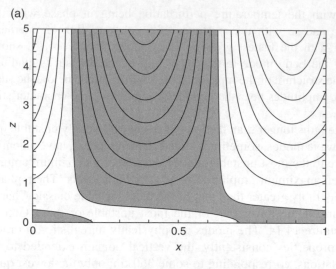

(b)

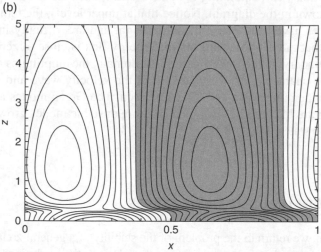

Figure 14.14 The structure of the most unstable Green mode. Notice that the vertical scale is different from that of Figure 14.12. (a) Poleward velocity perturbation and (b) temperature perturbation. As in Figure 14.12, the fields have been arbitrarily normalized so that the maximum value is 1, with negative values shaded

$$\psi' = \psi_k(y,z)e^{ikx}, \quad q' = q_k(y,z)e^{ikx} \tag{14.47}$$

The potential vorticity and geostrophic streamfunction perturbations are related by Equation 14.6c, which may be written as

$$q_k = -k^2\psi_k + \frac{\partial^2\psi_k}{\partial y^2} + \frac{f_0^2}{N^2}\frac{\partial^2\psi_k}{\partial z^2} + \frac{1}{\rho_R}\frac{\partial}{\partial z}\left(\frac{\rho_R f_0^2}{N^2}\right)\frac{\partial\psi_k}{\partial z} \tag{14.48}$$

The (y, z) domain must first be discretized in some way. This may be accomplished by setting up a grid in (y, z) space, on the nodes of which values of the flow variables are given. Alternatively, the flow might be represented by some appropriate spectral series which is truncated after a sufficient number of terms. For simplicity, we shall consider the gridpoint approach. Divide the domain into N_y points in the y-direction and N_z points in the z-direction, making $N_y \times N_z$ gridpoints in total. The $N_y \times N_z$ values of q_k and ψ_k are conveniently written as column vectors $\mathbf{q}$ and $\boldsymbol{\psi}$, in which case, Equation 14.48 can be written more compactly as

$$\mathbf{q} = \mathbf{E}\boldsymbol{\psi}$$

The elements of the matrix $\mathbf{E}$ are generated by writing the various derivatives of ψ_k in finite-difference approximations, for example,

$$\frac{\partial^2 \psi_k}{\partial y^2} \simeq \frac{\psi_k(y+\delta y, z) - 2\psi_k(y, z) + \psi_k(y-\delta y, z)}{\delta y^2}$$

In this same matrix notation, the potential vorticity throughout the fluid interior evolves according to

$$\frac{\partial \mathbf{q}}{\partial t} = -ik\bar{u}\mathbf{q} - ik\frac{\partial \bar{q}}{\partial y}\boldsymbol{\psi} = -ik\left(\bar{u}\mathbf{E} + \frac{\partial \bar{q}}{\partial y}\right)\boldsymbol{\psi} \qquad (14.49)$$

Operating on the entire equation with the matrix $\mathbf{E}^{-1}$ gives an equation for the evolution of the streamfunction

$$\frac{\partial \boldsymbol{\psi}}{\partial t} = -ik\left(\bar{u} + \frac{\partial \bar{q}}{\partial y}\mathbf{E}^{-1}\right)\boldsymbol{\psi} \equiv \mathbf{A}\boldsymbol{\psi} \qquad (14.50)$$

The boundary conditions 14.9 can also be written in terms of the geostrophic streamfunction ψ:

$$\left(\frac{\partial}{\partial t} + ik\bar{u}\right)\frac{\partial \psi}{\partial z} + ik\left(\frac{1}{f_0}\frac{\partial \bar{b}}{\partial y}\right)\psi = 0 \text{ for } z = 0, H \qquad (14.51)$$

This equation will contribute further rows, corresponding to the levels $z = 0, H$, to the matrix $\mathbf{A}$ in Equation 14.50. Seeking solutions of the form $\psi = \Psi e^{\sigma t}$ generates a classic matrix eigenvalue problem:

$$(\mathbf{A} - \sigma\mathbf{I})\boldsymbol{\psi} = 0 \qquad (14.52)$$

Standard methods are available to determine the eigenvalues σ and corresponding eigenfunctions of the matrix $\mathbf{A}$. The former give the growth rate and phase speed of each normal mode, while the corresponding eigenvectors give the structure of the normal mode. In general, the matrix $\mathbf{A}$ has complex elements and, depending upon the number of gridpoints $N_y \times N_z$ needed to represent the jet adequately, may be rather large. Consequently, the computational effort required to determine the eigenvalues and eigenvectors can be very large indeed: typically, the computer time required increases as the sixth power of the number of gridpoints $N_y \times N_z$. Furthermore, only a handful of the $N_y \times N_z$ normal modes generated will be of interest, generally those which are the most unstable.

For the primitive equations on a spherical domain, the difficulty in handling the huge matrices and also in producing a code for the linearized version of the model may often mean that a more straightforward initial value technique is preferable to the matrix technique. In the initial value technique, the nonlinear primitive equation model is initialized with the zonal flow plus a perturbation in the zonal number of interest. The amplitude of the perturbation is taken to be small enough that terms that are quadratic in it are negligible but large enough that the calculations are not compromised by the accuracy of representing the numbers in the computer. The structure of the perturbation is not crucial, but if a good guess at the most unstable wave structure is available, then this should be used. The model is integrated for a period during which there is some growth, and then the amplitude of the perturbation to the zonal flow is reduced back to a value comparable with its initial amplitude. This process is repeated until the structure hardly changes and the growth is almost exponential. This is a sign that the most unstable wave at this wavenumber has emerged from the initial structure, which could be envisaged as being composed of a large number of neutral, decaying and growing modes. This technique provides only the most unstable wave at each wavenumber and works well provided that there are no two modes with almost equal growth rates.

Figure 14.15a shows a zonal mean flow, which is an idealization of the observed midlatitude jet in winter, having a maximum westerly wind near the tropopause of $45\,\mathrm{m\,s^{-1}}$; the jet is centred at 45°N and confined to latitudes 30°N–60°N. A normal-mode calculation has been carried out using the initial value technique. The growth rate and phase speeds of the most unstable waves at zonal wavenumbers 1–11 are given in Figure 14.15b. The maximum growth rate is at wavenumber 8, a wavelength of about 3500 km at 45°N. There is a weak drop-off towards higher wavenumbers and a smooth drop-off towards 0 at lower wavenumbers. This behaviour is similar to that in the Charney model in Figure 14.12a except that there is no sign of the longwave-cut-off and very low-wavenumber behaviour. The phase speed shows little variation with wavenumber. At the latitude of the jet, this corresponds to the jet speed close to 70 kPa, so that this is the steering level. Again the behaviour is like that in the Charney model at medium and higher wavenumbers (Figure 14.12b) but not at low wavenumbers. This can be understood by noting that for the Charney model, the growth rate is $\sigma = kc_i(K)$. Figure 14.12 is drawn for a particular choice of

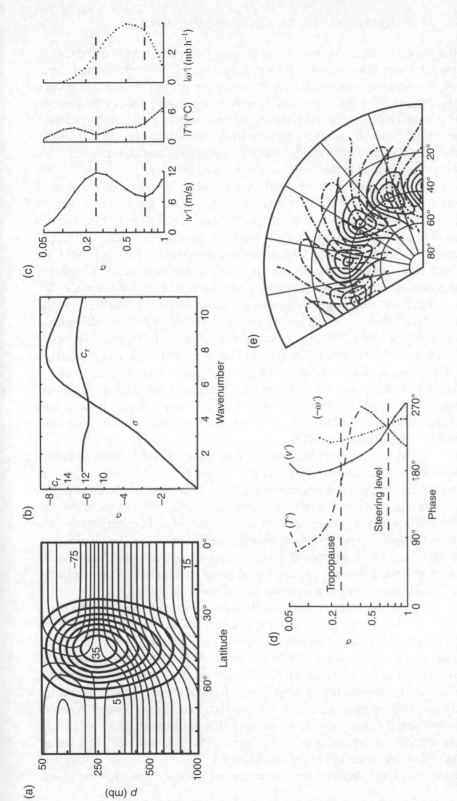

Figure 14.15 Instability of a realistic zonal jet based on the primitive equations with full spherical geometry. (a) The basic jet, with zonal wind contours shown every 5 m s⁻¹ and potential temperature contours every 5 K. (b) The growth rate (unit day⁻¹) and phase speeds (m s⁻¹ at 45°N) of the most unstable waves as a function of zonal wavenumber are shown. (c–e) Properties of the most unstable mode at zonal wavenumber 6 are shown. The vertical structure of the wave amplitude (c) and phase (d) of perturbation variables are given at 52°N, the latitude of maximum amplitude. The steering level and tropopause at this latitude are indicated by dashed lines. (e) The horizontal structure of the streamfunction at σ=0.97 and 0.19 (dashed); two wavelengths of the growing wave are plotted

meridional wavenumber, l. However, if l is allowed to vary, this gives that the most unstable wave at very low wavenumbers will have $\sigma = kc_i(K_m)$, where the total wavenumber K_m is a constant such that the low-wavenumber growth rates would be on a straight line that is from the origin and tangent to the main growth rate curve in Figure 14.12a, and the phase speed of these low-wavenumber modes would all have the value at $k = K_m$. The Green modes become irrelevant, and the structure is always that of the main Charney/Eady modes but with decreasing meridional scale. This agrees with what is found in the realistic case discussed here.

The structure of the most unstable zonal wavenumber 6 normal mode is shown in the remainder of Figure 14.15. This wavenumber 6 mode has a growth rate of $0.706\,\mathrm{day}^{-1}$ and a phase speed of 11.9° of longitude per day. Its vertical structure is summarized in terms of the phase and amplitude of variables in Figure 14.15c and d. The amplitude of the mode is largest in the troposphere, with decay into the stratosphere. v has maxima at the surface and also near the tropopause as expected given the potential vorticity gradients there. The main maximum for T is at the surface, and ω has a mid-tropospheric maximum. The vertical tilts of the fields in the troposphere show behaviour which is very similar to that in the Eady model, with ω tilting westward less than v, T tilting eastwards and the phases being nearly identical at the steering level. The horizontal structure of the streamfunction near the surface and the tropopause is shown in Figure 14.15e. The mode decays away from the central latitude of the jet and tilt with the jet. The upper-level wave is somewhat broader meridionally than the low-level wave. At shorter wave lengths, the structure becomes shallower, and the near-tropopause maximum in v is not found, again consistent with the ideas discussed earlier.

In common with the most unstable Charney mode, the poleward temperature flux is largest between the lower boundary and the steering level, and falls away in the upper troposphere. There is convergence of the temperature flux poleward of the jet and divergence equatorward. The unstable mode acts to reduce the temperature gradient in the midlatitudes. The correlation of vertical motion and temperature perturbation means that there is also an upward temperature flux with a similar distribution. The unstable eddies transport internal energy polewards and upwards. An association between poleward thermal advection and rising motion is of course exactly what our discussion of the omega equation in Chapter 13 predicted.

These qualitative properties of the unstable normal-mode feedback on the mean flow are very similar to those of the unstable Charney or Eady modes. However, the horizontal shears of the jet and tilts of the wave seen in Figure 14.15e introduce a new element which is not demonstrated by the simple modes of baroclinic instability. That new element is the presence of meridional fluxes of westerly momentum. These are directed equatorwards on the poleward flank of the jet, with rather weaker poleward fluxes on the equatorward flank and zero fluxes near the jet centre. Therefore there is convergence of the westerly momentum flux towards the centre of the jet, so that the eddies induce westerly acceleration in the midlatitudes. The largest momentum fluxes are again below the steering level. So not only do the eddies tend to accelerate the jet, but they also tend to reduce the vertical shear. However, these linear momentum fluxes are very weak compared to those observed, and their distribution is far from that observed. Linear theory does a reasonable job of predicting the observed temperature fluxes but a rather inadequate job of predicting the observed momentum fluxes.

In conclusion, the linear instability analysis of a realistic zonal jet yields unstable normal modes not dissimilar in growth rate, phase speed and structure to those predicted by simpler models such as the Eady and Charney models. The temperature fluxes carried by these waves have much the same pattern as that observed. However, the momentum fluxes are weak and unrealistic compared to those observed. To improve on them, we shall have to follow the development of the growing mode into the nonlinear regime, when feedbacks between the growing mode and the zonal mean state start to become appreciable. This discussion will be resumed in Chapter 16.

14.7   Initial value problem

All the models of baroclinic instability discussed so far involve a paradigm in which small-amplitude, constant-structure, wave-like disturbances grow on an appropriate sheared zonal flow. Although disturbances with a range of different wavenumbers can grow, after a time, one would expect the flow to be dominated by the fastest growing modes. In fact, purely zonal flows on which wave-like disturbances develop are rarely observed in the atmosphere. Rather, the flow is always disturbed, with a variety of vorticity patches of different scale and shape to be seen. Such patches sometimes intensify, sometimes weaken and sometimes are simply shredded by the flow to lose their identity. So, as an alternative to the approach taken in previous sections, in this section, we explore the evolution of isolated disturbances in a sheared flow, the so-called initial value problem. We shall retain the assumptions of quasi-geostrophic dynamics and linearity.

The distinction between the initial value problem and the normal-mode problem is neatly illustrated by a problem discussed by Orr in 1912, which we shall call the 'Venetian blind' problem.

For simplicity, consider the basic state of uniform shear with $\bar{u} = \Lambda z$ and constant stratification used for the Eady model. But now assume the fluid is infinitely deep, so that the effects of upper and lower boundaries are ignored. Because the basic state has zero-potential vorticity gradient and there are no boundary temperature gradients, the flow is stable to normal-mode disturbances according to the Charney–Stern criteria, Equation 14.18. However, this does not mean that disturbances cannot amplify transiently. Suppose there are small-amplitude potential vorticity perturbations which take the form of plane waves with zonal wavenumber k and vertical wavenumber m:

$$q' = Qe^{i(kx+mz)} \qquad (14.53)$$

Scaling z by the quasi-geostrophic aspect ratio factor f/N, the perturbations have the following tilt:

$$\alpha = \frac{f}{N}\left(\frac{m}{k}\right) \qquad (14.54)$$

When m is positive, the phase lines tilt towards the east with height. When m is negative, they tilt towards the west. The question we address is how such a disturbance evolves as it is sheared by the zonal flow. Since the potential vorticity gradient is zero, and there are no effects of boundaries in the vertical, the potential vorticity perturbations are simply advected with the flow at each level. The amplitude and zonal wavenumber of the potential vorticity perturbation remain unchanged, but shear changes the vertical wavenumber. Substituting the form Equation 14.53 with m as a function of time into the potential vorticity advection equation gives $dm/dt = -\Lambda k$ so that

$$m = m_0 - \Lambda kt$$

It follows that the tilt changes at a constant rate as a result of the shear:

$$\frac{d\alpha}{dt} = -\frac{f\Lambda}{N} \qquad (14.55)$$

The lines of constant phase rotate under the influence of the shear. A wave which initially had phase lines tilting to the west with height would become first vertical and then would begin to tilt to the east.

Given the relationship between potential vorticity and streamfunction,

$$q' = \frac{\partial^2 \psi'}{\partial x^2} + \frac{f^2}{N^2}\frac{\partial^2 \psi'}{\partial z^2} \qquad (14.56)$$

the perturbation streamfunction is related to α by

$$\psi' = -\frac{q'}{k^2\left(1+\alpha^2\right)} \qquad (14.57)$$

This relationship shows that the amplitude of the streamfunction perturbation will change as α changes in response to the shear. The total kinetic energy of the disturbance per unit area is

$$KE = \frac{1}{XZ} \int_0^X \int_0^Z \frac{\rho}{2} \left(\frac{\partial \psi'}{\partial x} \right)^2 dxdy$$

(14.58)

$$= \frac{\rho Q^2}{4k^2 (1+\alpha^2)^2}$$

where X represents a distance equal to the x-wavelength of the disturbance and Z the vertical wavelength. The magnitude of the kinetic energy is arbitrary, depending upon the amplitude Q of the potential vorticity perturbation. But Equation 14.58 shows that the kinetic energy is a maximum for zero tilt and is always smaller for $|\alpha|>0$. Imagine that a disturbance initially tilts to the west with height, that is, $\alpha<0$. Then as the pattern is sheared by the background flow, the phase lines become more vertical and kinetic energy increases to its maximum when $\alpha=0$. As the phase lines begin to tilt towards the east, the kinetic energy decreases.

The rate of change of kinetic energy is

$$\frac{d(KE)}{dt} = -\frac{\rho \alpha Q^2}{k^2 (1+\alpha^2)^3} \frac{d\alpha}{dt}$$

Substituting from Equation 14.55,

$$\frac{d(KE)}{dt} = \frac{Q^2}{k^2} \frac{\rho f \Lambda}{N} \frac{\alpha}{(1+\alpha^2)^3}$$

(14.59)

Since the amplitude Q of the potential vorticity disturbances is arbitrary, the kinetic energy and its rate of change are also arbitrary. A more useful result comes from normalizing Equation 14.59 by the kinetic energy itself to obtain an effective kinetic energy growth rate:

$$\sigma \equiv \frac{1}{KE} \frac{d(KE)}{dt} = 4 \frac{f\Lambda}{N} \frac{\alpha}{(1+\alpha^2)}$$

(14.60)

The kinetic energy growth rate is a maximum of $2(f\Lambda/N)$ when $\alpha=1$. Compare this with the kinetic energy growth rate of the most unstable Eady wave, which is $0.62(f\Lambda/N)$.[1] For a time, this disturbance can grow nearly three times more rapidly than an unstable normal mode. But as α becomes negative, growth is replaced by decay. Generation of kinetic energy in the first phase is balanced by decay in the second phase, and so there is no net growth. This is consistent with the predictions of the Charney–Stern theorem. We are dealing with a state which is baroclinically stable. The variations of KE and σ with $-\alpha$, and hence with time, are shown in

Figure 14.16. Note that the total energy is like Equation 14.58 but with just $(1+\alpha^2)$ in the numerator. Hence, the total energy also has a maximum growth rate at $\alpha = 1$, but it is now $f\Lambda/N$. Therefore, there is a conversion of energy from the mean flow to the perturbation for a westward tilt and in the opposite direction for an eastward tilt with height.

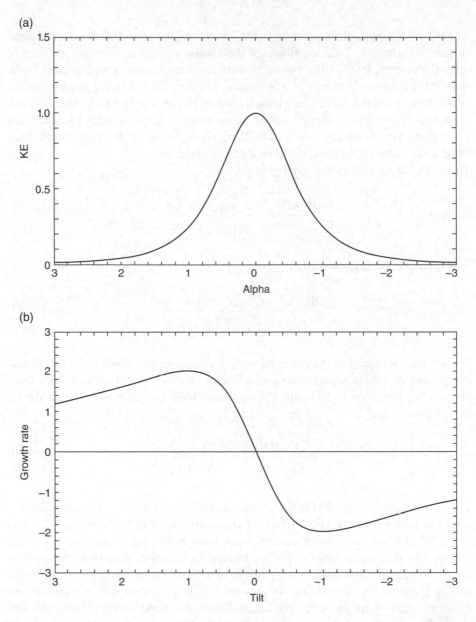

Figure 14.16 Showing the variation of KE (a) and kinetic energy growth rate (b) versus tilt, or equivalently, time, for the Orr calculation

An alternative approach to the wave-based analysis just carried out is to imagine isolated anomalies of potential vorticity. Assume a similar basic state of uniform vertical shear, zero-potential vorticity gradients and with boundary removed to infinity. In the point vortex limit, point potential vorticity anomalies can be taken as δ-functions. A single isolated vortex has the following potential vorticity:

$$q' = Q\delta\left(x - x_0, Z - Z_0\right) \tag{14.61}$$

Here, as earlier, Z is related to the geometrical height z by $Z = (N/f)z$. Q denotes the strength of the vortex. The geostrophic streamfunction corresponding to such a potential vorticity anomaly is obtained by solving the Poisson equation, Equation 14.56. The result may be written as follows:

$$\psi' = -\frac{Q}{4\pi R} \text{ where } R^2 = \left(x - x_0\right)^2 + \left(Z - Z_0\right)^2 \tag{14.62}$$

From this distribution of ψ', geostrophic wind anomalies, proportional to the horizontal gradient of ψ', and temperature anomalies, proportional to the vertical gradient of ψ', are implied. Figure 14.17 is a schematic illustration of these anomalies. A positive or cyclonic vortex induces cyclonic winds around its centre, which die away with distance from the centre. Above the vortex is a warm temperature anomaly and below a cold temperature anomaly. Again, these temperature anomalies die away as distance from the centre of the vortex increases. The temperature and wind anomalies are linked through the thermal wind relationship.

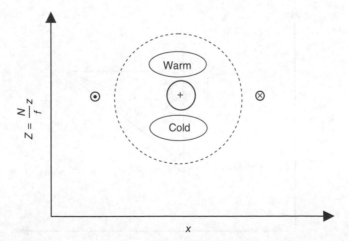

Figure 14.17 Schematic illustration of the wind and temperature anomalies associated with a point potential vorticity anomaly in a deep fluid of uniform potential vorticity. The dashed line indicates the associated perturbation geostrophic streamfunction

Now suppose two such vortices, with the same magnitude but opposite signs, are stacked vertically, forming a dipole structure, as shown in Figure 14.18. The geostrophic streamfunction induced by each vortex is given by Equation 14.62. When the distance from the centre of the dipole R is large compared to the separation of the vortices s, the streamfunction due to the combined effect of the two point vortices is given by superposing the streamfunctions for two vortices of opposite sign, using Equation 14.62; Figure 14.18 illustrates the geometry:

$$\psi' = -\frac{Q}{4\pi}\frac{s}{R^2}\cos(\theta) \qquad\qquad (14.63)$$

At locations where $R \gg s$, the streamfunction anomalies excited by the two vortices nearly cancel out, and so the disturbance falls away more rapidly, as R^{-2}, than that induced by a single monopole vortex. It is as if the second vortex effectively shields the fluid from the circulations induced by the first vortex and *vice versa*. Stronger circulations are concentrated in the region between and close to the vortices; if s is large compared with the distance from one or other vortex, then the streamfunction falls off as R^{-1} in accordance with Equation 14.62.

Now if the dipole is situated in a sheared flow with zero-potential vorticity gradients, each component vortex will simply be advected by the flow at that level. So, starting from an initial configuration with the two vortices stacked vertically, the spacing between the dipoles increases, and the axis of the dipole tilts towards the horizontal. The effect is that the mutual shielding of the vortices weakens, and a larger volume of the fluid is set into motion. Just as in the Venetian blind problem, the eddy kinetic energy increases as the plane bisecting the dipole becomes more

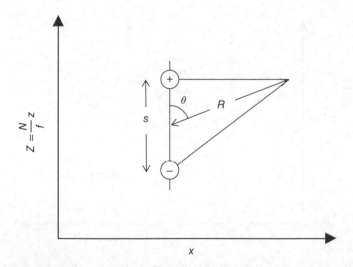

Figure 14.18 Schematic illustration of a pair of potential vortices of opposite sign and the circulation induced by them

vertical. Indeed, the Venetian blind problem is an example of potential vorticity unshielding applying to plane waves rather than to point vortices.

The previous calculations ignored the effects of boundaries. A more realistic problem includes upper and lower boundaries as in the Eady instability problem. Figure 14.19 shows the meridional wind in a calculation with a confined vortex as its initial state. In terms of potential vorticity, this state actually corresponds to a tripole structure, with patches of negative potential vorticity anomaly above and below a stronger patch of positive potential vorticity anomaly. When the potential vorticity anomalies are lined up vertically, the shielding effect is very strong, and there is little circulation outside the immediate vicinity of the vortices. Figure 14.19b shows the circulation after the ambient shear has begun to separate the vortex anomalies. The shielding effect is reduced, and a series of boundary-trapped waves propagates to the west along the upper boundary and to the east along the lower boundary. A strong westward tilting vortex has formed near the centre of the anomaly. Its structure closely resembles that of the unstable Eady modes, Figure 14.10a. The boundary-trapped wave fringes propagate at the speed of the ambient zonal flow near the upper and lower boundaries, giving rise to downstream development of disturbances near the upper boundary and upstream development near the lower boundary.

In this example, the horizontal scale of the anomaly was smaller than the scale of the shortwave cut-off in the Eady problem. So it should be stable in the normal-mode sense. Yet there was a substantial, if transient, increase in kinetic energy associated with the anomaly during its early evolution, as the vortex unshielding took place. This kind of mechanism, in which an upper-level potential vorticity feature comes into an appropriate alignment with a low-level potential vorticity or temperature anomaly, is implicated in many examples where a depression system deepens unusually rapidly.

Similar results are found for a localized disturbance to a zonal jet on the sphere using the nonlinear primitive equations. Figure 14.20 shows the surface pressure, 90 kPa temperature and 50 kPa streamfunction 17 days after a localized surface pressure perturbation was inserted at longitudes between 60° and 90° from the top of the plots. A succession of low- and high-pressure systems developed downstream from the initial perturbation with the fringe of the development moving at a typical jet speed and proceeding in the manner of the downstream propagation of Rossby wave activity trapped at the tropopause. This is consistent with the analysis in

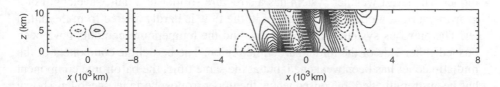

Figure 14.19 Development of the meridional wind anomaly associated with an isolated potential vorticity anomaly in a sheared zonal flow. Left: the initial state. Right: a subsequent state. From Badger and Hoskins (2001)

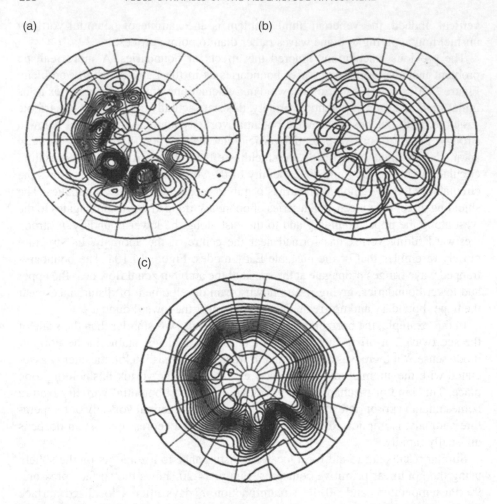

Figure 14.20   An example of downstream development in a spherical primitive equation model. (a) Surface pressure, (b) temperature at 90 kPa and (c) streamfunction at 50 kPa. From Simmons and Hoskins (1979)

Section 14.3 of the group velocity for waves trapped at a boundary in a much simpler model. The latest weather system has a structure similar to a half wavelength of a growing normal mode. Downstream of it, the flow is hardly altered from the initial jet. The previous systems have occluded, and the temperature contours have been pushed either side of the initial baroclinic region. The baroclinic component of the midlatitude jet has been weakened, but at the same time, the barotropic component has been strengthened. In consequence, the upper tropospheric jet speed has shown only a small increase, but the surface westerlies are strengthened considerably. The nonlinear wave structure and the behaviour of baroclinic waves are the topic for Chapter 16.

The initial jet has zero flow at the ground, and consistent with the simpler model in Section 14.3, new disturbances are also created successively in this region. However, consistent with an asymmetry that would be given in the Charney set-up, the scale of these systems in the horizontal and vertical directions is much smaller.

In the real atmosphere, the downstream development of Rossby waves and synoptic weather systems is very commonly observed.

Note

1. Since the kinetic energy depends upon the square of the perturbation velocity or streamfunction, the kinetic energy growth rate is twice the streamfunction growth rate, which was calculated in Section 14.4.

15
Frontogenesis

15.1 Frontal scales

One of the significant features of midlatitude depression systems is their propensity for developing fronts. Fronts are extended features across which there are rapid changes of velocity and temperature. Associated with them, and of overwhelming practical importance are often clouds and rain. The ubiquity of frontal structures in midlatitude flow was the basis of the Norwegian frontal model of cyclones discussed in Chapter 1.

Elementary kinematic considerations suggest circumstances where gradients of conserved quantities, such as potential temperature, might increase. Figure 15.1 illustrates two of these. Consider an initially gentle gradient of potential temperature θ. Then persistent deformation flow, as discussed in Sections 2.3 and 13.3, shown in Figure 15.1a, in which the axis of dilation is roughly parallel to the θ-contours, will concentrate the gradient of θ into an increasingly narrow zone. The magnitude of the gradient will increase indefinitely as long as the deformation persists, approaching a discontinuity only as time tends towards infinity. Of course, this simple two-dimensional picture is inadequate. As the temperature gradients tighten, so the wind field must adjust if thermal wind balance is to be retained. Following the discussion of Section 13.3, tightening of the temperature gradients implies that meridional circulations must develop to bring the wind and temperature fields into balance. This chapter explores the nature and consequences of such circulations. Similar considerations apply to the other frontogenetic situation, shown in Figure 15.1b. Here, a persistent sheared wind, blowing roughly perpendicular to isotherms, causes a tightening of the gradient around the zero wind line. Again, the gradient will only approach a discontinuity after an infinite time, and again, maintenance of thermal wind balance requires meridional circulations to develop.

The quasi-geostrophic theories developed in Chapters 12 and 13 give a qualitatively realistic picture of the meridional circulation in the vicinity of fronts. However, a typical frontal region might be only of order 100 km across, although the relevant scale in the long-front direction will be much larger, typically of order 1000 km.

Fluid Dynamics of the Midlatitude Atmosphere, First Edition. Brian J. Hoskins & Ian N. James.
© 2014 John Wiley & Sons, Ltd. Published 2014 by John Wiley & Sons, Ltd.

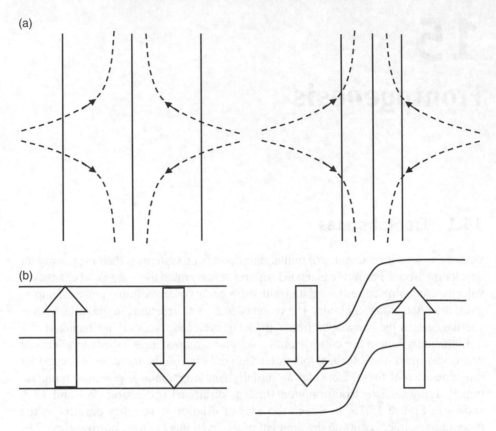

Figure 15.1 Examples of frontogenetic flow. (a) Persistent deformation, (b) persistent shear. The solid contours represent isopleths of potential temperature, which is advected conservatively in the absence of heating or cooling

A Rossby number based on the cross-frontal scale is therefore Ro = $V/fl \sim 10/$ $(10^{-4}10^5) \sim 1$. Such a large Rossby number implies that the quasi-geostrophic theory developed in earlier chapters is likely to be inadequate to represent the essential dynamics of frontal zones.

   The different scales in the across-front and along-front directions suggest that a more careful scale analysis of the momentum equations, looking at the balance of terms in these directions separately, might be fruitful. Figure 15.2 explains the notation we shall use. Consider a frame of reference which is fixed relative to the front. The y-axis is parallel to the front, while the x-axis is perpendicular to the front. The along-front scale is denoted L, while the smaller cross-front scale is denoted l. The corresponding velocity scales are V parallel to y and U parallel to x.

   Observations suggest that the component of velocity across the front, U, is small compared with the velocity component along the front, V. This is consistent with there being little interaction between the air on the two sides of the front, and so is consistent with the traditional concept of air masses. However, at the same time, the

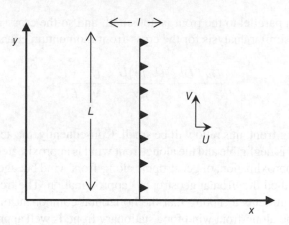

Figure 15.2   Notation used to describe a front

length scale for changes across the front is much smaller than that along the front. So assume that advection in the along-front direction is comparable to, or smaller than, that in the cross-front direction, that is,

$$v \frac{\partial}{\partial y} \sim u \frac{\partial}{\partial x}.$$ (15.1)

In terms of the characteristic length and velocity scales we have introduced

$$\frac{V}{L} \sim \frac{U}{l}.$$ (15.2)

Assume also that the temporal rate of change is not larger than the advection terms, so that

$$\frac{D}{Dt} \sim \frac{U}{l}.$$ (15.3)

Now apply this scaling to the two components of the momentum equation. Comparing the acceleration and Coriolis term in the along-front momentum equation gives

$$\frac{Dv/Dt}{f_0 u} \sim \frac{(U/l)V}{f_0 U} \sim \frac{V}{f_0 l}.$$ (15.4)

This is a Rossby number involving the large wind speed and the small-length scale, and as we have already shown, it is of order 1. This implies that for a strong front, the acceleration parallel to the front is not small, and so the cross-front flow is not geostrophic. A similar analysis for the cross-front momentum equation gives

$$\frac{Du/Dt}{fv} \sim \frac{(U/l)U}{f_0V} \sim \frac{U^2}{V^2}\frac{V}{f_0l} \tag{15.5}$$

Even for a strong front, this ratio will be small. Consequently, the acceleration of the cross-front flow is negligible and the along-front wind is approximately in geostrophic balance. This approximation, of geostrophic along-front wind but ageostrophic cross-front wind, is called the 'frontal geostrophy' approximation. The frontal geostrophic approximation amounts to noting that the horizontal component of wind, which is dominated by the along-front wind for a stationary front, is well approximated by the geostrophic wind. However, because of the large gradients in the cross-frontal direction, the full, rather than geostrophic, wind has to be used in the cross-front advection terms. The advecting wind will, of course, include any vertical velocity.

15.2   Ageostrophic circulation

Having established a consistent scaling for an idealized front, we shall now discuss the meridional circulation associated with a frontal structure. This is a necessary first step to a discussion of how a front might develop and evolve. We use the hydrostatic, z-coordinate equations, with the anelastic approximation, as in Equation 7.36. The variation of f is considered secondary and will be ignored, although it might play a role in a front which is sufficiently extended in the meridional direction. The equations will be written making the frontal geostrophy approximation. As in the previous section, the coordinates are orientated so that the y-axis is parallel to the front. The cross-frontal momentum equation is

$$\frac{Dv}{Dt} + f_0u + \frac{\partial\phi}{\partial y} = 0 \tag{15.6}$$

and the thermodynamic equation is

$$\frac{D\theta}{Dt} = 0. \tag{15.7}$$

The set of equations is completed with a number of diagnostic relationships. First, anelastic continuity equation

$$\nabla\cdot(\rho_R\mathbf{u}) = 0, \tag{15.8}$$

second, geostrophy for the along-front flow

$$f_0 v = \frac{\partial \phi}{\partial x}, \tag{15.9}$$

and, finally, the hydrostatic relationship

$$\frac{g}{\theta_0} \theta = \frac{\partial \phi}{\partial z}. \tag{15.10}$$

The advection terms are

$$\frac{D}{Dt} = \frac{\partial}{\partial t} + \mathbf{u} \cdot \nabla \equiv \frac{\partial}{\partial t} + u \frac{\partial}{\partial x} + v \frac{\partial}{\partial y} + w \frac{\partial}{\partial z} \tag{15.11}$$

From Equations 15.9 and 15.10, the equation of thermal wind balance is derived:

$$f_0 \frac{\partial v}{\partial z} = \frac{g}{\theta_0} \frac{\partial \theta}{\partial x} \tag{15.12}$$

The logic proceeds in a parallel way to the derivation of the Q-vector form of the quasi-geostrophic omega equation given in Section 13.3. That is, a cross-frontal circulation is derived such that thermal wind balance, Equation 15.12, continues to hold as the flow evolves.

It is convenient to partition the velocity field into its geostrophic and ageostrophic parts:

$$\mathbf{u} = \left(u_g \mathbf{i} + v_g \mathbf{j} + 0\mathbf{k} \right) + \left(u_a \mathbf{i} + 0\mathbf{j} + w\mathbf{k} \right) \tag{15.13}$$

First, consider the thermodynamic equation, Equation 15.7. Taking the x-derivative and multiplying by g/θ_0 gives

$$\frac{g}{\theta_0} \frac{\partial}{\partial x} \left[\frac{\partial \theta}{\partial t} + \left(u_g + u_a \right) \frac{\partial \theta}{\partial x} + v_g \frac{\partial \theta}{\partial y} + w \frac{\partial \theta}{\partial z} \right] = 0.$$

The differentiation acting on each of the advection terms yields two terms, and so we obtain

$$\frac{D}{Dt} \left[\frac{g}{\theta_0} \frac{\partial \theta}{\partial x} \right] = Q_1 - \frac{\partial u_a}{\partial x} \frac{g}{\theta_0} \frac{\partial \theta}{\partial x} - \frac{\partial w}{\partial x} \frac{g}{\theta_0} \frac{\partial \theta}{\partial z} \tag{15.14}$$

where

$$Q_1 = -\frac{\partial u_g}{\partial x}\frac{g}{\theta_0}\frac{\partial \theta}{\partial x} - \frac{\partial v_g}{\partial x}\frac{g}{\theta_0}\frac{\partial \theta}{\partial y}. \tag{15.15}$$

This expression for Q_1 is just the same expression as in Equation 13.19, which was derived in Section 13.3 for a quasi-geostrophic scaling. Equation 15.14 should be compared with its quasi-geostrophic form, Equation 13.18. Second, take the z-derivative of the momentum equation, equation and multiply by f_0. This gives

$$\frac{D}{Dt}\left[f_0\frac{\partial v}{\partial z}\right] = -f_0\frac{\partial u_g}{\partial z}\frac{\partial v_g}{\partial x} - f_0\frac{\partial v_g}{\partial z}\frac{\partial v_g}{\partial y} - f_0\frac{\partial u_a}{\partial z}\frac{\partial v_g}{\partial x} - f_0\frac{\partial w}{\partial z}\frac{\partial v_g}{\partial z} - f_0^2\frac{\partial u_a}{\partial z}$$

or, making use of the thermal wind equation, Equation 15.12

$$\frac{D}{Dt}\left[f_0\frac{\partial v}{\partial z}\right] = -Q_1 - f_0\left(f_0 + \frac{\partial v_g}{\partial x}\right)\frac{\partial u_a}{\partial z} - \frac{g}{\theta_0}\frac{\partial \theta}{\partial x}\frac{\partial w}{\partial z}. \tag{15.16}$$

which should be compared with the quasi-geostrophic form, Equation 13.21. As before, the geostrophic term Q_1 acts to modify both sides of the thermal wind Equation 15.12 with equal magnitude but with opposite signs. In other words, the geostrophic forcing, if non-zero, continually acts to destroy thermal wind balance. Compensating meridional circulations are required to maintain a balanced state.

Although Equations 15.14 and 15.16 resemble their quasi-geostrophic counterparts derived in Section 13.3, there are some important differences. First, the Lagrangian rates of change in both equations follow the full three-dimensional flow, not simply the horizontal geostrophic flow. In the region of fronts, where strong meridional circulations form, this is a significant difference. Second, the static stability term, $(g/\theta_0)\partial\theta/\partial z$, on the right-hand side of Equation 15.14 represents the local instantaneous static stability rather than a reference static stability. Again, this is particularly important in the region of front where there can be substantial changes of stratification between the different air masses. In the upper troposphere, a particular θ- surface can cross from the low static stability troposphere into the high static stability stratosphere across a frontal region. Thirdly, new terms involving the ageostrophic flow have been introduced. These are the first ageostrophic term in Equation 15.14 and the last ageostrophic term in Equation 15.16. Finally, the coefficient of $\partial u_a/\partial z$ in Equation 15.16 is more complicated, involving the relative vorticity as well as the planetary vorticity.

The frontal circulation required to maintain thermal wind balance follows immediately if the Lagrangian time derivatives are eliminated between Equations 15.14 and 15.16, using the thermal wind relationship, Equation 15.12. The result is

$$\frac{g}{\theta_0}\frac{\partial\theta}{\partial z}\frac{\partial w}{\partial x} - \frac{g}{\theta_0}\frac{\partial\theta}{\partial x}\left(\frac{\partial w}{\partial z} - \frac{\partial u_a}{\partial x}\right) - f_0\left(f_0 + \frac{\partial v_g}{\partial x}\right)\frac{\partial u_a}{\partial z} = 2Q_1. \qquad (15.17)$$

Define the coefficients:

$$N^2 = \frac{g}{\theta_0}\frac{\partial\theta}{\partial z}, \quad S^2 = f_0\frac{\partial v}{\partial z} \equiv \frac{g}{\theta_0}\frac{\partial\theta}{\partial x}, \quad F^2 = f_0\left(f_0 + \frac{\partial v_g}{\partial x}\right) \qquad (15.18)$$

each of which is the square of a frequency. Then Equation 15.17 can be rewritten as follows:

$$N^2\frac{\partial w}{\partial x} - S^2\left(\frac{\partial w}{\partial z} - \frac{\partial u_a}{\partial x}\right) - F^2\frac{\partial u_a}{\partial z} = 2Q_1 \qquad (15.19)$$

This can be compared with the quasi-geostrophic version in Equation 13.22. There the S^2 term is omitted and N^2 is taken to be a function of z only, while F^2 is replaced by the constant f_0^2.

The ageostrophic cross-frontal circulation is $(u_a\mathbf{i}+0\mathbf{j}+w\mathbf{k})$. From the continuity equation, Equation 15.8, it follows that

$$\frac{\partial}{\partial x}\left(\rho_R u_a\right) + \frac{\partial}{\partial z}\left(\rho_R w\right) = 0.$$

The cross-frontal circulation can be summarized by a stream function ψ, such that

$$u_a = \frac{1}{\rho_R}\frac{\partial\psi}{\partial z}, \quad w = -\frac{1}{\rho_R}\frac{\partial\psi}{\partial x}. \qquad (15.20)$$

Substituting in Equation 15.19 gives an equation for the cross-frontal stream function:

$$N^2\frac{\partial^2\psi}{\partial x^2} - S^2\left(\rho_R\left(\frac{\partial}{\partial z}\frac{1}{\rho_R}\frac{\partial\psi}{\partial x}\right) + \frac{\partial^2\psi}{\partial x\partial z}\right) + F^2\left(\rho_R\frac{\partial}{\partial z}\frac{1}{\rho_R}\frac{\partial\psi}{\partial z}\right) = -2\rho_R Q_1.$$

For situations in which the vertical scale of the motion is much less than the scale-height, this can be approximated as follows:

$$N^2\frac{\partial^2\psi}{\partial x^2} - 2S^2\frac{\partial^2\psi}{\partial x\partial z} + F^2\frac{\partial^2\psi}{\partial z^2} = -2\rho_R Q_1. \qquad (15.21)$$

This celebrated equation is often called the Sawyer–Eliassen equation, after the two scientists who originally and independently derived it. Again, this equation would be identical in quasi-geostrophic theory except that S^2 term would be omitted, and N^2 taken to be a function of z only, while F^2 would be replaced by f_0^2.

The character of the Sawyer–Eliassen equation depends upon the coefficients N, S and F. Notice first that if S is zero or small enough to be ignored, then Equation 15.21 starts to look very much like the Q-vector form of the omega equation discussed in Section 13.3. This could be solved for the meridional stream function, provided suitable boundary conditions on ψ were specified. If S is not small, but $(N^2F^2 - S^4)$ is nevertheless positive, then Equation 15.21 is elliptic and its solution is analogous to that of the omega equation: given boundary conditions on ψ, values of ψ throughout the domain can be calculated.

It is worth discussing the physical significance of the coefficients in more detail. From the equations with the frontal geostrophy approximation, Equations 15.6 and 15.9 with 15.8, the vector vorticity equation can be written in its full form:

$$\frac{D}{Dt}\left(\frac{1}{\rho}\varsigma_g\right) = \left(\frac{1}{\rho}\varsigma_g.\nabla\right)\mathbf{u} - \mathbf{k}\times\nabla\frac{g}{\theta_0}\theta, \tag{15.22}$$

where the relative vorticity is now only determined from the component v so that the geostrophic absolute vorticity is now

$$\varsigma_g = f\mathbf{k} + \nabla\times\left(0\mathbf{i} + v_g\mathbf{j} + 0\mathbf{k}\right) = \left(-\frac{\partial v}{\partial z}, 0, f + \frac{\partial v}{\partial x}\right),$$

and the vertical component is

$$\varsigma_g = f_0 + \frac{\partial v_g}{\partial x}.$$

Hence, the coefficient $F^2 = f_0\varsigma$. From the vorticity equation and the potential temperature equation, an equation for conservation of potential vorticity can be derived, but now

$$P = \frac{1}{\rho_R}\varsigma\cdot\nabla\theta = \frac{1}{\rho_R}\left(\varsigma_g\frac{\partial\theta}{\partial z} - \frac{\partial v_g}{\partial z}\frac{\partial\theta}{\partial x}\right). \tag{15.23}$$

With a little manipulation, this expression can be related to N, S and F so that

$$\mathcal{N}^2 \equiv \frac{g}{f_0\theta_0}\rho_R P = \frac{1}{f_0^2}\left(N^2F^2 - S^4\right). \tag{15.24}$$

Since $\mathcal{N}^2$ is proportional to the potential vorticity, conservation of potential vorticity implies that if the Sawyer–Eliassen equation, Equation 15.21, is elliptic initially, it will remain elliptic as the flow evolves. However, as gradients increase in a frontal region and N^2, F^2 and S^2 all become large, then the constancy of $\mathcal{N}^2$ means that there is strong cancellation between the terms in it and the Sawyer–Eliassen equation becomes close to parabolic. The implications of this will become clear in the following.

15.3   Description of frontal collapse

The form of the Sawyer–Eliassen equation can be simplified by a transformation of coordinates. This frontal transformation maps the horizontal distance x onto X where

$$X = x + \frac{v_g}{f_0} \qquad (15.25)$$

Note that

$$\delta X = \left(1 + \frac{1}{f_0} \frac{\partial v_g}{\partial x} \right) \delta x$$

As in the previous section, the bracketed term on the right-hand side is related to ζ_g and it follows that

$$\delta X = \frac{\zeta_g}{f_0} \delta x \qquad (15.26)$$

The implication of Equation 15.26 is that surfaces of constant X cluster towards regions of large absolute vorticity. Figure 15.3 illustrates the transformation for an idealized flow in which there is a concentration of low level cyclonic vorticity around $x = 0$.

We now rewrite the Sawyer–Eliassen equation, Equation 15.21 in (X, Z) space where X is given by the transformation 15.25. The vertical co-ordinate Z is simply z. Note that

$$\frac{\partial}{\partial x} = \frac{\partial X}{\partial x} \frac{\partial}{\partial X} = \frac{\zeta_g}{f_0} \frac{\partial}{\partial X}; \quad \frac{\partial}{\partial z} = \frac{\partial X}{\partial z} \frac{\partial}{\partial X} + \frac{\partial}{\partial Z} \qquad (15.27)$$

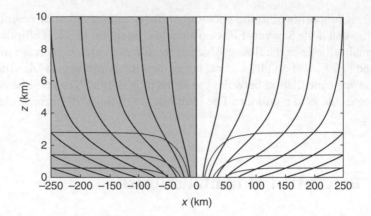

Figure 15.3   Illustrating the frontal transformation, Equation 15.25, albeit for an idealized artificial situation. A geostrophic wind blowing into the plane of the cross section decays with height
and has surface values of $-20\,\mathrm{m\,s^{-1}}$ changing to $+20\,\mathrm{m\,s^{-1}}$ across a zone 100km wide. The heavy
contours are lines of constant X, contour interval 100km. The light contours show the geostrophic
wind, contour interval $5\,\mathrm{m\,s^{-1}}$, with negative values shaded

Note that even though $z = Z$, the derivative in this direction holding X constant
$\partial/\partial Z \neq \partial/\partial z$ because surfaces of constant X do not coincide with surfaces of constant x. In fact, surfaces of constant X have slope $-(\partial v_g/\partial z)/(f + \partial v_g/\partial x)$ and are therefore parallel to the absolute vorticity vector ζ_g. Also, from the definitions 15.18, it
follows that

$$f_0^2 \frac{\partial X}{\partial x} = F^2; \ f_0^2 \frac{\partial X}{\partial z} = S^2 \tag{15.28}$$

Then the transformed Sawyer–Eliassen equation is

$$\frac{\partial}{\partial X}\left(N^2 \frac{\partial \psi}{\partial X}\right) + \frac{\partial^2 \psi}{\partial Z^2} + \frac{f_0^2}{H_\rho}\frac{\partial \psi}{\partial Z} = -2\rho_R \frac{f_0}{\zeta_g} Q_1$$

or, equivalently, but more elegantly,

$$\frac{\partial}{\partial X}\left(N^2 \frac{1}{\rho_R}\frac{\partial \psi}{\partial X}\right) + \frac{\partial}{\partial Z}\left(\frac{f_0^2}{\rho_R}\frac{\partial \psi}{\partial Z}\right) = -2\frac{f_0}{\zeta_g} Q_1.$$

For the case when $D \ll H_\rho$ this becomes

$$\frac{\partial}{\partial X}\left(N^2 \frac{\partial \psi}{\partial X}\right) + f_0^2 \frac{\partial^2 \psi}{\partial Z^2} = -2\rho_R \frac{f_0}{\zeta_g} Q_1 \tag{15.29}$$

The mixed derivative term has gone, and Equation 15.29 is identical to the quasi-geostrophic form of the cross-frontal circulation equation except that it is in (X, Z) space, the Brunt–Väisälä frequency N is replaced by $\mathcal{N}$, which is closely related to the potential vorticity, as shown by Equation 15.24, and the right-hand side includes the factor f_0/ζ_g. Indeed, in the limit $\xi \to 0$ so that the factor $f_0/\zeta_g \to 1$ and $\partial/\partial X \to \partial/\partial x$, Equation 15.29 reduces to the quasi-geostrophic form. It is clear that in transformed space there is no special behaviour as frontal gradients become large.

Having established the transformed Sawyer–Eliassen equation, its implications for frontogenesis can now be explored. Consider first a frontogenetic deformation flow such as that illustrated schematically in Figure 15.1a. As discussed in Chapter 13, in this case, $\nabla_H \theta$ is perpendicular to the dilation axis, which is taken to be the y-axis. The large-scale variation of the geostrophic wind is

$$\frac{\partial u_g}{\partial x} < 0, \quad \frac{\partial v_g}{\partial x} = 0$$

Consequently, Q_1 is positive; the Q-vector is parallel to the x-axis, and points from cold to warm air. A similar Q-vector arises from shear frontogenesis, illustrated in Figure 15.1b. Here, in the initial large-scale field, $\nabla_H \theta$ is parallel to the y-axis, with

$$\frac{\partial \theta}{\partial y} < 0.$$

The u_g component of the geostrophic wind is zero, but

$$\frac{\partial v_g}{\partial x} > 0.$$

Once again, this implies that Q_1 is positive in the frontal region. The implied meridional circulation, therefore, has rising motion at large x, that is, in the warm air, and sinking motion in the colder air, at small x.

Making the frontal geostrophy approximation, so that the Sawyer–Eliassen equation describes the meridional circulation, the meridional circulation in transformed (X, Z) space is similar to that derived from quasi-geostrophic theory. Near the surface, there will be rising motion in the warm air and sinking motion in the cold air. The circulation is represented schematically in Figure 15.4a. Now consider the effect of the co-ordinate transform. Figure 15.4b shows the same situation, but this time in physical (x, y) space. Assuming that the strong shears are confined to the lower troposphere, the surfaces of constant X are strongly distorted near the surface, but less so at upper levels. The meridional circulation is therefore distorted in the lower troposphere, as shown. The shrinking of the x- and z-scales in the vicinity of the front implies that all the second derivatives of ψ, but especially

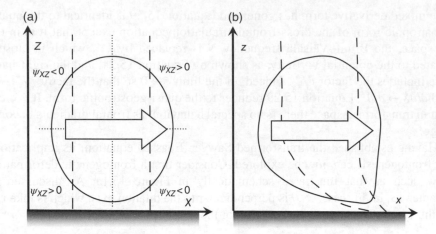

Figure 15.4 Schematic illustration of a frontal circulation in (a) the transformed (X, Z) space and (b) in the physical (x, z) space. The heavy dashed lines are lines of constant X

$\partial^2\psi/\partial x\partial z$, will have large values here. The stronger the front, so that the collapse of the x-scale is more pronounced, the stronger the distortion of the meridional circulation will be.

The implication of this distortion of the meridional circulation in the vicinity of regions of strong horizontal shear leads to an important feedback which provides the primary mechanism for frontogenesis. Consider Equation 15.22 for the vertical component of vorticity with the cross-frontal geostrophy approximation. On a boundary at which $w = 0$, the equation becomes

$$\frac{D\zeta_g}{Dt} = \frac{\zeta_g}{\rho_R}\frac{\partial}{\partial z}\left(\rho_R w\right). \tag{15.30}$$

This equation differs in two respects from the quasi-geostrophic vorticity equation. Both are related to the fact that the Rossby number is not necessarily small in the frontal region. First, the advecting velocity in the Lagrangian time derivative is the total velocity, not simply the geostrophic velocity. Second, the geostrophic vorticity is not necessarily small compared to the planetary vorticity, and so the stretching term includes the total absolute vorticity, not merely the planetary vorticity. From Equation 15.20, the stretching is given by

$$\frac{1}{\rho_R}\frac{\partial}{\partial z}\left(\rho_R w\right) = -\frac{\partial u}{\partial x} = -\frac{1}{\rho_R}\frac{\partial^2\psi}{\partial x\partial z}$$

This is related to the geostrophic stretching by

$$\frac{1}{\rho_R}\frac{\partial}{\partial z}(\rho_R w) = -\frac{1}{\rho_R}\frac{\partial^2 \psi}{\partial x \partial z} = -\frac{1}{\rho_R}\frac{\zeta_g}{f_0}\frac{\partial}{\partial X}\left(\frac{\partial X}{\partial z}\frac{\partial \psi}{\partial X} + \frac{\partial \psi}{\partial Z}\right). \qquad (15.31)$$

The second equality follows from application of the transformation formulae, Equation 15.27, and using the fact that ψ and its X derivatives are zero on $Z=0$. Near the ground, at $Z=0$, where ψ and $\partial\psi/\partial X$ are zero, this reduces to

$$\frac{1}{\rho_R}\frac{\partial}{\partial z}(\rho_R w) = -\frac{1}{\rho_R}\frac{\zeta_g}{f_0}\frac{\partial^2 \psi}{\partial X \partial Z} \qquad (15.32)$$

Because ψ satisfies equation 15.29, an equation which has no singular behaviour as the front becomes stronger and N^2, S^2 and F^2 become larger, it is clear from Equation 15.31 that the stretching becomes larger with the vorticity. For the short time of a frontal collapse and material movement in that time, we can take

$$\gamma = -\frac{1}{\rho_R f_0}\frac{\partial^2 \psi}{\partial X \partial Z}$$

to be constant. Then the vorticity equation, Equation 15.30, can be written:

$$\frac{D\zeta_g}{Dt} = \gamma\zeta_g^2 \qquad (15.33)$$

Now suppose an air parcel remains for a sufficient length of time near the same location in (X, Z) space. Then Equation 15.33 may be integrated following a fluid parcel to give

$$\left(A - \frac{1}{\zeta_g}\right) = \gamma t,$$

A being a constant of integration. This constant can be determined by specifying an initial condition. Suppose that when $t=0$, the absolute vorticity is ζ_0, then

$$\zeta_g = \frac{\zeta_0}{(1 - \zeta_0 \gamma t)} \qquad (15.34)$$

Persistent stretching, which becomes larger and larger as the circulation in (x, y) space becomes progressively more distorted means that the absolute vorticity increases from ζ_0 at $t=0$ and becomes infinite at $t=(\zeta_0\gamma)^{-1}$. For example, suppose the front forms in an already well-developed depression where $\zeta_0 = 2f_0$. A reasonable estimate for γ might be 0.2. Then the time to frontal collapse would be about 7 hours. If the initial vorticity were larger, the time to collapse would be correspondingly shorter. The crucial ingredient in the collapse to a discontinuity is that as the vorticity gets larger, so the stretching gets larger. This leads to the quadratic term on the right-hand side of Equation 15.33.

Another perspective on the fact that the convergence in the frontal region increases with the vorticity considers a state independent of one horizontal direction, taken to be y, and perturbations to it. In the first panel of Figure 15.5, frontal gradients are weak, with θ-surfaces almost horizontal and X-surfaces almost vertical. In the second panel, a strong front is present and the angle between the surfaces is much smaller. That this must be the case can be seen by writing Equation 15.24:

$$\frac{\theta_0}{g}\mathcal{N}^2 \equiv \frac{1}{f_0}\rho_R P = \frac{\partial\theta}{\partial z}\frac{\partial X}{\partial x} - \frac{\partial\theta}{\partial x}\frac{\partial X}{\partial z} \tag{15.35}$$

Dividing by $\partial\theta/\partial z$ and $\partial X/\partial z$ gives

$$\frac{f\dfrac{\theta_0}{g}\mathcal{N}^2}{X_z\theta_z} = \frac{\rho_R P}{X_z\theta_z} = \frac{X_x}{X_z} - \frac{\theta_x}{\theta_z}$$

which is simply the difference between the slopes of the θ and X surfaces. As the gradients in X and θ become larger, the numbers on the left-hand side of this equation must become smaller and the slopes must become more similar.

It can be shown that for perturbations to this state, the highest frequency oscillations, frequency σ_{max}, are almost buoyancy oscillations in the vertical direction, and the lowest frequency oscillations, with frequency σ_{min}, are almost inertial oscillations along the isentropes. A rather beautiful result is that

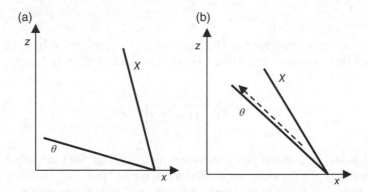

Figure 15.5 Slopes of X and θ surfaces in a cross-front, vertical section. (a) Weak frontal region, (b) strong frontal region, with the direction of the upglide indicated by the arrow

$$\sigma_{max}^2 \times \sigma_{min}^2 = f \frac{g}{\theta_0} \rho_R P \qquad (15.36)$$

For frictionless, adiabatic motion P is conserved when moving with the fluid, and so the product of the two frequencies is therefore also conserved.

For weak gradients, the Sawyer–Eliassen circulation is like that in Figure 15.4a. However, as the vertical stability and therefore σ_{max} increases during frontogenesis, so the inertial stability to motion along θ-surfaces must decrease, as indicated in Figure 15.5b. Consequently, the cross-frontal circulation distorts in the manner indicated in Figure 15.4b so that the stretching near the surface front becomes stronger and stronger. This is the crucial ingredient in the analysis of the frontal collapse in a finite time.

15.4 The semi-geostrophic Eady model

The Eady problem, discussed in detail in Section 14.4, envisages a Bousinesq, anelastic fluid with constant N^2 and vertical shear $\Lambda = \partial u / \partial z$, confined between rigid boundaries a distance H apart, with constant potential vorticity in the fluid interior. For the current argument, we confine our attention to the case of a perturbation with no y-variation, that is, to the $l = 0$ case. In the fluid interior, from Equation 14.32, conservation of quasi-geostrophic potential vorticity can be written in the following form:

$$\left(\frac{\partial}{\partial t} + U \frac{\partial}{\partial x} \right) \left(\frac{1}{f^2} \frac{\partial^2 \phi'}{\partial x^2} + \frac{1}{N^2} \frac{\partial^2 \phi'}{\partial z^2} \right) = 0. \qquad (15.37)$$

Here, ϕ' is the perturbation geopotential. The boundary conditions at the top and bottom are

$$\left(\frac{\partial}{\partial t} + U \frac{\partial}{\partial x} \right) \frac{\partial \phi'}{\partial z} + \Lambda \frac{\partial \phi'}{\partial x} = 0. \qquad (15.38)$$

A similar problem can be set up, but using the extension of the cross-front geostrophy approximation to be discussed in Chapter 16, and making the co-ordinate transforms introduced in the previous sections:

$$X = x + \frac{v_g}{f}$$

$$Y = y; \quad Z = z; \quad T = t.$$

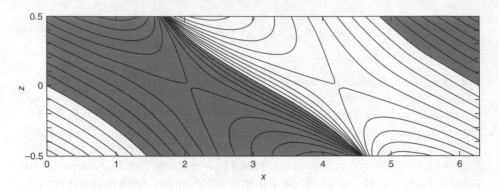

Figure 15.6 The fastest growing semi-geostrophic Eady mode, showing the perturbation poleward velocity component, negative values shaded. The amplitude has been chosen so that frontal discontinuities at the upper and lower boundaries are about to form

Then the so-called semi-geostrophic equations parallel to Equation 15.37 can be written in the following form:

$$\left(\frac{\partial}{\partial T} + U \frac{\partial}{\partial X}\right)\left(\frac{1}{f^2}\frac{\partial^2 \Phi'}{\partial X^2} + \frac{1}{\mathcal{N}^2}\frac{\partial^2 \Phi'}{\partial Z^2}\right) = 0 \qquad (15.39)$$

subject to boundary conditions at $z = 0, H$:

$$\left(\frac{\partial}{\partial T} + U \frac{\partial}{\partial X}\right)\frac{\partial \Phi'}{\partial Z} + \Lambda \frac{\partial \Phi'}{\partial X} = 0. \qquad (15.40)$$

Here, ϕ' is a modified geopotential

$$\Phi' = \phi' + \frac{1}{2} v_g^2$$

What is immediately apparent is that Equations 15.37 and 15.38 have exactly the same form as Equations 15.39 and 15.40. The differences are that the semi-geostrophic version uses the transformed co-ordinate set, and that $\mathcal{N}^2$, which is proportional to the potential vorticity P, plays the role of N^2 in the advected quantity in the fluid interior. It follows that the solutions to Equations 15.39 and 15.40 in transformed space are nearly the same as the solutions to the quasi-geostrophic form shown in Figure 14.11. However, in physical x-y space, the scale of regions of positive perturbation vorticity is compressed, while the anti-cyclonic region of negative perturbation vorticity expands. Temperature gradients in the cyclonic regions increase compared to those in the anti-cyclonic regions. The wave becomes increasingly asymmetric as its amplitude increases. This stretching and compressing of the x-coordinate is greatest near the upper and lower boundaries and small in the fluid interior. In a finite time, just as for the simple frontogenetic flow discussed in the previous section, infinite temperature and velocity gradients will develop adjacent to the boundaries and so sharp fronts will have formed.

Figure 15.6 illustrates the solution to the semi-geostrophic Eady problem in x-y space, using the fastest growing $l=0$ mode. The diagram may be compared with the quasi-geostrophic solution shown in Figure 14.11. The amplitude of the Eady wave in Figure 15.6 was chosen so that frontal discontinuities were about to form at the upper and lower boundaries. The mechanism for the frontal formation is similar to the idealized frontogenesis described in Section 15.3: secondary circulations that develop to maintain balance distort as the gradients become stronger. This leads to the formation of discontinuities in a finite time.

15.5   The confluence model

The Eady model demonstrates frontogenesis in the manner of the shear mechanism shown in Figure 15.1b. The first mechanism, confluence has been used as an example in Chapter 13 also. As there, consider a simple confluence field:

$$u = -\alpha x, v = \alpha y$$

with a balancing pressure field. Then the impact of this flow on a potential temperature field that varies in x but not y, as in Figure 13.6, gives a problem where the deviations from the confluence flow remain independent of y. Making the cross-front geostrophy approximation and using the transformation of coordinates leads to the conservation of potential vorticity. Since the potential vorticity is initially uniform, then

$$\frac{1}{f^2}\frac{\partial^2 \Phi'}{\partial X^2} + \frac{1}{N^2}\frac{\partial^2 \Phi'}{\partial Z^2} = 0$$

as in the Eady model. In this model it can be shown that the conditions on horizontal boundaries above and below the domain are

$$\left(\frac{\partial}{\partial t} - \alpha X \frac{\partial}{\partial X}\right)\theta = 0.$$

In X-space, θ is simply advected by the confluence flow. Again the equations are identical to those given by a quasi-geostrophic analysis of the problem but here they apply in X-space.

For an initially broad, smooth temperature contrast, which is symmetrical about $x=0$ and independent of height, the solution to this confluence model at a time of large absolute vorticity ($\sim$5 f) at the surface front is illustrated in Figure 15.7. Only

(a)

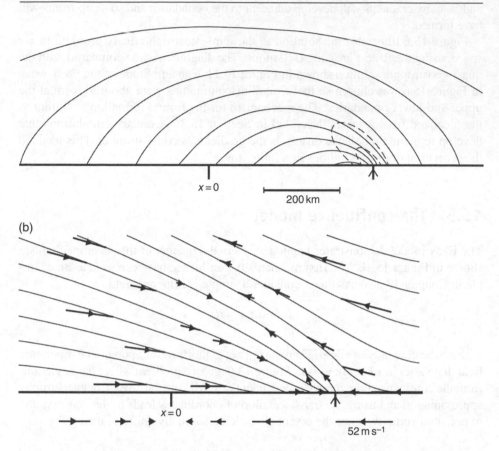

Figure 15.7 Solution of the uniform potential vorticity confluence model at a time when the maximum absolute vorticity has reached 5 f. Only the top half of the domain is shown. Top panel: the along-front wind (into the section) with contour interval $4\,m\,s^{-1}$. Also shown in dashed lines are contours of Ri at 0.5 and 1.0. Lower panel: contours of potential temperature with interval $2.4\,K$. Also shown are air parcel displacements from a previous time, which can be compared with the displacements associated with the basic confluence flow alone, shown below the surface. In both panels, an arrow indicates the location of the surface vorticity maximum. From Hoskins (1971)

the lower half of the domain is shown. Large gradients in v (upper panel) and θ (lower panel) are seen near the surface on the warm side of the temperature contrast. In both panels the location of the largest vorticity at the surface is indicated by an arrow and in the upper panel low Richardson numbers (1.0 and 0.5) by dashed contours. Also shown in the lower panel is the full motion in the x, z section, and the basic confluence motion which is indicated below the lower boundary. The warm air rises almost along isentropic surfaces above the sloping frontal region with significant ascent occurring even close to the lower boundary. Again the frontogenesis

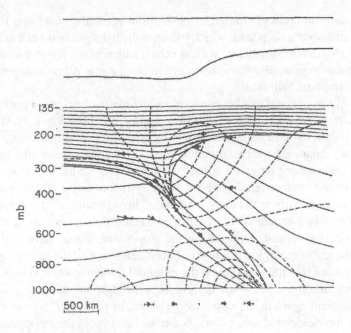

Figure 15.8 A model showing the formation of an upper-air front. A troposphere with uniform potential vorticity is bounded above by a tropopause and a layer of four times larger uniform potential vorticity. A basic confluence motion, indicated at the bottom, has acted on the initially broad temperature contrast. At the time of the solution, the tropopause would have looked like that shown at the top of the figure if only the confluence field had acted on it. The potential temperature field is shown by continuous contours drawn every 7.8 K, and the velocity into the section by dashed contours every 10.5 m s⁻¹. Fluid displacement vectors from a previous time are shown by vectors. From Hoskins (1972)

mechanism is as discussed in Section 15.3. There is weaker descent on the cold side of the temperature contrast.

15.6   Upper-level frontogenesis

The focus so far in this chapter has been on frontogenesis near the surface. However, as discussed in Section 1.5 and shown in Figure 1.14, frontogenesis also occurs near the tropopause with relatively thin 'tongues' of stratospheric air sometimes descending deep into the troposphere. The two-dimensional Eady shear frontal model and the confluence frontal model both have rigid lids at which frontogenesis occurs much as it does near the surface. The tropopause, with its high potential vorticity stratospheric air above it, does act somewhat like a rigid boundary for the troposphere, but this is clearly not a good model to represent the observed upper-air frontogenesis.

The cross-frontal circulation equation is quite generally valid and if the geostrophic motion over a deep layer which includes the tropopause is such as to increase horizontal temperature gradients, we can expect a direct circulation that will include the tropopause region. However, without studying some actual solutions it is not clear what response will occur.

The confluence model can be extended to include two regions, a uniform potential vorticity troposphere with a uniform higher potential vorticity stratosphere above it. Solutions have to be determined numerically with matching across the 'tropopause' separating the two regions. Figure 15.8 shows the results for a case in which the tropospheric potential vorticity is the same as in the previous confluence model while the stratospheric potential vorticity is four times larger. The variation of the reference density with height is included here because of the height range of interest. The x-component of the basic confluence field is shown at the bottom. The solution is shown at a time when the tropopause would have had the shape shown at the top of the figure if it had not been deformed by the ageostrophic circulation. The wind into the section and the potential temperature field indicate strong near-surface frontogenesis, a weaker gradient through the mid-troposphere and a stronger gradient again in the upper troposphere. In particular, a tongue of stratospheric air has descended along some isentropes that cross the tropopause down to the mid-troposphere. The tongue is a region of large cyclonic vorticity (about $8f$) and temperature gradients. The jet maximum sits on the high tropopause on the warm side of the temperature contrast.

The vectors on the figure show fluid parcel displacements from a previous time. They exhibit signatures of both the basic confluence flow and the cross-frontal circulation with ascent on the warm side and descent on the cold side. Of particular interest here is the tropopause folding motion with the descent almost along sloping isentropes in the region of the tongue. This can be understood from consideration of the stability to displacements in direction across and along isentropes, discussed at the end of Section 5.3. The high potential vorticity of the stratosphere means that it is very stable to displacements across isentropes. However, its stability is no larger than that for the troposphere to motion along isentropes. Therefore, provided there are isentropes crossing the tropopause, the response of the tropopause in the cross-frontal circulation forced by the tendency of the geostrophic motion to increase temperature gradients will tend to be descent along those sloping isentropes. The similarity of the model solution in Figure 5.8 with the observed upper-air front in Figure 1.13 is marked, though the latter is more intense in the depth of the descent of the tongue of stratospheric air and in the mid-tropospheric gradients. Without a rigid boundary there is no tendency for the model to give a discontinuity in a finite time. The even more intense descent and mid-tropospheric gradients in the observed case can only be achieved with fully three-dimensional flow.

16

The nonlinear development of baroclinic waves

16.1 The nonlinear domain

In Chapter 14, the development of baroclinic waves was discussed by means of a classical perturbation analysis. The flow was partitioned into a basic state, which was zonally invariant, and a wave-like perturbation. A set of linearized equations for the flow resulted by neglecting all terms involving the products of perturbation quantities. Such an approximation is valid provided the perturbation remains sufficiently small. However, for a range of wavelengths, perturbations grow exponentially in time. So no matter how small the initial perturbation amplitude might be, there comes a time when it is no longer small, and consequently the neglect of products of perturbation quantities is no longer justified. The purpose of this chapter is to explore the consequences of the breakdown of the linear assumption as the amplitude of the waves becomes larger.

The linear normal mode analysis of Chapter 14 implies a strict partitioning between the basic state and the developing wave: the basic state remains fixed, and the wave grows. In fact, the normal modes discussed in Chapter 14 have the property that although their amplitude increases exponentially with time, their shape remains constant. Both these features break down as the amplitude of the perturbation becomes appreciable. The shape of the waves begins to change, and at the same time, the perturbations begin to modify the background flow. When the amplitude becomes sufficiently large, indeed, the distinction between basic state and perturbation becomes more arbitrary. The effect of nonlinearity can be thought of in terms of scale interactions: on the one hand, the growing disturbance begins to develop small-scale structures, manifested in the changing shape of the wave, particularly in developing small-scale frontal structures. On the other hand, the disturbance also begins to modify the larger-scale flow in which it is embedded. In what follows, a convenient approach is to consider first the development of small-scale structures, and then the ways in which the growing waves begin to modify the larger scale.

Fluid Dynamics of the Midlatitude Atmosphere, First Edition. Brian J. Hoskins & Ian N. James.
© 2014 John Wiley & Sons, Ltd. Published 2014 by John Wiley & Sons, Ltd.

16.2   Semi-geostrophic baroclinic waves

In discussing the generation of small-scale structures within a developing wave, we shall make use of similar scaling arguments to those deployed in Chapter 15. The basis is the recognition that the horizontal velocity comes into the basic equations in two different ways, the advection of fluid parcels and the momentum carried. The core scaling result is that if the Lagrangian time rates of change are longer than f^{-1}, then the momentum of an air parcel is well approximated by its geostrophic value. In Section 7.4, this was seen as the criterion for the geostrophic velocity to be a good approximation to the full velocity. However, here we consider the further implications of this criterion for the momentum equations. We use the basic equations, Equation 7.36, with the momentum equation in the following form:

$$\frac{D\mathbf{v}}{Dt} + f\mathbf{k} \times (\mathbf{v} - \mathbf{v}_g) = 0 \qquad (16.1)$$

For simplicity, take f to be constant. Taking $\mathbf{k} \times$ this equation and rearranging gives

$$\mathbf{v} = \mathbf{v}_g + \mathbf{k} \times \frac{1}{f}\frac{D\mathbf{v}}{Dt} \qquad (16.2)$$

Substituting for this expression for $\mathbf{v}$ into the second term on the right-hand side gives

$$\mathbf{v} = \mathbf{v}_g + \mathbf{k} \times \frac{1}{f}\frac{D\mathbf{v}_g}{Dt} - \frac{1}{f^2}\frac{D^2\mathbf{v}}{Dt^2} \qquad (16.3)$$

Note that no approximation beyond the basic set momentum equation with f constant has been made at this point Comparing the left-hand side and the last term on the right-hand side, it is clear that if the Lagrangian timescale for changing the velocity is much longer than f^{-1}, then the latter term may be neglected to give

$$\mathbf{v} = \mathbf{v}_g + \frac{1}{f}\frac{D\mathbf{v}_g}{Dt} \qquad (16.4)$$

Finally, taking the vector product of $\mathbf{k}$ with this equation and rearranging gives

$$\frac{D\mathbf{v}_g}{Dt} + f\mathbf{k} \times (\mathbf{v} - \mathbf{v}_g) = 0 \qquad (16.5)$$

Note that here

$$\frac{D}{Dt} = \frac{\partial}{\partial t} + \mathbf{u} \cdot \nabla \equiv \frac{\partial}{\partial t} + \mathbf{v} \cdot \nabla + w \frac{\partial}{\partial z} \tag{16.6}$$

so that the ageostrophic horizontal velocity is neglected in the momentum but not in the advecting velocity. In this way, the 'geostrophic momentum' approximation is analogous to the hydrostatic approximation in which w is neglected in the momentum but not in the advecting velocity.

The basic equations, Equation 7.35, with the geostrophic momentum approximation, are

$$\frac{D\mathbf{v}_g}{Dt} + f\mathbf{k} \times (\mathbf{v} - \mathbf{v}_g) = 0$$

$$\mathbf{v}_g = \frac{1}{f}\mathbf{k} \times \nabla\Phi \tag{16.7}$$

$$\frac{\partial\Phi}{\partial z} = \left(\frac{g}{\theta_0}\right)\theta$$

$$\nabla \cdot (\rho_R \mathbf{u}) = 0$$

For simplicity we will assume that the vertical scale of the motion is much smaller than H_ρ, so that ρ_R will be taken as a constant ρ_0 and the atmosphere is effectively incompressible.

From the momentum equations, it can be shown with some effort that the vorticity equation has the usual form:

$$\frac{D\varsigma_g}{Dt} = (\varsigma_g \cdot \nabla)\mathbf{u} - \mathbf{k} \times \frac{g}{\theta_0}\nabla\theta \tag{16.8}$$

However, the three-dimensional vorticity is

$$\varsigma_g = \left(-\frac{\partial v_g}{\partial z}\mathbf{i} + \frac{\partial u_g}{\partial z}\mathbf{j} + \left(f + \frac{\partial v_g}{\partial x} - \frac{\partial u_g}{\partial y} \right)\mathbf{k} \right)$$

$$+ \left(\frac{1}{f}\frac{\partial(u_g,v_g)}{\partial(y,z)}\mathbf{i} + \frac{\partial(u_g,v_g)}{\partial(z,x)}\mathbf{j} + \frac{\partial(u_g,v_g)}{\partial(x,y)}\mathbf{k} \right) \tag{16.9}$$

Here

$$\frac{\partial\left(u_g,v_g\right)}{\partial\left(x,y\right)}=\frac{\partial u_g}{\partial x}\frac{\partial v_g}{\partial y}-\frac{\partial u_g}{\partial y}\frac{\partial v_g}{\partial x}$$

and for the other two similar terms x and y are replaced by y and z, and by z and x. The first term in Equation 16.9 is the expected vorticity determined from the geostrophic velocity alone. In conditions in which the geostrophic convergence is much smaller than f and the rotation of the geostrophic flow with height is small, it can be shown that the additional term is small.

It follows from the vorticity and potential temperature equations, Equations 16.8 and 7.35), that there is also a potential vorticity conservation:

$$\frac{DP_g}{Dt}=0, \quad \text{where } P_g=\frac{1}{\rho_0}\varsigma_g\cdot\nabla\theta \tag{16.10}$$

As in Section 15.2 and Equation 15.23, it is useful to use as the conserved potential vorticity-like variable

$$N^2\equiv\frac{g}{f_0\theta_0}\rho_0 P=\frac{g}{f_0\theta_0}\varsigma_g\cdot\nabla\theta. \tag{16.11}$$

The equations are simplified with a coordinate transformation, which is simply a two-dimensional generalization of the coordinate transformation Equation 15.25 which was introduced in the study of fronts. The transformation may be written as follows:

$$\left(X,Y,Z,T\right)=\left(x+\frac{v_g}{f},y-\frac{u_g}{f},z,t\right) \tag{16.12}$$

The horizontal velocity components in transformed space are

$$\frac{DX}{Dt}=u+\frac{1}{f}\frac{Dv_g}{Dt}, \quad \frac{DY}{Dt}=v+\frac{1}{f}\frac{Du_g}{Dt}$$

With the geostrophic momentum approximation, the momentum equation, Equation 16.5, gives

$$\frac{DX}{Dt}=u_g, \quad \frac{DY}{Dt}=v_g \tag{16.13}$$

Therefore, the horizontal motion in transformed space is geostrophic. Consequently, X and Y are referred to as geostrophic coordinates. The ageostrophic horizontal motion is implicit in the coordinate transformation. Note, however, that there is still vertical motion in the transformed space.

The transformation of coordinates is

$$\frac{\partial}{\partial x} = \left(1 + \frac{1}{f}\frac{\partial v_g}{\partial x}\right)\frac{\partial}{\partial X} - \frac{1}{f}\frac{\partial u_g}{\partial x}\frac{\partial}{\partial Y}$$

$$\frac{\partial}{\partial y} = \frac{1}{f}\frac{\partial v_g}{\partial y}\frac{\partial}{\partial X} + \left(1 - \frac{1}{f}\frac{\partial u_g}{\partial y}\right)\frac{\partial}{\partial Y} \qquad (16.14)$$

$$\frac{\partial}{\partial z} = \frac{1}{f}\frac{\partial v_g}{\partial z}\frac{\partial}{\partial X} - \frac{1}{f}\frac{\partial u_g}{\partial z}\frac{\partial}{\partial Y} + \frac{\partial}{\partial Z}$$

Therefore, the Jacobian of the transformation is

$$J = \frac{\partial(X,Y)}{\partial(x,y)} = \frac{1}{f}\mathbf{k}\cdot\varsigma_g \qquad (16.15)$$

A physical interpretation of the Jacobian is that it represents the ratio of the volume of a cell in transformed space, $\delta X\delta Y\delta Z$, to its corresponding volume $\delta x\delta y\delta z$ in ordinary Cartesian (x, y, z) space. So, Equation 16.15 shows that a uniform mesh of points in transformed space clusters together in regions of (x, y, z) space where the vertical component of the absolute vorticity is large, and becomes more widely spaced where it is small. This is, of course, exactly the two-dimensional horizontal coordinate transformation generalization of the one-dimensional transformation introduced in Section 15.3.

Eliminating between the coordinate transformation equations, Equation 16.14, gives

$$J\frac{\partial}{\partial Z} = \frac{1}{f}\varsigma_g\cdot\nabla \qquad (16.16)$$

Therefore, from Equation 16.11, the conserved variable potential vorticity can be written as follows:

$$\mathcal{N}^2 \equiv \frac{g}{f_0\theta_0}\rho_0 P = \frac{g}{\theta_0}J\frac{\partial\theta}{\partial Z} \qquad (16.17)$$

Also, from the coordinate transformations it can be shown that

$$\frac{1}{J} = 1 - \frac{1}{f}\frac{\partial v_g}{\partial X} + \frac{1}{f}\frac{\partial u_g}{\partial Y} + \frac{1}{f^2}\frac{\partial(u_g,v_g)}{\partial(X,Y)} \qquad (16.18)$$

The last term on the right-hand side is analogous to the additional term in the absolute vorticity with the geostrophic momentum approximation, Equation 16.9, and is in general small.

The three-dimensional gradient of ϕ in physical space determines the geostrophic velocity components and the potential temperature. If we define

$$\Phi = \phi + \frac{1}{2}\left(u_g^2 + v_g^2\right) \tag{16.19}$$

it can be shown that it plays the same role in transformed space:

$$\frac{\partial \Phi}{\partial X} = f v_g, \quad \frac{\partial \Phi}{\partial Y} = -f u_g, \quad \frac{\partial \Phi}{\partial Z} = \frac{g}{\theta_0}\theta \tag{16.20}$$

Substituting into Equation 16.18 and neglecting the last term on the right-hand side of it gives

$$\frac{1}{J} = 1 - \frac{1}{f^2}\left(\frac{\partial^2 \Phi}{\partial X^2} + \frac{\partial^2 \Phi}{\partial Y^2}\right)$$

Finally, substituting this and the third expression of Equation 16.10 into Equation 16.17 gives that the conserved potential vorticity variable in transformed space can be written as

$$\mathcal{N}^2 = \frac{\partial^2 \Phi}{\partial Z^2} \Big/ \left(1 - \frac{1}{f^2}\left(\frac{\partial^2 \Phi}{\partial X^2} + \frac{\partial^2 \Phi}{\partial Y^2}\right)\right)$$

or

$$\frac{1}{f^2}\left(\frac{\partial^2 \Phi}{\partial X^2} + \frac{\partial^2 \Phi}{\partial Y^2}\right) + \frac{1}{\mathcal{N}^2}\frac{\partial^2 \Phi}{\partial Z^2} = 1 \tag{16.21}$$

On horizontal boundaries, the boundary conditions are

$$w = 0$$

and from the conservation of potential temperature

$$\left(\frac{\partial}{\partial T} - \frac{1}{f}\frac{\partial \Phi}{\partial Y}\frac{\partial}{\partial X} + \frac{1}{f}\frac{\partial \Phi}{\partial X}\frac{\partial}{\partial Y}\right)\left(\frac{\partial \Phi}{\partial Z}\right) = 0 \text{ at } Z = 0, H \tag{16.22}$$

The conservation of $\mathcal{N}^2$ (PV) and θ moving with velocity (u_g, v_g, w) in transformed space and the Equations 16.21 and 16.20 relating $\mathcal{N}^2$ and θ to Φ, with the boundary conditions, Equations 16.22 form a complete set of equations for Φ and w that are known as the semi-geostrophic equations. They are closely similar to the corresponding quasi-geostrophic equations, which are derived under more restrictive assumptions, with the difference that the semi-geostrophic equations are expressed in 'semi-geostrophic' (X, Y, Z, T) space, while the quasi-geostrophic set are in real (x, y, z, t) space. The similarity is exact when the potential vorticity is uniform, so that the coefficients in Equation 16.21 are all constant. For such an initial state, the potential vorticity remains a uniform constant for all subsequent times and so the only time dependence in the problem enters through the boundary conditions, Equation 16.22. The case in which the basic state has no Y-variation but simply increases with height satisfies this constant potential vorticity condition. This is simply the Eady problem which was discussed in Section 14.4. The difference from the quasi-geostrophic Eady problem is that when the solutions are transformed from semi-geostrophic to real space, the scale of cyclonic regions shrinks and the scale of the anti-cyclonic regions expands; in other words, as the amplitude increases, the shape of the growing mode changes. Eventually, the shrinking of coordinates in the regions of largest absolute vorticity leads to frontogenesis, that is, to infinite gradients of conserved quantities such as θ in a finite time. This is the frontogenesis model that was discussed in Section 15.4.

A more realistic example of the solution of this semi-geostrophic equation set was given by Hoskins and West (1979). Rather than a simple linear shear with height, it is possible to construct a jet-like basic but still with constant potential vorticity, that is, for a zonal flow that satisfies

$$\frac{\partial^2 \bar{u}}{\partial Y^2} + \frac{f^2}{\mathcal{N}^2} \frac{\partial^2 \bar{u}}{\partial Z^2} = 0$$

An example of such a jet is a simple shear in the vertical, plus a flow that is zero at the lower boundary, $Z=0$, and sinusoidally varying at an upper boundary, $Z=H$. Defining non-dimensional variables $\tilde{Z} = Z/H$, and $\tilde{Y} = Y/L_R$, where $L_R = \mathcal{N}H/f$, consider

$$\bar{u} = U\left[Z - \frac{\mu}{2}\left(Z + \frac{\sinh lZ}{\sinh l}\cos(lY)\right)\right] \qquad (16.23)$$

This is zero at $Z=0$, and at $Z=H$ $\bar{u} = U$ at the central latitude ($Y=\pi L_R/l$) and it reduces to $(1-\mu)U$ at $Y=0$ and $Y=2\pi L_R/l$. Therefore, the total vertical shear in the wind at the central latitude is always the same, but for increasing μ it becomes more jet-like. For illustrating solutions, we use the parameter values given in Hoskins and West (1979) with $H=9$ km, $\mathcal{N}^2 \approx 1.3 \times 10^{-4}$ s^{-2}, $U \approx 29.4$ m s^{-1}, and a wavelength in the meridional

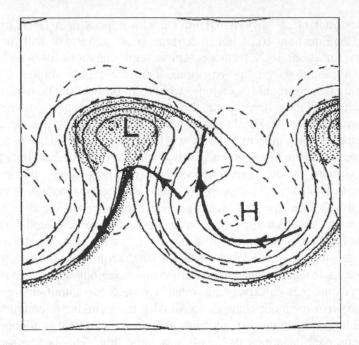

Figure 16.1 Surface map at day 6.3 for a perturbation to a strong jet zonal flow. Potential temperature contours are continuous and drawn every 4 K, Φ' contours are dashed, and the region of relative vorticity greater than $f/2$ is dashed. The relative vorticity in the cold front has a maximum of $5f$. The bold lines indicate two trajectories relative to the system from day 3. From Hoskins(1982)

direction of about 5600 km. For each zonal flow, the initial conditions comprise the basic zonal flow plus the most unstable wave for that flow with a small amplitude.

First we take the strong jet case with $\mu = 1$. Figure 16.1 shows the surface development after 6.3 days of integration. The plot shows the surface pressure, potential temperature and vorticity fields, along with the geopotential field. There is large vorticity in the cyclone, and strong temperature gradients and a streak of large vorticity along a strong cold front on the warm side of the temperature contrast to the southwest of the surface low. There is also a weaker warm front feature ahead of the surface low pressure. The surface temperature gradients and vorticity associated with the cold front would become infinite some 5 hour after this time.

The structure of the cold front is very similar to that in the two-dimensional frontogenesis cases discussed in Chapter 15. Pursuing this similarity, the transform space, $Z = 2.8$ km **Q**-vector and vertical velocity distributions together with θ are shown in Figure 16.2. In the cold-front region, the **Q**-vectors are oriented down the temperature gradient, indicating a frontogenetic region there. The vertical velocity field shows ascent ahead and descent behind. In three-dimensional motion the trajectories of the air become very important in determining the extent to which frontogenesis actually occurs. One of the trajectories in Figure 16.1 shows that air

(a)   (b)

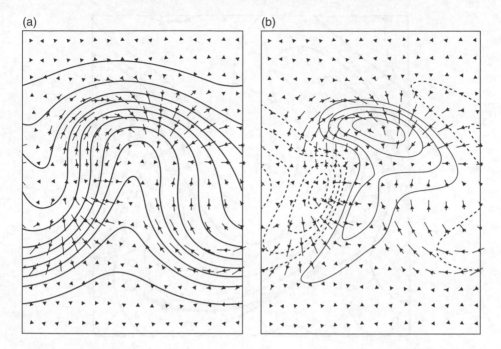

Figure 16.2 Some day 6.3 fields in transform space at 2.8 km. **Q**-vectors are shown together with (a) potential temperature (contour interval 4 K) and (b) vertical velocity (contour interval 0.5 cm s^{-1}, negative values dashed. From Hoskins and Pedder (1980), permission of the Royal Meteorological Society

parcels tend to move along the cold-front region and so strong frontogenesis occurs. However, in the weak warm-front region to the north-east of the low, air parcels move rapidly around the top of the low pressure system, forming large gradients in a structure often referred to as a bent-back occlusion.

Figure 16.2 shows that, in the region of the temperature contrast ahead of the warm sector region, the Q-vectors are almost parallel to the temperature contours and so there is no frontogenesis there. This is associated with the tendency of the wave to tilt in a southwest–northeast direction on the southern side of the domain in sympathy with the basic jet shear. However, this can be changed by a simple adjustment to the basic zonal like that of Equation 16.23 but with the latitudinal variation having a maximum at the surface and zero at the lid. For a value of μ of 0.3, the surface flow is -4.4 m s^{-1} at the central latitude and $+4.4$ m s^{-1} to north and south, and there is a uniform flow of 25.0 m s^{-1} at the lid. The strong ambient cyclonic shear near the surface and the westerly winds to the south of the strongest baroclinic region influence the tilt of the developing baroclinic wave as is seen in Figure 16.3. Consistent with this, there is now a very strong warm-front south-east of the low and ahead of the warm sector. Q-vectors (not shown) indicate that this is now the frontogenetic region, and the trajectory shown in Figure 16.3 indicates that air parcels move along this region.

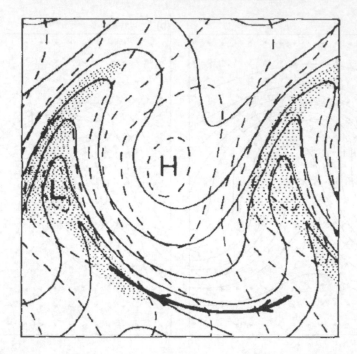

Figure 16.3   Surface map at day 5.5 for a perturbation to a flow with surface easterly winds in the middle and westerly winds on either flank. Potential temperature contours are continuous and drawn every 8 K (double that in Figure 16.1), Φ' contours are dashed, and the region of relative vorticity greater than $f/2$ is dashed. The relative vorticity in the warm front has a maximum of $5f$. The bold line indicates a trajectory relative to the system from day 3. From Hoskins (1982)

16.3   Nonlinear baroclinic waves on realistic jets on the sphere

The calculations discussed in Section 16.2 are based on the assumption of uniform potential vorticity, which in turn requires a very specific zonal flow profile. These assumptions, together with the geostrophic momentum, f-plane and constant density approximations are relaxed in the calculations discussed in this section, in which the results were obtained using the primitive equations on the sphere. The technique is essentially the same as in the previous section: specify a zonal flow, determine a most unstable normal mode, and start a nonlinear integration with the zonal flow plus a small amplitude normal mode. Here the latitudinal scale is not specified but the zonal wave number is.

The basic state zonal flow in the case that will be known as LC1 flow is the same jet whose stability was discussed in Section 14.6. We will mostly consider the development of the most unstable wave number 6, whose structure was discussed earlier and shown in Figure 14.15. Figure 16.4 shows the surface pressure and near-surface temperature for days 4, 6 and 9. The structure at day 4 is still close to that of the

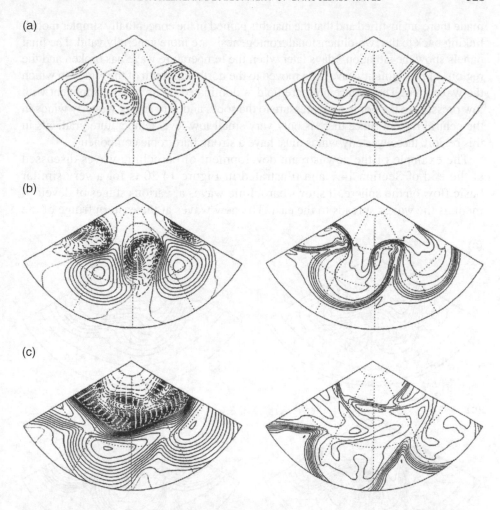

Figure 16.4 LC1 surface pressure and near surface ($\sigma = 0.967$) temperature at days (a) 4, (b) 6 and (c) 9. The contour interval for pressure is 0.4 kPa, with the 100 kPa contour dotted and contours for values below that dashed, and for temperature is 4 K. Sectors are shown for two wavelengths between latitude 20°N and the pole. Lines of longitude and latitude are drawn every 30° and 20°, respectively. From Thorncroft *et al.* (1993), with permission of the Royal Meteorological Society

normal mode, though the lows are slightly stronger than the highs. This is mainly because they have drifted polewards and therefore occupy a small area than the highs that have drifted southwards. However, nonlinear stretching is also involved to a small extent. By day 6, a very strong surface low has formed and the temperature wave has steepened. The gradients in temperature and the curvature of the pressure field make it clear that a very strong cold front had formed. Also, there is again a 'bent-back occlusion'. In fact, the similarity with the first semi-geostrophic, nonlinear baroclinic wave shown in Figure 16.1 is striking. This strongly suggests that the approximations made there are justified and that the insights gained in the conceptually simpler model, linking back to the two-dimensional frontogenesis are more generally valid. The final panels show the situation 3 days later when the temperature wave has broken and the regions of baroclinicity have been moved to the north and south of the region in which the wave grew. The surface pressure field is dominated by its zonal component with low pressure to the north, high pressure to the south and very strong westerly winds in the central region. Since there is only very small low-level temperature gradients in this region, these westerly winds must have a strong barotropic component.

The example of the downstream development of baroclinic waves discussed at the end of Section 14.7 and illustrated in Figure 14.20 is for a very similar basic flow on the sphere. It shows baroclinic waves at various stages of development as the waves spreads to the east. The new waves at the eastern fringe of the

(a)

(b)

(c)

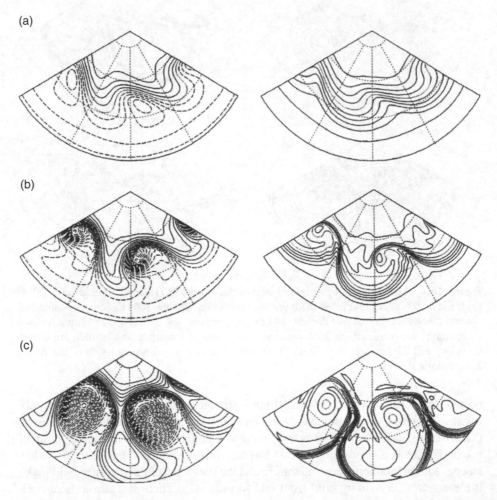

Figure 16.5   As Figure 6.4, but for the LC2 life cycle at days (a) 4, (b) 6 and (c) 9. From Thorncroft *et al.* (1993), with permission of the Royal Meteorological Society

wave-packet are like the early stages of the life cycle shown in Figure 16.4, and the older ones are like the mature stage.

We now show results for another similar basic zonal flow, but to which a cyclonic shear has been added on the equatorward flank. This was originally inspired by the semi-geostrophic results of Hoskins and West (1979) shown in the previous section. The basic thermal field is identical to that in LC1, but in this case, which will be referred to as LC2, barotropic easterly winds are added at 50°N and westerlies at 20°N, both with amplitude $10\,\mathrm{m\,s^{-1}}$. In Figure 16.5, the LC2 surface structure is also shown for days 4, 6 and 9. Again the day 4 structure shows almost linear growth of the wave. Its SE-NW slope is clear, as are the ambient easterlies to the north and westerlies to the south. The strong weather system at day 6 does indeed show the development of a very strong surface warm front ahead of the warm sector, as anticipated by the semi-geostrophic investigation. The day 9 picture indicates a finite amplitude behaviour which is very different from that of LC1. The low pressure system is very strong and almost circular. The baroclinic region has not separated into two regions to north and south. Rather it is organized around the periphery of the lows which are cold core vortices. This very strong eddy structure continues to exist over many days.

It is evident that the development of the near surface weather systems and the fronts in them can be generally understood from the quasi-geostrophic and semi-geostrophic theory given in Chapters 14 and 15 and Section 16.1. However, the changes in the mean flow and the life cycles of the baroclinic waves clearly need more analysis. This is the subject of the rest of this chapter.

16.4   Eddy transports and zonal mean flow changes

The basis of the linear theory described in Chapter 14 was a strict partitioning between a constant zonal background flow, and a small amplitude perturbation. The feedback of the perturbations onto the background flow is proportional to the square of the amplitude, and is therefore very small as long as the amplitude remains small. But as the amplitude of a growing mode increases, this assumption is no longer accurate. Eddies then systematically transport fluid properties, such as internal energy, momentum or potential vorticity, from place to place in the meridional plane. These transports depend upon the square of the perturbation amplitude, and so rapidly become important as the amplitude exceeds some significant value. The transports of tracers, such as water vapour or chemical constituents, by the growing eddies are important in other contexts. As the background flow changes, the development of the perturbations themselves begins to change. Feedbacks between the mean flow and the eddies develop and these feedbacks become increasingly important as the amplitude increases. The feedbacks may be either negative, reducing the growth of the eddy, or else positive, increasing the eddy growth rate.

Once such feedbacks become important, linear theory breaks down, and the wave is said to be 'nonlinear'. The threshold amplitude for substantial feedbacks depends upon the details of the background flow and the perturbation. But generally, when the perturbation quantities become comparable in magnitude to the background variation of the same quantity, we expect linear theory to begin to break down. So, for example, if the temperature perturbation associated with a growing baroclinic wave becomes comparable to the temperature difference across the jet, linear theory is of doubtful validity. Similarly, if the eddy velocities become comparable to the typical zonal flow, again linear theory is likely to breakdown. Another interpretation of such criteria is that linearity becomes invalid when isotherms or streamlines of the total flow make appreciable angles with latitude circles.

Using the basic equation set (7.35), if Q is any quantity, the mass conservation equation enables us to write the material rate of change of Q in its so-called flux form:

$$\frac{DQ}{Dt} = \frac{\partial Q}{\partial t} + \left(\frac{\partial(uQ)}{\partial x} + \frac{\partial(vQ)}{\partial y} + \frac{1}{\rho_R} \frac{\partial(\rho_R wQ)}{\partial z} \right) \tag{16.24}$$

Taking the zonal average, the x-derivative term becomes 0. Analogous to the procedure in Section 11.5, we separate variables into their zonally averaged components and the departure from the zonal average. For example,

$$u = \bar{u} + u', \quad \text{where } \overline{u'} = 0$$

Then, in the zonal average of Equation 16.24, the products of zonal mean and eddy terms become zero, so that

$$\frac{\overline{DQ}}{Dt} = \left(\frac{\partial}{\partial t} + \frac{\partial \bar{v}\bar{Q}}{\partial y} + \frac{1}{\rho_R} \frac{\partial \rho_R \bar{w}\bar{Q}}{\partial z} \right) + \left(\frac{\partial}{\partial y} \overline{v'Q'} + \frac{1}{\rho_R} \frac{\partial}{\partial z} \rho_R \overline{w'Q'} \right) \tag{16.25}$$

The first bracketed term on the right-hand side of this equation is the zonal mean flow advection of the zonal mean Q; we denote this by $\tilde{D}\bar{Q}$. The zonal average of the mass conservation equation is

$$\frac{\partial \bar{v}}{\partial y} + \frac{1}{\rho_R} \frac{\partial(\rho_R \bar{w})}{\partial z} = 0 \tag{16.26}$$

and so we can write $\tilde{D}\bar{Q}$ as a simple meridional advection by the zonal mean flow:

$$\tilde{D}\bar{Q} = \left(\frac{\partial}{\partial t} + \bar{v} \frac{\partial}{\partial y} + \bar{w} \frac{\partial}{\partial z} \right) \bar{Q} \tag{16.27}$$

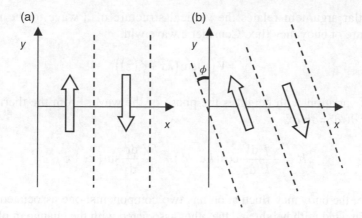

Figure 16.6   A plane wave in the horizontal plane. (a) Troughs and ridges parallel to the y-axis. (b) Troughs and ridges making an angle ϕ with the y-axis

The impact of the eddy terms in Equation 16.24 is through terms involving the correlations of the departures from the zonal averages of a velocity and the Q. They are referred to as the poleward or upward eddy fluxes or transports of Q.

Consider first the poleward eddy flux of westerly momentum, u, associated with a plane wave in the horizontal, shown in Figure 16.6a. In this case, the troughs and ridges are parallel to the y-axis. Let the y-component of the eddy velocity be

$$v' = V \sin(kx)$$

The x-component u' is 0, and consequently $\overline{u'v'} = 0$; there are zero momentum fluxes. Now consider the situation with the same plane wave, but where the troughs and ridges make an angle ϕ with the y-axis, shown in Figure 16.6b. The components of eddy velocity on the x-axis are now

$$v' = V \sin\left(k \cos(\phi) x\right)\cos(\phi)$$
$$u' = -V \sin\left(k \cos(\phi) x\right)\sin(\phi)$$

so that

$$\overline{u'v'} = -\frac{V^2}{4} \sin(2\phi) \tag{16.28}$$

If $0 \le \phi \le \pi/2$, so that the troughs and ridges run from south-east to north-west, as shown in Figure 16.4b, the momentum flux is negative, that is, westerly momentum is transported southwards. When the troughs and ridges run from south-west to north-east, the westerly momentum flux is northwards.

A similar argument relates the vertical structure of a wave to the poleward temperature or buoyancy flux. Consider a wave with

$$v' = V(z)\sin\big(kx + \phi(z)\big)$$

The only variation with height is the phase of the wave. From the thermal wind relation $f_0 \partial v/\partial z = \partial b/\partial x$,

$$b' = -\frac{f}{k}\frac{dV}{dz}\cos(kx + \phi) + \frac{fV}{k}\frac{d\phi}{dz}\sin(kx + \phi)$$

Therefore, the buoyancy fluctuation has two components, one associated with the increase in wind with height and the other associated with the change in phase with height. However, the first term is out of phase with v' and so it is only the second that contributes to the poleward buoyancy flux:

$$\overline{v'b'} = \frac{fV^2}{2k}\frac{d\phi}{dz} \tag{16.29}$$

If the meridional wind field does not tilt with height, then there is zero buoyancy flux. However, if $d\phi/dz$ is positive, so that the wave tilts westward with height, then the temperature flux is poleward.

To conclude, the net transport of westerly momentum and temperature by the waves depends upon their shape. Waves that tilt in the horizontal from south-west to north-east transport westerly momentum northwards. Waves that tilt to the west with height transport temperature polewards. Of course, these transports depend upon the square of the wave amplitude. This means they are negligible in the limit of small amplitude. In that case, consistent with the linear assumption, the waves have negligible effect on the mean flow. However, as the perturbation amplitude increases, the eddy transports become appreciable and eventually begin to modify the mean flow.

We apply these ideas to the baroclinic waves we have examined already. A poleward buoyancy flux associated with a westward tilt with height of the pressure and meridional velocity waves was discussed in Section 14.1 as being the essence of baroclinic instability. The poleward eddy transport of buoyancy implies a mean cooling on the equatorial side and warming on the poleward side, and so a tendency to reduce the baroclinicity in the zonal flow. The Eady wave has this westward tilt with height. In fact for the most unstable Eady wave, the poleward buoyancy flux is positive and independent of height. However, the Eady wave has no tilt in the horizontal and hence it has zero poleward transport of westerly momentum. Note that the eddy upward flux of buoyancy is also an essential ingredient in baroclinic instability. It is formally negligible to order Rossby number according to quasi-geostrophic theory but in reality is important in determining the static stability of the atmosphere. The impact of the eddy vertical flux of westerly momentum is also formally small in quasi-geostrophic theory. It is zero in the Eady wave and is in general small for realistic waves.

For the LC1 wave number 6 normal mode, discussed in Section 14.6, the structures of the poleward eddy transports of temperature and westerly momentum are shown in Figure 16.5. There is poleward temperature transport throughout the troposphere, though the values are largest at low levels. As we have noted before, the wave tends to tilt with the shear in the jet, southwest–northeast to the south and northwest–southeast to the north. Consistently, the momentum transports are poleward to the south and equatorward to the north, that is, they converge into the jet, acting to strengthen it. So the LC1 normal mode acts in the sense of reducing the zonally averaged baroclinicity, but increases the westerly momentum in the latitude of the jet.

The day 4–14 average fluxes for the LC1 nonlinear integration are also shown in Figure 16.7. The emphasis here is on structure rather than magnitude, which is arbitrary for the linear normal mode. All measures of the nonlinear development of the wave show that amplitudes increase preferentially in the upper troposphere. The temperature flux has a second maximum in the upper troposphere and is broader in latitude than the normal mode. The westerly momentum flux occurs in the upper

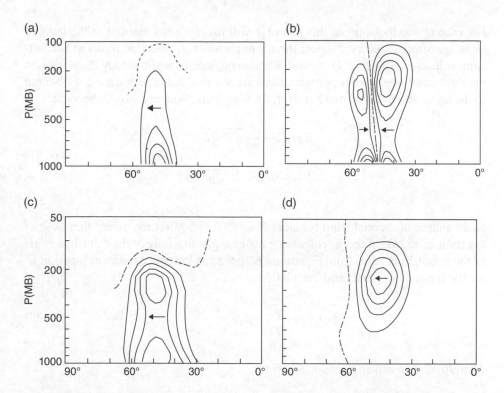

Figure 16.7 The poleward eddy fluxes of (a) temperature and (b) westerly momentum for the LC1 normal mode, and (c) and (d) the same for the average of days 4–14 in the nonlinear integration. The directions of the fluxes are indicated by the arrows. The contour interval in each case is 0.2 of the maximum value. The zero contour is drawn only at latitudes at which the value at some height exceeds 0.1 of the contour interval. From Simmons and Hoskins (1977, 1978), with permission from the American Meteorological Society

troposphere and is now dominated by the poleward flux, acting to shift the jet poleward. The vertical fluxes are not shown but have a very similar structure to the poleward fluxes, with upward fluxes of temperature and westerly momentum, consistent with the strong correlation of v and w throughout the development.

To gain a complete picture of the feedback of the baroclinic waves on to the mean flow, we need to take account of the fact that thermal wind balance will be approximately maintained in the zonal mean flow by the mean meridional circulation. The technique for doing this is totally analogous to that used in the frontal circulations and omega equation discussions. From Equation 16.25, the zonal mean zonal wind and buoyancy equations can be written as follows:

$$\tilde{D}\bar{u} = f\bar{v} - \frac{\partial}{\partial y}\overline{u'v'} + \frac{1}{\rho_R}\frac{\partial}{\partial z}\rho_R\overline{w'u'} + \bar{\ddot{u}}$$

$$\tilde{D}\bar{b} = -N^2\bar{w} - \frac{\partial}{\partial y}\overline{v'b'} + \frac{1}{\rho_R}\frac{\partial}{\partial z}\rho_R\overline{w'b'} + \bar{\ddot{b}}$$

(16.30)

For ease of manipulation, at this point we will make approximations at the level of quasi-geostrophic theory. Neglect the mean meridional advection terms in the left-hand side, approximating $\tilde{D}$ by $\partial/\partial t$. We also neglect the vertical eddy flux terms on the right-hand side. These approximations are not very accurate and are not essential to the analysis. However, making them for simplicity, Equations 16.30 become

$$\frac{\partial}{\partial t}\bar{u} = f\bar{v} - \frac{\partial}{\partial y}\overline{u'v'} + \bar{\ddot{u}}$$

$$\frac{\partial}{\partial t}\bar{b} = -N^2\bar{w} - \frac{\partial}{\partial y}\overline{v'b'} + \bar{\ddot{b}}$$

(16.31)

Maintenance of thermal wind balance $f\partial\bar{u}/\partial z = -\partial\bar{b}/\partial y$ then means that $f\partial/\partial z$ of the right-hand side of the $\bar{u}$ equation must be equal to $-\partial/\partial y$ of the right-hand side of the $\bar{b}$ equation. Also, from Equation 16.26 we can introduce a stream function $\bar{\psi}$ for the mean meridional mass circulation:

$$\rho_R\bar{v} = \frac{\partial\bar{\psi}}{\partial z}, \quad \rho_R\bar{w} = -\frac{\partial\bar{\psi}}{\partial y}$$

(16.32)

Therefore, we reach an equation for $\bar{\psi}$:

$$N^2\frac{1}{\rho_R}\frac{\partial^2\bar{\psi}}{\partial y^2} + f^2\frac{\partial}{\partial z}\frac{1}{\rho_R}\frac{\partial\bar{\psi}}{\partial z} =$$

$$-f\frac{\partial^2}{\partial y\partial z}\overline{u'v'} + \frac{\partial^2}{\partial y^2}\overline{v'b'} + f\frac{\partial}{\partial z}\bar{\ddot{u}} - \frac{\partial}{\partial y}\bar{\ddot{b}}$$

(16.33)

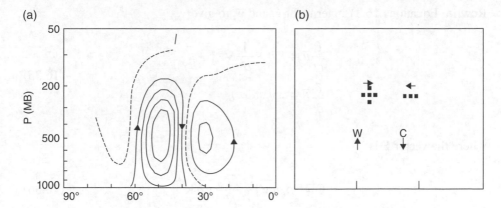

Figure 16.8 (a) Mean meridional circulation for LC1. (b) The corresponding forcing. From Simmons and Hoskins (1978), with permission from the American Meteorological Society

The form of the operator on the left-hand side is quite familiar. The first term describes adjustment of the buoyancy through adiabatic temperature changes and the second adjustment of the westerly wind through the Coriolis term.

The actual meridional circulation for LC1 averaged over days 4–14 is shown in the left-hand panel of Figure 16.8. There is a strong 'indirect' cell with rising motion near 55° poleward of sinking near 40°, and a weaker direct cell on the equatorward side of it. The driving of this circulation can be understood using the theory and the right-hand panel in Figure 16.8b. The strengthening of the westerlies near 50° by the eddy momentum flux, indicated by a dashed +, can be compensated by the Coriolis force associated with the equatorward motion indicated, and the reduction in the westerlies near 37° (dashed-) by poleward motion. Similarly, the eddy poleward heat transport leads to warming near 60° which can be compensated by ascent and cooling near 40° by descent as indicated. The net result is clearly consistent with the actual meridional circulation. It is the meridional circulation that drives the development of the very strong low-level mid-latitude westerly winds noted by day 9 in Figure 16.4, and also the subtropical easterly winds. The circulations can also be understood from Equation 16.33 with the temperature and momentum fluxes shown in Figure 16.7.

It should be noted that the removal of the mid-latitude temperature contrast and the development of the strong surface westerly winds are also seen in the region 'processed' by the baroclinic waves in the example of the downstream development of baroclinic waves shown in Figure 14.20.

A different perspective emerges by defining a residual circulation $(\bar{v}_r, \bar{w}_r)$ where $\bar{w}_r$ incorporates both the mean circulation and the eddy terms in the buoyancy equation in Equation 16.31. At the same time, $\bar{v}_r$ is defined so that mass conservation is satisfied:

$$\bar{v}_r = \bar{v} - \frac{1}{\rho_R}\frac{\partial}{\partial z}\left(\frac{\rho_R}{N^2}\overline{v'b'}\right), \quad \bar{w}_r = \bar{w} + \frac{\partial}{\partial y}\left(\frac{1}{N^2}\overline{v'b'}\right). \tag{16.34}$$

Rewrite Equations 16.31 in terms of $\bar{v}_r$ and $\bar{w}_r$ to give

$$\frac{\partial}{\partial t}\bar{u} = f\bar{v}_r + \frac{1}{\rho_R}\nabla\cdot\mathbf{F} + \bar{\dot{u}}$$

$$\frac{\partial}{\partial t}\bar{b} = -N^2\bar{w}_r + \bar{\dot{b}}$$

(16.35)

where the vector $\mathbf{F}$ is

$$\mathbf{F} = -\rho_R\overline{u'v'}\mathbf{j} + \frac{f\rho_R}{N^2}\overline{v'b'}\mathbf{k} \qquad (16.36)$$

The vector $\mathbf{F}$ is called the 'Eliassen-Palm' (E-P) flux. The divergence of the E-P flux measures the total effect of the eddies on the zonal mean wind. It includes both the direct effect of eddy momentum fluxes, and also the indirect effect of the eddy temperature fluxes. In a steady state, any buoyancy source, that is, heating, has to be balanced directly by the residual circulation. As before, requiring the maintenance of thermal wind balance we can form an equation, analogous to Equation 16.33, for a stream function for the residual circulation $\bar{\psi}_r$:

$$N^2\frac{1}{\rho_R}\frac{\partial^2\bar{\psi}_r}{\partial y^2} + f^2\frac{\partial}{\partial z}\frac{1}{\rho_R}\frac{\partial\bar{\psi}_r}{\partial z} = -\frac{\partial}{\partial z}\left(\frac{f}{\rho_R}\nabla\cdot\mathbf{F}\right) + f\frac{\partial}{\partial z}\bar{\dot{u}} - \frac{\partial}{\partial y}\bar{\dot{b}} \qquad (16.37)$$

The simplification given by the residual circulation and the Eliassen-Palm flux perspective does come with a price. Because of the definition of the residual vertical velocity, Equation 16.34, the lower boundary becomes more complicated

$$\bar{w}_r = \frac{\partial}{\partial y}\left(\frac{1}{N^2}\overline{v'b'}\right) \text{ at } z = 0 \qquad (16.38)$$

This also provides the lower boundary condition for the residual mean circulation, Equation 16.37.

Before looking at the Eliassen-Palm flux for LC1, we note that the Eliassen-Palm flux can be interpreted in terms of quasi-geostrophic potential vorticity. From the conservation of quasi-geostrophic potential vorticity for frictionless, adiabatic motion, the zonal mean potential vorticity evolves according to

$$\frac{\partial\bar{q}}{\partial t} + \frac{\partial\overline{v'q'}}{\partial y} = 0 \qquad (16.39)$$

where

$$q' = +\frac{\partial v'}{\partial x} - \frac{\partial u'}{\partial y} + \frac{1}{\rho_R}\frac{\partial}{\partial z}\left(\frac{\rho_R f}{N^2}b'\right)$$

Terms representing the zonal mean sources and sinks of potential vorticity resulting from heating, friction can be added to the right-hand side of Equation 16.39 but here the emphasis is on the eddy effect on the zonal mean potential vorticity. The zonal mean potential vorticity changes through the divergence of the poleward eddy flux of potential vorticity. A little manipulation, using the continuity equation and thermal wind balance equations, enables this flux divergence to be written in the following form:

$$\overline{v'q'} = \frac{1}{\rho_R}\left\{\frac{\partial}{\partial y}\left(-\rho_R\overline{u'v'}\right) + \frac{\partial}{\partial z}\left(\frac{\rho_R f_0}{N^2}\overline{v'b'}\right)\right\} = \frac{1}{\rho_R}\nabla\cdot\mathbf{F} \qquad (16.40)$$

Therefore, the divergence of the Eliassen-Palm flux, which can be viewed as the eddy driving of westerly zonal momentum, is actually the poleward eddy flux of potential vorticity. Conversely, the zonal mean distribution of potential vorticity is determined by the Eliassen-Palm fluxes, since Equation 16.39 may be rewritten in the compact form as follows:

$$\rho_R\frac{\partial\overline{q}}{\partial t} = -\frac{\partial}{\partial y}\nabla\cdot\mathbf{F} \qquad (16.41)$$

We shall now look at pictures of the Eliassen-Palm flux and its divergence through the development of LC1, given in Figure 16.9. The normal mode Eliassen-Palm flux is dominated by the contribution of the low-level buoyancy flux, with convergence in the lower troposphere. By day 5 this buoyancy flux, and hence the Eliassen-Palm flux, is much deeper. There is now divergence near the surface and convergence in the upper troposphere. Thereafter, the upper tropospheric poleward momentum flux becomes important and even dominant. At day 8, the upper tropospheric Eliassen-Palm flux arrows are strongly oriented towards the tropics, and where at day 5 there was convergence in the upper troposphere there is now divergence. The time average picture then shows Eliassen-Palm flux arrows rising through the troposphere and bending towards the tropics, with low-level divergence in mid-latitudes indicating zonal mean westerly forcing there and convergence in the subtropical upper troposphere, indicating easterly forcing there.

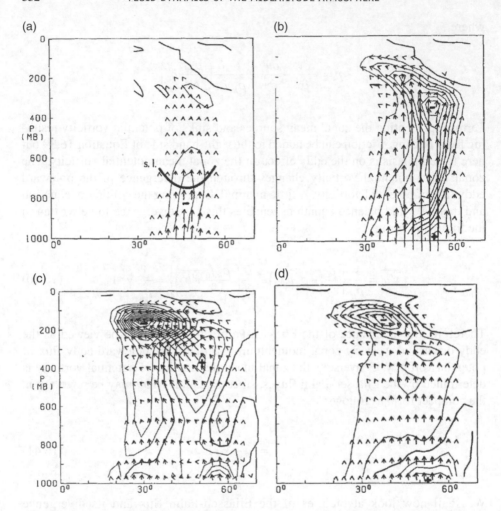

Figure 16.9 Eliassen-Palm fluxes and their divergence for LC1. (a) Shows the normal mode, (b) day 5, (c) day 8 and (d) the time average. The scales in (a) are arbitrary, but the scale for the arrows is the same in the other panels. The contour interval in (d) is 3/8 that in (b) and (c). From Edmon *et al.* (1980), with permission from the American Meteorological Society

16.5 Energetics of baroclinic waves

In Section 7.6, we showed that a positive correlation between vertical motion and temperature anomalies leads to the generation of kinetic energy of atmospheric flow. Here, we will extend those ideas to the development and decay of baroclinic waves. We may split the kinetic energy and available potential energy into energy associated with the zonal flow and into energy associated with the deviations from the zonal mean, that is, with the eddies. The kinetic energy is

$$\langle K \rangle = K_Z + K_E = \left\langle \rho_R \frac{\bar{u}^2}{2} \right\rangle + \left\langle \rho_R \frac{\overline{u'^2 + v'^2}}{2} \right\rangle \tag{16.42}$$

and the available potential energy is

$$\langle A \rangle = A_Z + A_E = \left\langle \frac{\rho_R}{2N^2} \bar{b}^2 \right\rangle + \left\langle \frac{\rho_R}{2N^2} \overline{b'^2} \right\rangle \tag{16.43}$$

Starting from the momentum equation and the thermodynamic equation, a similar analysis to that given in Section 7.6, but splitting the variables into their eddy parts, leads to expressions for the conversion of energy from one form to another. The details of the analysis are omitted here, but will be found in various textbooks, such as that of James (1994). For example, the conversion from A_E to K_E is

$$C_E = \left\langle \rho_R \overline{w'b'} \right\rangle \tag{16.44}$$

This expression implies that upward eddy buoyancy fluxes are required to convert eddy available potential energy into eddy kinetic energy. For balanced flows, the vertical eddy motion is given by the omega equation of Chapter 13: it is the vertical velocity demanded to keep the velocity and buoyancy fields in balance with each other.

So, the context of the baroclinic life cycles discussed in this chapter, the eddy available potential energy and the eddy kinetic energy are not independent reservoirs of energy; they are linked together by the requirements of balance. The total eddy energy is

$$E = A_E + K_E = \left\langle \frac{\rho_R}{2} \left(\overline{u'^2} + \overline{v'^2} + \frac{\overline{b'^2}}{N^2} \right) \right\rangle \tag{16.45}$$

Eddy energy changes either by conversion of zonal available potential energy A_Z to E, denoted C_A, or by the conversion of zonal kinetic energy K_Z to E, denoted C_K. These conversions are

$$C_A = \left\langle -\frac{\rho_R}{N^2} \frac{\partial \bar{b}}{\partial y} \overline{v'b'} \right\rangle \tag{16.46}$$

and

$$C_K = \left\langle -\rho_R \frac{\partial \bar{u}}{\partial y} \overline{u'v'} \right\rangle \tag{16.47}$$

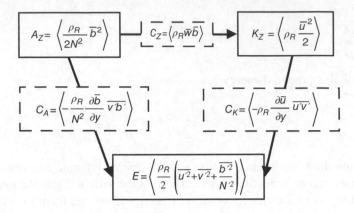

Figure 16.10   Summary of energetics for the baroclinic life cycle calculations

In the Eady and Charney linear models of baroclinic instability, the poleward fluxes of zonal momentum are zero and so C_K is 0. The buoyancy fluxes are directed down the gradient of zonal mean buoyancy and so C_A is positive, generating E. In more realistic models of linear instability, in which $\partial \bar{u} / \partial y \neq 0$, the momentum fluxes are weak but generally directed up the gradients of $\bar{u}$, the conversion C_K is weakly negative and so presents a sink of eddy energy. Figure 16.10 is a diagram summarizing this energetic scheme. Omitted here for clarity, and because they are not relevant to the frictionless, adiabatic life cycles described in this chapter, the scheme should additionally include source-and-sink terms for each of the forms of energy due to heating and friction. In the time mean global circulation, available potential energy is generated by differential solar heating, transformed into zonal and eddy kinetic energy by atmospheric circulations, and dissipated by friction. Ultimately, generation and dissipation must balance.

Turning now to the nonlinear regime, Figure 16.11 shows the variation of the energy conversions C_A and C_K through the LC1 life cycle. The total eddy energy grows, initially exponentially, and reaches its maximum around day 7. The conversion C_A follows the same pattern, converting energy between A_Z and E. During this growing phase, the conversion C_K remains small, but it becomes large and negative as the wave approaches its maximum amplitude, representing a drain on E. This conversion remains important during the decay of the wave until day 10, after which both the eddy energy and the conversions are small. So, buoyancy fluxes down the temperature gradient drive the growth of the wave, while momentum fluxes up the gradient of zonal wind drive its subsequent decay.

The conversion C_A, Equation 16.46, relates to the zonal wind shear by virtue of the thermal wind relationship:

$$C_A = \left\langle \frac{\rho_R f}{N^2} \overline{v'b'} \frac{\partial \bar{u}}{\partial z} \right\rangle \tag{16.48}$$

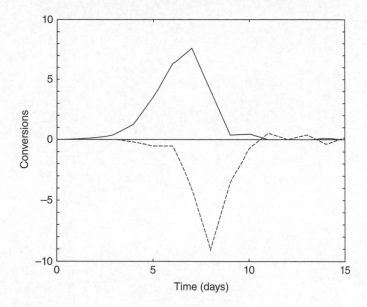

Figure 16.11 Variation of the conversions C_A (solid line) and C_K (dashed line) in units of $\mathrm{W\,m^{-2}}$ during the LC1 life cycle

It follows that the total generation of eddy energy is

$$\frac{\mathrm{d}E}{\mathrm{d}t} = C_A + C_K = \left\langle \rho_R \left(-\overline{u'v'}\frac{\partial \bar{u}}{\partial y} + \frac{f}{N^2}\left(\overline{v'b'}\frac{\partial \bar{u}}{\partial z}\right)\right)\right\rangle$$
$$= \left\langle \mathbf{F}\cdot\nabla\bar{u}\right\rangle \qquad (16.49)$$

where $\mathbf{F}$ is the Eliassen-Palm flux introduced in the previous section, Equation 16.36. So, eddy energy is generated if, on the average, the Eliassen-Palm flux is directed up the zonal wind shear, from low wind to higher wind, and is dissipated if the Eliassen-Palm flux is directed down the zonal wind shear. During the early stages of the LC1 life cycle, the Eliassen-Palm flux is predominantly vertical, directed up the wind shear, from low values of zonal wind near the ground to larger zonal winds in the upper troposphere, thereby generating eddy energy. In the later decay phase, the Eliassen-Palm flux is largest in the upper troposphere, directed mainly horizontally from the upper tropospheric jet towards the tropics where the zonal wind becomes small. Such a configuration is associated with *CZ* rapidly draining eddy energy and returning it to the zonal flow.

17

The potential vorticity perspective

17.1 Setting the scene

In many previous chapters, potential vorticity has formed the basis of the analytical development. The object of this chapter is to tie together and extend this potential vorticity perspective to midlatitude weather systems, including those in the real atmosphere. We start with a recapitulation of some of the results of the earlier chapters.

In Section 10.1, we derived the conservation of $\zeta_n/\rho\delta h$ for a small cylinder between two neighbouring isentropic surfaces with distance δh apart, where ζ_n is the component of absolute vorticity normal to the isentropic surfaces. This was shown to be the essence of the material conservation of the Rossby–Ertel potential vorticity:

$$P = \frac{1}{\rho}\zeta \cdot \nabla\theta \qquad (17.1)$$

This is referred to as potential vorticity, sometimes abbreviated as PV. In Chapter 12, the quasi-geostrophic approximations were introduced, and in Section 12.3, the conservation moving with the horizontal geostrophic motion was derived for the quasi-geostrophic potential vorticity:

$$q = f_0 + \beta y + \nabla_H^2 \psi_g + \frac{1}{\rho_R}\frac{\partial}{\partial z}\left(\rho_R \frac{f_0^2}{N^2}\frac{\partial \psi_g}{\partial z}\right) \qquad (17.2)$$

In Section 12.4, we showed that quasi-geostrophic potential vorticity is not an approximation to potential vorticity, but that the rate of change of q and its horizontal advection based on horizontal surfaces are proportional to approximations to the

Fluid Dynamics of the Midlatitude Atmosphere, First Edition. Brian J. Hoskins & Ian N. James.
© 2014 John Wiley & Sons, Ltd. Published 2014 by John Wiley & Sons, Ltd.

rate of change and horizontal advection of potential vorticity based on isentropic surfaces.

The necessity to pose a basic static stability independent of the horizontal position and also the use of geostrophic advection are major caveats to the validity of the application of the conservation of quasi-geostrophic potential vorticity to the nonlinear development of weather systems and in the region of a sloping tropopause. In Chapter 15 and Section 16.2, we showed that, provided the timescale for the material change in wind speed and direction is much longer than f^{-1}, an analytical approximation to the conservation of potential vorticity is possible, and that, with a transformation of horizontal coordinates, the equations became similar, and, in simple cases, identical, to the quasi-geostrophic potential vorticity equation. This enabled analytic solutions for frontogenesis and simple numerical simulations of nonlinear baroclinic waves. This semi-geostrophic theory therefore provides a link between the full conservation of potential vorticity and the geostrophic conservation of quasi-geostrophic potential vorticity.

The two fundamental properties of potential vorticity are its material conservation in the absence of friction and heating and its invertibility. The implication of the change in potential vorticity due to friction and heating is the topic of Section 17.6. As discussed in more detail in Section 10.3, invertibility means that, provided the total mass between isentropic surfaces is known and boundary conditions specified, all the fields associated with a balanced motion can be determined from the distribution of potential vorticity on isentropic surfaces. The quasi-geostrophic potential vorticity theory demonstrates this invertibility in that Equation 17.1 may be solved for ψ given suitable boundary conditions on ψ. Thus, the horizontal velocity and buoyancy are determined. Taking a very simple example, in Figure 17.1 is a sketch of the isentropes and circulation associated with a positive quasi-geostrophic potential vorticity 'δ-function', an infinitesimally small region of very large quasi-geostrophic potential vorticity. The isentropes bow towards the positive quasi-geostrophic potential vorticity region and there is cyclonic circulation around it. The positive vorticity anomaly above and below is compensated by reduced static stability to give zero quasi-geostrophic potential vorticity anomaly in each region. For a finite-sized quasi-geostrophic potential vorticity anomaly, outside of the anomaly very similar circulations and isentropic behaviour are found. Similarly, a negative quasi-geostrophic potential vorticity 'δ-function' would lead to isentropes bowing away from it and an anti-cyclonic circulation. These same features were also seen for the idealized circularly symmetric potential vorticity anomaly distributions shown in Section 10.3. The positive potential vorticity anomaly on isentropes crossing the tropopause in Figure 10.4a led to isentropes bowed towards the anomalous potential vorticity region, low static stability in the troposphere and cyclonic circulation. The same occurs for the positive boundary temperature anomaly case shown in Figure 10.5a. Conversely, in the case of negative potential vorticity anomalies on isentropic surfaces crossing the tropopause and for a negative surface temperature anomaly, the isentropes bow away from the anomaly, there is high tropospheric static stability and there is anti-cyclonic

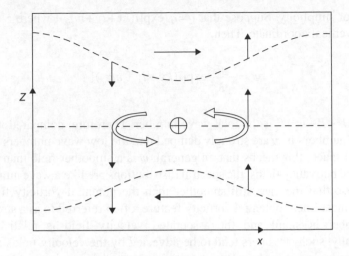

Figure 17.1 A vertical section showing the circulation and θ contours (dashed) for a positive δ-function in quasi-geostrophic potential vorticity. Aspects of the associated vertical motion are discussed in Section 17.2. Relevant to this, for a basic shear flow and in a coordinate system moving with the potential vorticity δ-function, the sense of motion along isentropes and the implied vertical motion are as indicated by arrows. From Hoskins *et al.* (2003), with permission from the Royal Meteorological Society

circulation. The way in which the vertical velocity is also given by the potential vorticity distribution and its rate of change will be discussed in the quasi-geostrophic context in Section 17.2.

Considering Equation 17.2, in the region of a finite positive potential vorticity anomaly, there will be both positive relative vorticity (the first two terms) and increased static stability (third term). It is evident that the relative strength of the two effects is given by the Burger number

$$\mathrm{Bu} = \frac{N^2 D^2}{f^2 L^2} \tag{17.3}$$

For $\mathrm{Bu} \gg 1$, that is, for small horizontal scales or large vertical scales, the circulation component dominates the quasi-geostrophic potential vorticity and its conservation. Conversely, for $\mathrm{Bu} \ll 1$, that is, for long horizontal scales or short vertical scales, the variation in the static stability component dominates. The presence of both cyclonic circulation and increased static stability (compared with values elsewhere on these surfaces) on the isentropes crossing the tropopause is evident in Figure 10.4a. In the region where there is a negative potential vorticity anomaly of a finite size, there will tend to be anti-cyclonic circulation and reduced static stability, as seen in Figure 10.4b.

It is useful to consider some scales for potential vorticity and the corresponding fields using quasi-geostrophic potential vorticity theory with f, ρ_R and N taken to be constant for simplicity. Suppose that $q=q_0\exp[i(kx+ly+mz_g)]$ where $z_g=Nz/f$ is a stretched vertical coordinate. Then,

$$\psi = -\frac{q_0}{K^2}\exp[i(kx+ly+mz_g)]\tag{17.4}$$

Here, $K=(k^2+l^2+m^2)^{1/2}$ is the total wave number. Therefore, compared with q, the high wave numbers in ψ are strongly damped and the low wave numbers dominate. In physical space, this means that, in general, ψ is a smoother field than q. For the velocity and buoyancy fields, the spectral distributions are like a wave number times that of ψ, so that they are still smoother than the potential vorticity field. For a shallow, almost linear potential vorticity feature, often referred to as a streamer, the wave number is small and the associated velocity field is relatively weak. Consequently, such streamers tend to be advected by the velocity fields associated with the nearby larger-scale potential vorticity anomalies.

For a boundary temperature or buoyancy anomaly $b=f\partial\psi/\partial z=b_0\exp[i(kx+ly)]$,

$$\psi = -\frac{b_0}{NK_h}\exp\left[i(kx+ly)-K_hz_g\right]\tag{17.5}$$

where $K_h=(k^2+l^2)^{1/2}$ is the horizontal wave number. The higher wave numbers are not as damped as they are for a q distribution, and in this case, the velocity field tends to have the same scale as the buoyancy field.

17.2   Potential vorticity and vertical velocity

In a potential vorticity perspective, the circulations that perform the adjustment to balance are implicit. However, it gives insight if the relationship is considered explicitly. This will be done in the quasi-geostrophic context and the analysis will produce yet another form of the omega equation beyond those given in Chapter 13.

We start by again considering a positive potential vorticity 'δ-function', by which we mean an infinitesimally small region of very large potential vorticity, as illustrated for the Northern Hemisphere in Figure 17.1. The potential vorticity δ-function and the structure associated with it will move with the basic flow in which it is embedded, whatever the nature of that flow. If this basic flow is uniform, the whole pattern is stationary and there will be no vertical motion. However, this is not the case if the potential vorticity δ-function is embedded in a westerly flow that increases with height. Referring to Figure 17.1, for simplicity, we take a frame of reference moving with the potential vorticity δ-function. The basic shear must be accompanied by isentropic surfaces that slope up towards the pole. For these surfaces not to move, the cyclonic circulation must be on these surfaces and so it must be accompanied by descent to the west and ascent to the east. This vertical motion is illustrated in Figure 17.1. Above the δ-function, the westerly flow must move down the isentrope to the west and up the isentrope to the east. Therefore, in this region, both components of the flow require descent to the west and ascent to the east if the isentropes are to be stationary. Similarly, below the δ-function both the relative easterly flow and the circulation require descent to the west and ascent to the east.

In this example, in which the potential vorticity field and isentropes do not change in time, the vertical motion is given entirely by that implied by the horizontal motion up and down fixed isentropic surfaces. We refer to this as 'isentropic upgliding', w_{IU}. In this example, $w=w_{IU}$. More generally, the potential vorticity field and the associated isentropic surfaces will change. Using the adiabatic buoyancy equation in a frame of reference that can be chosen arbitrarily,

$$w = w_{IU} + w_{ID} = -\frac{1}{N^2}\mathbf{v}\cdot\nabla b - \frac{1}{N^2}\frac{\partial b}{\partial t} \qquad (17.6)$$

where w_{ID} is the vertical motion associated with the fact that isentropes are not fixed, that is, with isentropic displacement. To relate w_{ID} to the potential vorticity, we use an identity relating q and b which can be quite simply verified:

$$L\left(\frac{b}{N^2}\right) = f\frac{\partial q}{\partial z}$$

Here, L is the operator

$$L(X) = N^2\nabla^2 X + f^2 \frac{\partial}{\partial z}\left(\frac{1}{\rho_R}\frac{\partial}{\partial z}(\rho_R X)\right) \qquad (17.7)$$

the same operator that acts on w in the omega equation in Chapter 13. Therefore,

$$Lw_{ID} = -f\frac{\partial}{\partial z}\frac{\partial q}{\partial t} = -f\frac{\partial}{\partial z}(\mathbf{v}\cdot\nabla q) \qquad (17.8)$$

On horizontal boundaries, the boundary condition $w=0$ means that

$$w_{ID} = -w_{IU} = \frac{1}{N^2}\mathbf{v}\cdot\nabla b \qquad (17.9)$$

Therefore, the vertical motion can be considered to be the sum of an isentropic upgliding, w_{IU}, and an isentropic displacement component, w_{ID}. The latter is

associated with changes in the interior potential vorticity and boundary buoyancy, which in the absence of heating or friction must be associated with interior potential vorticity or boundary buoyancy advection.

The analysis carries through in any frame of reference. The natural one to use is
the one which minimizes the changes in the interior potential vorticity and boundary temperature and the consequent magnitude of w_{ID}. In particular, it is natural to
use a frame of reference moving with the weather system or the major potential
vorticity anomaly of interest.

It can be verified by some manipulation that the form, of the omega equation
given by Equations 17.6, 17.8 and 17.9, is consistent with the other forms of the
omega equation given in Chapter 13. The advantage of this form for the discussion
here is that it again shows the primary roles played by interior potential vorticity and
boundary buoyancy. The frame of reference should normally be taken moving with
the weather system of interest. Isentropic upglide leads directly to no change in the
thermal structure of a system. It is the potential vorticity and boundary b advection
that leads to any changes. However, the isentropic upglide can often dominate the
vertical motion, as it did in the example illustrated in Figure 17.1, and can therefore
be very important for latent heat release.

The two-dimensional Eady wave was discussed in Section 14.4, and the vertical
velocity of the most unstable Eady wave is shown in Figure 14.11. In Figure 17.2,
the full Eady wave vertical velocity is also shown, along with the components w_{IU}
and w_{ID}. For the Eady wave, the latter is driven purely by the need to balance the
buoyancy advection on the two boundaries. At the mid-level, isentropic upglide
has the same phase as w but almost twice the magnitude. It is partially cancelled by
w_{ID} driven by the need to balance w_{IU} on the boundaries to make $w=0$ there. The
reduction enables the poleward ascending air to remain warm for its longitude and
height, consistent with arguments given in Chapter 14. At the short-wave cut-off,
as shown in the final panel in Figure 17.2, w_{IU} has a structure similar to that for the
most unstable wave. Indeed for the same amplitude surface v-wave, the wave at the
short-wave cut-off has a marginally larger amplitude. However, its w_{IU} now satisfies the boundary condition $w=0$ so that w_{ID} is zero and the structure does not
evolve in time.

17.3   Life cycles of some baroclinic waves

The synoptic development in the two nonlinear life cycles, which was discussed
previously in Section 16.5, will now be analysed from a potential vorticity
perspective. In particular, the development will be viewed in terms of the
interaction of the near-tropopause potential vorticity and the near-surface potential
temperature. As remarked before, the main real-world interest in these life cycles
is the range of different aspects that they illustrate rather than the complete
development over the period.

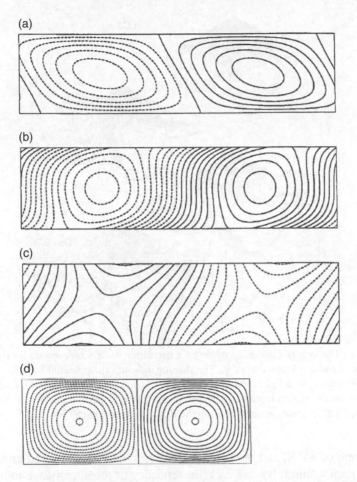

Figure 17.2 Vertical velocity analysis for two-dimensional Eady waves. (a)–(c) Most unstable mode w, w_{IU} and w_{ID}, respectively. (d) $w = w_{IU}$ for the mode at the short-wave cut-off. The most unstable and short-wave cut-off mode pictures are shown for the same amplitude in the boundary v field. Negative contours are dashed, and the phase is such that the middle level low pressure is in the centre of the plot. From Hoskins *et al*. (2003), with permission from the Royal Meteorological Society

Figure 17.3 shows the LC1 development between days 6.25 and 10.25 using contours of the potential vorticity field on the 300 K surface. Also shown is the near-surface potential temperature. By day 6.25, the wave amplitude is large. However, the basic vertical interaction is seen, with a positive (cyclonic) potential vorticity anomaly near the tropopause west of a warm (cyclonic) θ anomaly near the surface. By this stage, the θ field has become strongly distorted. The warm air has moved a long way polewards (L) but only in a very narrow strip with the tip of the occlusion further equatorwards (O), at the end of the strong cold front. The potential vorticity trough has extended equatorwards (U). Both fields show evidence of a cyclonic

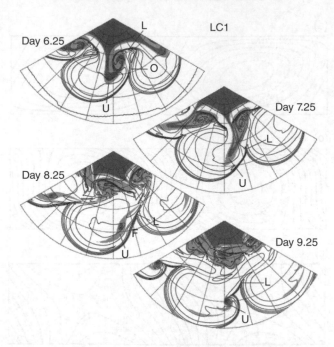

Figure 17.3 The synoptic evolution of the LC1 baroclinic wave. Contours are for θ on the lowest model level and are drawn every 5 K. The shading indicates the potential vorticity distribution on the 300 K surface, with light contours for PV > 1PVU and heavy shading for PV > 3PVU. Two wavelengths and the region north of 15°N are shown, and grid lines are drawn every 15°. From Methven *et al.* (2005), with permission from the Royal Meteorological Society

turning north of 45°N, and the potential vorticity has been wrapped around in the trough to form a spiral. By day 7.25, the tendency of the upper wave to break anti-cyclonically south of 45°N is apparent. The trough has extended (U) in a narrowing filament parallel to the cold front. Vertical sections show that this is part of an upper air front in this region, where the high potential vorticity air has locally moved down tilted isentropic surfaces. The occlusion process has continued to such an extent that the tip of the warm sector (L) is now much further south. Poleward of 45°N, the cyclonic spiral breaking is evident at both levels. By day 8.25, a vortex roll-up is evident in the filament of high potential vorticity. In this region of small large-scale strain, the high potential vorticity has started to create its own cyclonic circulation. This has wrapped additional high potential vorticity from the filament into it, and this leads to strengthening of the cyclonic circulation. This upper vortex causes a wave on the lower temperature field which is just apparent (F) at day 8.25. A coupled development follows, and by day 9.25 the wave is very strong. The warm air has been drawn in below and just to the east of the upper vortex (U) and has then cut off. Such a cut-off of warm sector air is referred to as a 'seclusion'.

Figure 17.4 shows similar fields for LC2 between days 5.75 and 10.25. This time, the potential vorticity contours are for the 320 K surface and so are at higher

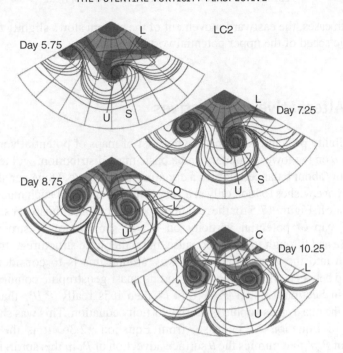

Figure 17.4 The synoptic evolution of the LC2 baroclinic wave. Contours are for θ on the lowest model level and are drawn every 5 K. The shading indicates the PV distribution on the 320 K surface, with light contours for PV > 2PVU and heavy shading for PV > 5PVU. From Methven *et al.* (2005), with permission from the Royal Meteorological Society

potential vorticity values. The finite amplitude θ wave shows a bent back λ structure bounded by the weak cold front and the strong warm front, this continuing on to the bent back occlusion. The structure is wrapping cyclonically and is producing a seclusion of the warm air (S). The upper potential vorticity trough is extending equatorwards (U) west of the warm air, again consistent with a westward tilting cyclone throughout the troposphere. By day 7.25, the seclusion (S) has become more evident and the warm air has extended further polewards (L). This continues until it is pinched off by the occlusion (O) at day 8.75. The upper potential vorticity trough (U) continues to break cyclonically. It forms a large-scale potential vorticity region over the warm seclusion (S) as seen at days 7.25 and 8.75, making a strong cyclonic vortex through the troposphere. By day 10.25, the positive upper potential vorticity region has pinched off (C) from its polar source region. Following this event, the vortex becomes more symmetric and persists for many days.

In both life cycles, as the primary low-level θ wave occludes, its induced meridional flow becomes weaker and its interactions with the upper potential vorticity wave become weaker. However, influenced by middle tropospheric potential vorticity anomalies, the upper potential vorticity wave continues to grow for several more days and dominates the large-scale development. Consistent with this, in both cases, the eastward movement of the system slows slightly to be nearer the intrinsic speed of the upper potential vorticity wave.

17.4 Alternative perspectives

The invertibility principle says that a whole set of maps of potential vorticity on θ, along with θ on the lower boundary and a basic mass distribution, will tell us all we need to know about balanced flow in a dry atmospheric state. However, there can be problems if one wishes to look only at a very limited number of such maps. Consider the schematic in Figure 17.5 for the middle/upper troposphere and lower stratosphere. Clearly, a map of potential vorticity on θ will show a lot of structure in the stratosphere and through to the tropopause, but rather little structure as the θ surface dives down into the troposphere. One way around this is to consider again the relationship between potential vorticity and its quasi-geostrophic counterpart, q. As discussed in Section 12.4, if $P=P_R(z)+P'$ then it is really P'/P_R that gives the essence of the quasi-geostrophic potential vorticity equation. This was shown by the derivation of Equation 12.33. Also, from Equation 12.36, it is the horizontal advection of $P_R q'$ that mimics the θ surface advection of P. In this spirit, Figure 17.6 shows the quantity P'/P_R for the most unstable baroclinic wave, wave number 7, on

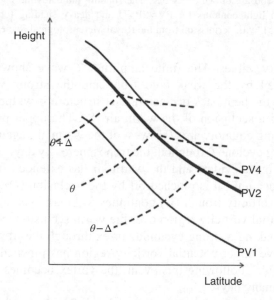

Figure 17.5 Schematic for the distribution of potential temperature (θ) and potential vorticity (PV) in a latitude–height section in the upper troposphere and lower stratosphere in middle latitudes. Three θ contours with equal spacing are shown in dashed lines. Potential vorticity contours at 1, 2 and 4 PVU are shown in continuous lines. The PV2 contour is heavier, symbolizing that this is also the dynamical tropopause

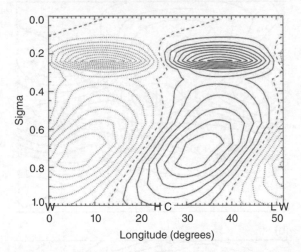

Figure 17.6 The potential vorticity structure of the most unstable wave number 7 on the LC1 basic flow at the latitude of maximum baroclinicity. The quantity shown is the perturbation potential vorticity divided by a basic potential vorticity distribution. Contour intervals are equal, and negative values are dashed. The positions of the maxima and minima in the surface pressure and temperature are indicated. From Methven *et al.* (2005), with permission from the Royal Meteorological Society

the LC1 flow. This shows nicely the perturbation in the troposphere, with magnitude peaking near the steering level, as well as that near the tropopause. Both are important in the structure of the wave.

The argument of the previous paragraph suggests that instead of the conservation of P, we might consider the conservation of $\ln(P)$. Equivalently, we could use a logarithmic contour interval for a map of potential vorticity on θ. Figure 17.7a shows a map of the 25 kPa geopotential height for a particular time and Figure 17.7b the corresponding potential vorticity on 315 K map with logarithmic contour intervals. This θ-level is chosen so that the picture does indeed show the low potential vorticity air displaced a long way poleward, giving an anti-cyclonic anomaly in the ridge in the height field, and high potential vorticity cyclonic anomalies in the troughs on either side. There is also a cut-off high region equatorward of the potential vorticity ridge and a corresponding cut-off low in the height field. However, this 315 K θ-level has been carefully chosen to capture all this structure, and it would not necessarily be suitable for showing the details in other events. Furthermore, there is no sign of the potential vorticity structure equatorward of 40°N. Here, the θ surface has descended into the lower troposphere and so there is no signature of the potential vorticity structure at higher levels.

One answer to this problem is to consider a number of θ-levels. However, if only one map is to be considered, Figure 17.5 suggests an alternative. Indeed, that alternative has its own advantages. Since potential vorticity and θ are both conserved in adiabatic, frictionless flow, we can equally well consider the motion of air on a potential vorticity surface and map the materially conserved property θ on that surface. Poleward movement of the air will lead to low potential vorticity on a θ

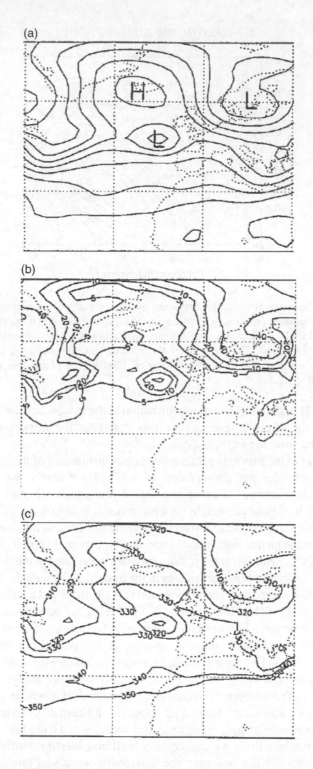

Figure 17.7 Maps for 12Z on 19 June 1996 from European Centre for Medium Range weather Forecasting (ECMWF) data. (a) 25 kPa height with contours every 100 m. (b) Potential vorticity on the 315 K surface, with contours at 0.5, 1, 2, and 4 PVU and marked by ten times these values. (c) θ on the 2 PVU surface with contours every 10 K up to 350 K. From Hoskins (1977), with permission of the Royal Meteorological Society

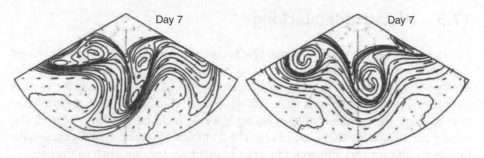

Figure 17.8 Maps of θ on the 2 PVU surface for day 7 of LC1 (left) and LC2 (right). Contours are drawn every 5 K from 290 to 350 K. Two wavelengths are shown. Here, the domain is given only from 20°N to the pole. From Thorncroft *et al.* (1993), with permission of the Royal Meteorological Society

surface, or equivalently to high θ on a potential vorticity surface. Both correspond to an anti-cyclonic, high static stability anomaly. Similarly high potential vorticity on a θ surface and low θ on a potential vorticity surface both correspond to a cyclonic, low static stability anomaly.

In some situations, such as the folding in an upper air front, there may not be a monotonic PV distribution in z or θ. However, in situations where this is not the case, it is clear that knowing the θ distribution on potential vorticity surfaces everywhere would also be a sufficient basis for invertibility. However, the focus is usually just on one potential vorticity surface, the 2 potential vorticity units (PVU) surface, shown by a thicker line in Figure 17.5. This is a very important surface as in middle latitudes it is co-located with the tropopause: it is often referred to as the dynamical tropopause. Therefore, we can consider motion of the air on the dynamical tropopause and follow it with maps of θ on a 2 PVU surface. This map is shown in Figure 17.7c for the same synoptic case. Clearly, the 315 K contour on this map is identical with the PV2 contour on PV on the θ map in Figure 17.7b. It captures all the middle/upper tropospheric and stratospheric structure seen on the 315 K PV map. However, unlike the 315 K potential vorticity map, it would do this for any synoptic situation, and in addition it would also show the subtropical structure. One caveat is that equatorwards of the subtropical jet, the 2 PVU surface can no longer be considered as a dynamical tropopause: the 2 PVU surface bends sharply upwards while the tropical tropopause is closer to a constant θ surface, such as the 380 K surface.

Figure 17.8 shows θ on potential vorticity surface maps for day 7 of both LC1 and LC2. Compare these with the day 7.25 potential vorticity on θ fields in the higher resolution calculations shown in Figures 17.3 and 17.4. There is much similarity in the higher latitude structures, but Figure 17.8 gives a much clearer picture of the anti-cyclonic wave breaking in the middle and lower latitude regions in LC1. This contrasts strongly with the cut-off cold cyclonic feature in LC2.

17.5 Midlatitude blocking

The synoptic case featured in Figure 17.7 shows an amplified ridge with a cyclonic circulation on its equatorward side. Between them, the midlatitude westerlies are locally replaced by easterlies. The midlatitude flow is blocked and so is the easterly passage of weather systems propagating from upstream. Such a phenomenon is referred to as a block. The emphasis often is on the blocking high, but the cut-off low is also usually part of such a system and can even be dominant. Blocking systems occur quite frequently and are very important in some midlatitude regions, such as Europe.

Figure 17.9 shows a schematic of a block. The contours could be those of upper tropospheric geopotential height. More relevant dynamically, they could be those of potential vorticity on a θ surface or θ on the 2 PVU surface. With the latter interpretation, the anti-cyclonic circulation corresponds to potentially warm air cut off from its subtropical source region and the cyclonic circulation to potentially cold air cut off from its polar source region. The cyclonic region will be one of low static stability. It will tend to be a region of convective instability, particularly as the underlying surface and any entrained ambient low-level air are likely to be warm and moist. The reverse is true for the anti-cyclonic region. The dipole structure in this schematic is dynamically stable and is an example of the sort of structure discussed in Figure 9.3b, but with signs reversed. The anti-cyclonic circulation around the warm pole acts to stop the cold pole being advected eastwards by the ambient westerly flow. Similarly, the cyclonic circulation around the cold pole helps keep the warm pole in position.

It is clear from this discussion that, unless heating processes are dominant, the formation of a block must come about through motions that lead to cut-offs of the warm and/or cold air. Figure 17.10 shows this happening in one event in the north-western North Atlantic, on the northern side of the jet. An oncoming weather system leads to poleward movement of the warm air and equatorward and eastward

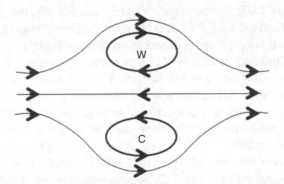

Figure 17.9 Schematic of a midlatitude blocking structure. The contours could be those of upper tropospheric height, potential vorticity on a θ surface or θ on the 2 PVU surface. In the latter case, the warm anomaly (W) gives anti-cyclonic circulation and the cold anomaly (C), cyclonic circulation. Both give an easterly anomaly on the central latitude, interrupting the westerly flow

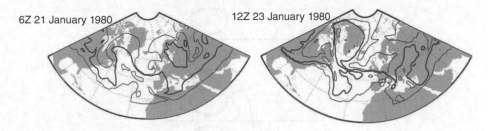

6Z 21 January 1980

12Z 23 January 1980

Figure 17.10 Theta on 2 PVU maps showing the development of a high-latitude block. The maps are for 6Z on 21 and 12Z on 23 January 1980, respectively, on the left- and right-hand sides. Theta contours are drawn at 10 K intervals. From Woolings *et al*. (2008), with permission from the American Meteorological Society

movement of the cold air. This is shown by the θ contours in the 21 January map. The warm air anomaly develops its own anti-cyclonic circulation, and the cold air develops its own cyclonic circulation. Both these circulations act to cut off their anomalies from their source region, leading to the situation seen on the 23rd January map. The whole process is a wave breaking that is visualized by θ on 2 PVU contours. The development of a structure like that in the schematic of Figure 17.9 is clear. In this case, the warm air/anti-cyclone dominates. Also the warm and cold anomalies move around one another in a cyclonic sense. In other cases, and in other regions, the cold air/cyclone can dominate. Also the sense of the wave breaking can be anti-cyclonic, particularly in the European region.

Blocks are so important because, once they have been established, they can persist for longer than the synoptic timescale. The dynamical perspective achieved through using potential vorticity and θ makes it clear why this is the case. Having set up such a dipole block, it can only be removed by the cut-off structures returning to their source regions or by frictional or heating processes. The latter will be discussed in the next section, but their timescale for removing a block is usually a week or more. Given the stability of the blocking dipole, the advection of the cut-off anomalies away from each other and back towards their source regions often requires the influence of a new, very strong, weather system. In fact, more routine weather systems are apt to reinforce the block. Ahead of a low-pressure system that slows as it reaches the block, the warm air extends polewards and then cuts off and joins the existing warm anomaly, while behind the low, the cold air extends equatorwards, cuts off and joins the cold anomaly. Figure 17.11 is a schematic illustration of this mechanism. In an incoming weather system, subtropical air is displaced polewards and polar air displaced equatorwards. As the weather system approaches the block, it extends meridionally. The subtropical air develops its own anti-cyclonic circulation which acts to produce a cut-off that reinforces the warm anti-cyclone in the block. Then, the polar air extends equatorwards and behaves in a similar manner, acting to reinforce the cold cyclone in the block. In this way, the blocking structure is strengthened and tends to extend westwards.

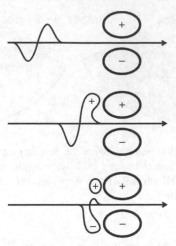

Figure 17.11 Schematic illustration of the mechanism for the reinforcement of a block by (a) an approaching synoptic weather system; (b) the tropical air is swept anti-cyclonically around the warm pool and develops its own anticyclonic circulation. (c) The warm anomaly cuts off and is effectively absorbed into the warm pool, while a similar process is repeated around the cold pool. The contours shown could be those of θ on a potential vorticity surface or potential vorticity on a θ surface

17.6   Frictional and heating effects

When heating and frictional torques are present, then potential vorticity is not materially conserved. For potential vorticity, Equation 10.6 gave

$$\frac{DP}{Dt} = \frac{1}{\rho}(\nabla \times \dot{\mathbf{u}}) \cdot \nabla\theta + \frac{1}{\rho}\boldsymbol{\zeta} \cdot \nabla\dot{\theta} \qquad (17.10)$$

For quasi-geostrophic potential vorticity, the derivation of the quasi-geostrophic potential vorticity equation in Section 12.3 can easily be extended to include non-conservative terms:

$$\frac{Dq}{Dt} = \mathbf{k} \cdot \nabla \times \dot{\mathbf{v}} + \frac{1}{\rho_R}\frac{\partial}{\partial z}\left(\rho_R \frac{f_0}{N^2}\dot{b}\right) \qquad (17.11)$$

The right-hand side of Equation 17.11 clearly has two terms that are approximations to those in Equation 17.10 when these are divided by $\rho_R d\theta_R/dz$. In Section 10.2, it was shown in Equations 10.8 and 10.9 that the mass integral over a region of potential vorticity can be written as a divergence and that

$$\rho\frac{DP}{Dt} = \nabla \cdot \mathbf{J} \qquad (17.12)$$

where

$$\mathbf{J} = [\nabla \times \dot{\mathbf{v}}]\theta + \zeta\dot{\theta} \qquad (17.13)$$

Note that the form of $\mathbf{J}$ is not unique since the curl of any vector that has zero divergence could be added to it. From Equation 10.9, the mass integral of potential vorticity in a region can only change because of non-zero sources on the boundary of the region. Internal sources can only lead to redistribution. Considering the quasi-geostrophic potential vorticity case and Equation 17.11, it is clear that horizontal integration turns the frictional torque term into a boundary integral while vertical integration does this for the buoyancy source term. Consequently, the same overall result holds for the quasi-geostrophic potential vorticity.

This integral result and the impact of localized heating and frictional torques will be considered in more detail later. However, first it is helpful to consider the timescales over which the material conservation of potential vorticity on isentropic surfaces may be a good approximation. Frictional torque is large in the boundary layer but generally very small in the free atmosphere. Consequently, the material conservation of potential vorticity is unlikely to be useful on synoptic timescales for air in the boundary layer but should be valid elsewhere. In the case of heating, there are two requirements for potential vorticity conservation:

1. The air on the given isentropic surface at a later time should not come from a significantly different part of the system.

2. The potential vorticity should not be changed by a significant proportion over the time.

The first implies that the change in θ over some time interval Δt, which is of the order $\dot{\theta}\Delta t$, should be much less than $D(\partial\theta_R/\partial z)$, D being the characteristic depth of the system. The second implies that the change in $\delta\theta$ of a small cylinder, $\delta h(\partial\dot{\theta}/\partial z)\Delta t$, should be much less than $\delta\theta$. If the height scale for the change in $\dot{\theta}$ is equal to D, the two criteria are equivalent. If it is smaller than D, the latter criterion will be more restrictive. In either case, the material conservation should be a good approximation for timescales:

$$\Delta t \ll \frac{\partial\dot{\theta}}{\partial z} \bigg/ \frac{\mathrm{d}\theta_R}{\mathrm{d}z} \qquad (17.14)$$

For radiative cooling, with a height scale of 3 km and a basic potential temperature gradient of $3 \mathrm{K\,km^{-1}}$, Equation 17.14 gives a timescale of 9 days. However, for heating by the release of latent heat, which can be three or more times larger and with a depth scale somewhat smaller, the timescale could be 3 days or less. Therefore, we conclude that when looking at maps of potential vorticity on θ over a few days, material conservation will be a good approximation, except in the boundary layer and near regions of significant latent heat release.

Unlike the resolved dynamical processes of earlier chapters, heating and friction are complicated processes which do not depend in any simple way upon the flow variables. For the sake of illustration, some idealized forms of heating and friction will sometimes be incorporated in the equations. But generally, these are so over-simplified that they are unlikely to be quantitatively correct. Consequently, the discussion here will be more descriptive and less quantitative than that elsewhere. A more realistic approach would involve simulating atmospheric flow with elaborate numerical models. Even then, there is great uncertainty about the best ways of representing many heating and friction mechanisms, especially when transport and mixing in a turbulent flow are central processes.

The response to heating and friction was discussed in Section 13.4 in terms of the ageostrophic circulation that is generated and its impact on the flow. However, from a potential vorticity perspective, we can consider frictional and heating effects in terms of a two-stage process:

1. Determine the material change in potential vorticity implied by the source.

2. Invert the potential vorticity to determine the balanced response.

The ageostrophic or higher-order circulation that achieves the balance is implicit in step 2 and need not be determined.

Consider first an isolated region of heating, illustrated in Figure 17.12. The heating causes the central isentrope in Figure 17.12 to dip down, as shown in the left panel. Therefore, for a cylinder above the heating, δh is increased for the same $\delta \theta$, implying that the potential vorticity is decreased, consistent with Equation 17.10. Conversely, below the heating, δh is decreased and the potential vorticity is increased. In an Eulerian sense, the total potential vorticity in the region has not changed, but it has been redistributed, consistent with Equation 17.12. In this panel, all the change is in static stability. However, as a result of the adjustment to balance process, step 2, shown in the right panel, the decreased potential vorticity above the heating is partitioned between low absolute vorticity, that is, anti-cyclonic

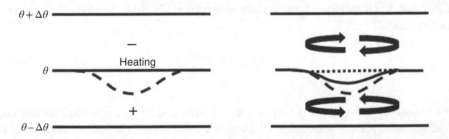

Figure 17.12 A vertical section schematic showing a potential vorticity interpretation of the response to heating. Left: initial θ surfaces (continuous), their perturbation by heating (dashed) and the consequent changes in potential vorticity. Right: the adjustment process in which the potential vorticity perturbation is shared between static stability and circulation, involving a relaxation of the perturbation of the θ surface

circulation, and low static stability. Similarly, the increased potential vorticity below the heating is partitioned between high static stability and high absolute vorticity, that is, cyclonic circulation. The adjustment process in step 2 involves an implicit ascent in the heating region with vortex shrinking above and stretching below, consistent with the discussion in Section 13.4. In the same way, isolated cooling would give a rise in the central isentrope, leading to high potential vorticity above the cooling and low potential vorticity below. In the balanced state, the deformation of the central isentrope would be reduced, and there would be high static stability and cyclonic circulation above, and low static stability and anti-cyclonic circulation below. In each case, the apportioning between static stability and absolute vorticity changes depends on the Burger number, Bu, with static stability changes dominating for low Bu and the vorticity changes for high Bu.

If we consider a steady heating situation in which the vortex lines are parallel and in a direction $\hat{s}$ which is close to vertical, then $\mathbf{J}$ is in the direction $\hat{s}$. For a parcel rising quasi-vertically along $\hat{s}$, if the density was constant, then the net change in PV on the parcel as it moved from below the heating region, given by

$$\Delta P = \int \frac{1}{\rho} \nabla \cdot \mathbf{J} dz$$

would be 0. Because its density actually decreases as the parcel ascends, the positive change below the heating will tend to be outweighed by the negative change above. Consequently, we can expect anomalously low potential vorticity to stream out from above a heating region in a weather system.

We now consider some additional aspects of three-dimensional cyclone development from the potential vorticity perspective, referring to Figure 17.13. As discussed earlier,

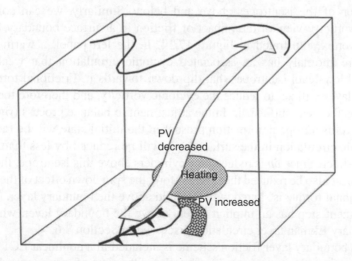

Figure 17.13 Schematic of the airflows and potential vorticity changes due to latent heat release in a cyclonic weather system

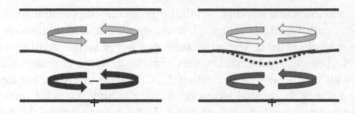

Figure 17.14 A schematic vertical section showing a potential vorticity interpretation of the impact of boundary friction on a low-level vortex. Left: the circulation and isentropic surfaces in a cyclonic vortex weakening with height associated with a warm boundary anomaly (+) and the change in potential vorticity associated with surface frictional reduction in the vortex (−). Right: the adjustment to balance

the latent heat release, which has a maximum in the low to mid-troposphere, implies a negative potential vorticity source above it and a positive one below. The air in the former region will generally be ascending and moving polewards and eastwards relative to the system, and the reduced potential vorticity in it will act to amplify the downstream ridge in the upper troposphere. In fact as argued earlier, even if air rises from below the heating region to above it, it is likely to have anomalously low potential vorticity. The increased potential vorticity in the air that remains in or below the heating maximum will mean that as this air tends to move polewards and westwards into and around the lower level cyclone, it will act to amplify the cyclone.

The impact of a frictional torque can be understood in a similar manner. Consider an initial state with a localized cyclonic vortex associated with a localized positive potential vorticity anomaly. A frictional torque will act to reduce the cyclonic circulation and therefore the potential vorticity anomaly. In the balanced state, there will be a smaller reduction in the cyclonic circulation but now also a reduction in the deformation of the isentropes above and below. Similarly, we can consider the potential vorticity view of the impact of friction in a surface boundary layer on a cyclonic vortex, illustrated in Figure 17.14. In the left panel, a warm boundary temperature anomaly has an associated cyclonic circulation that weakens with height and low-level isentropes that dip down towards it. Frictional torque in the boundary layer will act to reduce the cyclonic vorticity, and therefore the potential vorticity, in the near-surface air. In the adjustment to balance process (right panel), the low-level isentrope deformation present in the initial state will be reduced and the cyclonic circulation in the surface layer will be reduced by less than the direct frictional effect. Since δh is reduced for cylinders above this isentrope, the absolute vorticity must also be reduced there. Therefore, the spin down effect of the boundary layer frictional torque is shared with the fluid above the boundary layer. Of course, the adjustment step has an implicit ascent above the boundary layer, which is just the secondary Ekman layer circulation discussed in Section 8.6.

When a boundary layer friction scheme is included in a nonlinear LC1 baroclinic wave integration, the growth rate is reduced and the time of maximum eddy energy is delayed and the maximum reduced. This might be expected from the inclusion of

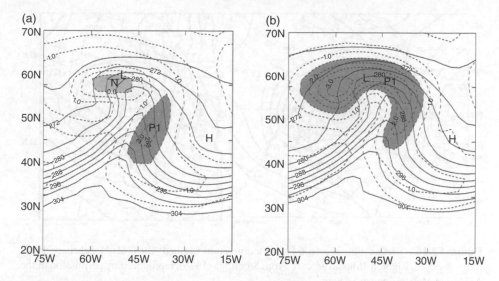

Figure 17.15 Potential vorticity distributions for day 6 at (a) $\sigma=0.98$ and (b) $\sigma=0.955$ for an LC1 plus boundary layer friction simulation, shown by dashed contours with an interval of 0.5PVU. Lighter shading indicates regions with potential vorticity less than zero and darker shading indicates regions with potential vorticity greater than 1.5PVU. Potential temperature contours (interval 4 K) are shown by continuous lines and the surface low- and high-pressure locations are indicated by L and H. From Adamson *et al.* (2006), with permission of the Royal Meteorological Society

the frictional effect, but the potential vorticity in the nonlinear wave also shows some unexpected aspects. Figure 17.15 shows the potential vorticity at day 6 on two boundary layer surfaces. In the lowest of these, the expected negative potential vorticity is seen near the low centre. However, at this level it is also clear that positive potential vorticity has been generated in the warm frontal region and advected around to the north of the low centre. At higher levels, this feature is dominant. This positive potential vorticity generation is associated with the generation of horizontal vorticity in the frontal region. Considering the relevant term, $((\nabla \times \dot{\mathbf{v}}) \cdot \nabla_H \theta)/\rho$, if we set $\dot{\mathbf{v}} = -\mathbf{v}/\tau_D$ and, for qualitative argument, assume approximate thermal wind balance, then

$$\frac{DP}{Dt} = -\frac{f\theta_0}{\tau_D \rho g} \mathbf{v} \cdot \frac{\partial \mathbf{v}}{\partial z} \qquad (17.15)$$

When the low-level wind is in the opposite sense to the shear, positive potential vorticity is generated. This is the case in this warm frontal region, where the flow relative to the system is almost 'backwards' along θ contours. The positive potential vorticity acts to intensify the warm front and 'bent-back occlusion'. However, it does not intensify the low as it is advected out of the boundary layer and forms a shallow layer of enhanced static stability above the surface low. In other idealized cyclone simulations, depending on the underlying surface conditions and the set-up used, vertical thermal fluxes can dominate the boundary layer potential vorticity generation.

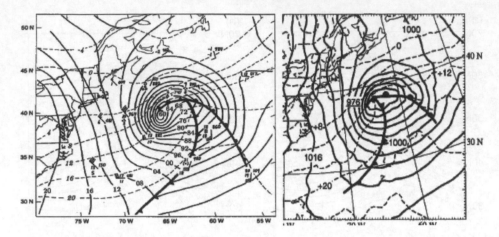

Figure 17.16   Surface pressure (solid) and temperature (dashed) for 00Z 24 February 1987. Left: observed. Right: a 36 hour forecast. From Stoelinga (1996), reproduced by permission of the American Meteorological Society

Finally, we discuss potential vorticity generation and its role using a simulation of a real case of explosive cyclogenesis. Figure 17.16 shows an analysis of surface pressure for a cyclone in the western North Atlantic that had deepened by some 4.5 kPa in the previous 36 hour and also for a model run in 1995 that had been initialized at the start of this period. The surface low in the model deepened by about 2/3 of the observed amount, making it realistic enough to form the basis for a potential vorticity analysis.

The model was run with an accumulation following the air motion of the potential vorticity changes due to the separate physical processes. Some of the results for the 95–70 kPa layer are shown in Figure 17.17. The positive lower tropospheric potential vorticity tendency due to large-scale latent heat release is again accumulated by the air moving poleward and rising ahead of the cold front, and then moving around the poleward side of the low. This is spread through the lower to middle troposphere, where the values are larger than in the lower tropospheric layer illustrated. Low-level heating associated with surface heat fluxes into the boundary layer in the cold air leads to reduced potential vorticity in the interior. Surface friction again leads to a predominantly positive potential vorticity anomaly.

The potential vorticity anomalies associated with the individual physical processes can be inverted to attribute portions of the observed structure to them. For this case, the potential vorticity associated with large-scale latent heat release makes by far the largest contribution to the surface pressure anomaly and circulation. In fact, it is even larger than that due to the surface temperature anomaly. However, it should be noted that the existence of the latent heat release depends upon the organized structure which has developed; this is largely a result of the dry dynamics in which the surface temperature anomalies play a crucial role.

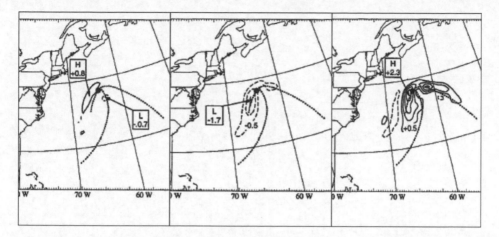

Figure 17.17 The potential vorticity changes accumulated over 36 hour and averaged for the 95–70 kPa layer for (left) latent heat release, (centre) surface heat fluxes and (right) surface momentum fluxes. The contour interval is 0.5PVU. Negative contours are dashed and the zero contour is not drawn. From Stoelinga (1996), reproduced by permission of the American Meteorological Society

The main contributions to large positive potential vorticity and associated large cyclonic circulation can come from three different origins: depression of the tropopause, large-scale latent heat release and poleward displacement of surface temperature contours. At the peak time of severe cyclonic systems, it is often found that these three forcings stack up in a 'potential vorticity tower' such that their circulations reinforce. The storm of October 1987, whose winds did great damage in southern England, was one example of a potential vorticity tower.

18
Rossby wave propagation and potential vorticity mixing

18.1 Rossby wave propagation

Rossby waves are a fundamental component of the dynamics of the synoptic and larger-scale flow in the midlatitudes. Their essential properties were discussed in Section 9.5 in the context of a highly simplified barotropic model of geophysical flows. This chapter begins by broadening the discussion to include more complicated flows, and, in particular, to include vertical as well as horizontal propagation. However, many of the results discussed earlier for the barotropic case carry over more or less directly to this baroclinic situation. We move on to discuss Rossby wave propagation when the gradients of potential vorticity are sudden rather than smooth. The chapter concludes with a comparison of the effects of individual vortices and their interactions on the large-scale mixing of potential vorticity.

We will use the quasi-geostrophic framework. Ignoring for the present the effects of friction and heating, the quasi-geostrophic potential vorticity equation is

$$\frac{D_g q}{Dt} = 0 \qquad (18.1)$$

where the potential vorticity q is related to the geostrophic stream function by the elliptic relationship

$$q = f_0 + \beta y + \nabla_H^2 \psi + \frac{1}{\rho_R} \frac{\partial}{\partial z} \left(\frac{\rho_R f_0^2}{N^2} \frac{\partial}{\partial z} (\psi) \right) \qquad (18.2)$$

The inversion of this relationship for a given distribution of q, which requires appropriate boundary conditions on ψ, yields the geostrophic stream function. As with the relationship between absolute vorticity and stream function, discussed in Chapter 9, the important feature of this relationship is that a localized change of q will affect ψ throughout the fluid.

Fluid Dynamics of the Midlatitude Atmosphere, First Edition. Brian J. Hoskins & Ian N. James.
© 2014 John Wiley & Sons, Ltd. Published 2014 by John Wiley & Sons, Ltd.

Partitioning q into its zonal mean and wavy parts,

$$q = \bar{q} + q'$$

where $\bar{q}$ denotes the zonal mean of q and q' the deviation of q from the zonal mean, leads to a linearized perturbation equation:

$$\frac{D_g q'}{Dt} + v' \frac{\partial \bar{q}}{\partial y} = 0 \tag{18.3}$$

We have assumed that the perturbation quantities are so small that their products are negligible. The perturbation potential vorticity is

$$q' = \frac{\partial^2 \psi'}{\partial x^2} + \frac{\partial^2 \psi'}{\partial y^2} + \frac{1}{\rho_R} \frac{\partial}{\partial z} \left(\frac{\rho_R f_0^2}{N^2} \frac{\partial \psi'}{\partial z} \right)$$

To simplify the discussion somewhat, we assume that the density drops off exponentially with height:

$$\rho_R = \rho_{R0} e^{-z/H}$$

This expression is exact for an isothermal atmosphere in height coordinates, but, more realistically, we can consider that we are now using the vertical coordinate $z = -H \log(p/p_0)$ as introduced in Section 6.3. We also take N^2 to be constant. If the perturbation is wave-like in the vertical as well as in the horizontal directions, an appropriate solution to Equation 18.3 is

$$\psi' = \Psi e^{z/2H} \exp\left(i\left(kx + ly + mz - \omega t \right) \right) \tag{18.4}$$

The factor $e^{z/2H}$ takes account of the variation of density with height. As the disturbance propagates in the vertical direction, from higher to lower density, its amplitude must increase if the energy per unit volume is conserved. With this form of solution substituted into Equation 18.3 and recalling that $v' = \partial \psi'/\partial x$, the perturbation potential vorticity equation leads to a relationship between wave number and frequency which is called the dispersion relationship:

$$\omega = \bar{u}k - \frac{\bar{q}_y k}{\left(k^2 + l^2 + \dfrac{f_0^2}{N^2} \left(m^2 + \dfrac{1}{4H^2} \right) \right)} \tag{18.5}$$

where $\bar{q}_y$ denotes $\partial \bar{q} / \partial y$. The dispersion relationship (18.5) is a generalization of the barotropic version given in Equation 9.18: new elements are the substitution of β by $\bar{q}_y$, and the new terms derived from the vertical variation of density in the denominator. The structure of $\bar{q}_y$ was discussed in Section 14.3: the principal result for present purposes is that midlatitude values of $\bar{q}_y$ are substantially larger than β, with the largest values in the upper troposphere and around the tropopause.

A useful re-arrangement of Equation 18.5 is

$$k^2 + l^2 + \frac{f_0^2}{N^2}\left(m^2 + \frac{1}{4H^2}\right) \equiv K_T^2 = \frac{\bar{q}_y}{\bar{u} - \omega/k} \tag{18.6}$$

Here, K_T is the effective total wave number of the Rossby wave. This expression makes it clear that the wave numbers k, l and m cannot be set arbitrarily. For example, if

$$k^2 + l^2 \geq \frac{\bar{q}_y}{\bar{u} - \omega/k} - \frac{f_0^2}{4N^2H^2},$$

then m would be imaginary and vertical propagation would not be possible. In the following discussion, we shall imagine a 'wavemaker' inserted into the fluid at some location and exciting oscillations with frequency ω. Under what circumstances can Rossby waves propagate away from the wavemaker, and what wave numbers are able to propagate? A similar discussion was given in Sections 9.6 and 9.7, but there the interest was only in horizontal propagation away from the wavemaker.

18.2   Propagation of Rossby waves into the stratosphere

A simple but suggestive calculation models the stratosphere as a uniform layer of air, with constant stratification N and constant zonal wind $\bar{u}$. The lower boundary is the tropopause and the rich dynamics of the underlying troposphere as weather systems develop, decay and propagate mean that the tropopause is continually deformed. So the tropopause acts as a wavemaker, the source of disturbances with a wide range of wave numbers and frequencies. Some of these disturbances will be confined to the vicinity of the tropopause, but when propagation is possible, Rossby waves may fill the stratosphere.

Re-arranging Equation 18.6 gives

$$m = \pm \frac{N}{f_0}\left(\frac{\bar{q}_y}{\bar{u} - \omega/k} - \frac{f_0^2}{4N^2H^2} - K^2\right)^{\frac{1}{2}} \tag{18.7}$$

Here, $K = \sqrt{k^2 + l^2}$ denotes the total horizontal wave number. For Rossby waves to propagate vertically, m must be real. This means that the first term in the square root must be positive and must exceed the sum of the remaining terms. Propagation is favoured by large $\bar{q}_y$, small but positive eastward motion relative to the basic flow, $\bar{u} - \omega/k$ and small K. Vertical propagation is inhibited by large K and prevented entirely if $\bar{u} - \omega/k < 0$. To make it easier to interpret these results, note that each of

(a)

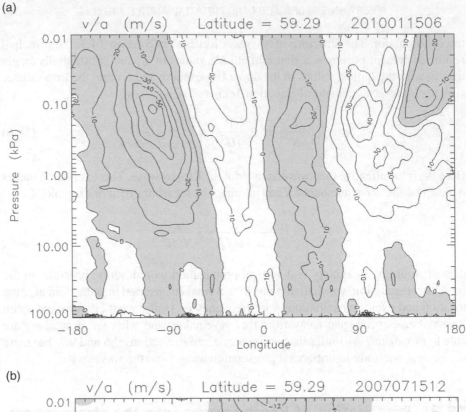

(b)

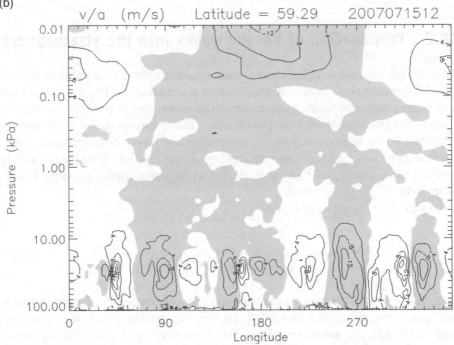

Figure 18.1 Longitude–height sections of the poleward wind in the troposphere and stratosphere at 60°N, for (a) typical winter conditions on 15 January 2010, contour interval 10 m s^{-1}, and (b) typical summer conditions on 15 July 2010, contour interval 5 m s^{-1}. Shading indicates negative values. Based on ERA-Interim data

the terms within the square root has the dimensions of m^{-2}, that is, wave number squared. For a given latitude ϕ, multiplying a wave number by $a\cos(\phi)$ gives a dimensionless wave number, namely, the number n of wavelengths which will fit around the latitude circle. Equation 18.7 can therefore be rewritten as

$$m = \pm \frac{N}{f_0 a \cos(\phi)} \left(n_q^2 - n_H^2 - n_K^2 \right)^{1/2}$$

(18.8)

$$\text{where } n_q^2 = \frac{\overline{q}_y a^2 \cos^2(\phi)}{\overline{u} - \omega/k}, n_H^2 = \frac{f_0^2 a^2 \cos^2(\phi)}{4N^2 H^2} \text{ and } n_K^2 = K^2 a^2 \cos^2(\phi).$$

Taking typical values for the mid-winter lower stratosphere, an estimate of n_q is around 4 or 5. A representative value of n_H is 1.4. It follows that long waves, with $n_K < 3$, can propagate vertically. In summer conditions, the midlatitude stratosphere is filled with easterly wind. In these conditions, no waves can propagate vertically, and so Equation 18.8 predicts that Rossby waves of all scales will be confined to the troposphere.

Figure 18.1 shows longitude–height sections of the poleward wind at 60°N for two representative days, one in the northern hemisphere summer and one in the summer. In both cases, there is vigorous activity in the troposphere, due to a succession of weather systems, with a typical zonal wave number of 5–7. However, these relatively short wavelengths are confined to the troposphere and die away rapidly above 20 kPa. The stratosphere, between 10 and 0.1 kPa, supports long waves of zonal wave number 1–2, whose amplitude increases with height, in January. In July, the stratosphere is essentially devoid of any significant waves. These diagrams provide a striking confirmation of the predictions of Equation 18.8.

18.3 Propagation through a slowly varying medium

The ideas introduced in the previous section can be extended to more realistic situations using a technique called 'ray tracing', a concept already introduced in a barotropic context in Section 9.7. We will consider the propagation of Rossby waves through an atmosphere in which $\overline{q}_y$, $\overline{u}$, etc., vary across the meridional plane.

Suppose that Rossby waves, with a variety of wave numbers and frequencies, are excited at some location within the atmosphere by a wavemaker. The nature of this wavemaker need not be specified: it might be provided by vortex shrinking and stretching as air flows over some features such as mountains, or it might be provided by the instability of the flow itself generating wave-like disturbances. The principle is that the wave front advances from the wavemaker at the group velocity of the Rossby waves. The disturbances excited by the wavemaker may be of any arbitrary structure, but they can be Fourier decomposed into a set of wave-like disturbances with various frequencies and wave numbers. The group velocity, now including a vertical component, is a vector given by

$$\mathbf{c}_g = \frac{\partial \omega}{\partial k}\mathbf{i} + \frac{\partial \omega}{\partial l}\mathbf{j} + \frac{\partial \omega}{\partial m}\mathbf{k} \tag{18.9}$$

As well as the zonal wind and potential vorticity gradient, the group velocity depends upon the frequency and wave number of the waves excited. It follows from the dispersion relationship, Equation 18.5, with a little manipulation that

$$\mathbf{c}_g = \left(\frac{\omega}{k} + \frac{2\bar{q}_y k^2}{K_T^4} \right)\mathbf{i} + \left(\frac{2\bar{q}_y kl}{K_T^4} \right)\mathbf{j} + \frac{f_0^2}{N^2}\left(\frac{2\bar{q}_y km}{K_T^4} \right)\mathbf{k} \tag{18.10}$$

This is a generalization to three dimensions of the two dimensional Rossby wave group velocity in Equation 9.26. The group velocity will depend upon the local values of $\bar{q}_y$, $\bar{u}$ and wave number. If the waves are of small amplitude and therefore linear, then the properties of the zonal flow will be fixed. In general, as the packet moves through the fluid, its wave number will evolve in response to local changes of $\bar{q}_y$, $\bar{u}$.

The critical assumption in ray tracing is that the background flow varies only slowly, that is, on a length scale which is long compared to the wavelength of the Rossby waves. In the original optical applications of ray tracing, this is often an extremely good assumption. It is a rather poor one for Rossby waves in the atmosphere, especially as the major focus is frequently upon rather long wavelength disturbances. However, the theory does give an illuminating conceptual model of how Rossby waves can propagate over large distances, and, qualitatively at least, proves surprisingly robust even when this 'slowly varying' assumption is not very good.

It remains to determine how the frequency and the wave number of a wave packet evolve as the waves propagate through the varying medium. For a dispersion relationship of the form $\omega = W(\mathbf{k}, \mathbf{r}, t)$, $\bar{u}$ or where the position and time dependence arises through variation of the basic state, such as $\bar{q}_y$, or the parameters of the problem, such as f, with position and time.

$$\frac{D_p \omega}{Dt} = -\frac{\partial W}{\partial t}, \frac{D_p k}{Dt} = \frac{\partial W}{\partial x}, \frac{D_p l}{Dt} = \frac{\partial W}{\partial y}, \frac{D_p m}{Dt} = \frac{\partial W}{\partial z} \tag{18.11}$$

where

$$\frac{D_p}{Dt} = \frac{\partial}{\partial t} + \mathbf{c}_g \cdot \nabla$$

is the rate of change following the wave packet.

Now suppose we confine our attention to a situation in which the mean flow varies neither in time nor in the zonal direction. Then in Equation 18.11, $\partial P/\partial t = \partial P/\partial x = 0$ and so

$$\frac{D_p \omega}{Dt} = \frac{D_p k}{Dt} = 0,$$

that is, both the frequency and zonal wave number of the packet are conserved. The meridional wave number l and the vertical wave number m will vary as the packet propagates. However, if the variation of one of these wave numbers is calculated by integrating Equation 18.11, the other can be diagnosed from the dispersion relationship, Equation 18.5. So extending the barotropic ray tracing argument of Section 9.7, the path of the ray can be calculated from the set of ordinary differential equations:

$$\frac{D_p X}{Dt} = c_{gx}; \quad \frac{D_p Y}{Dt} = c_{gy}; \quad \frac{D_p Z}{Dt} = c_{gz}$$

$$\frac{D_p l}{Dt} = -\frac{1}{2\left(\bar{q}_y / (\bar{u} - \omega/k) - K_T^2 + l^2\right)^{1/2}} \frac{\partial}{\partial y}\left(\frac{\bar{q}_y}{\bar{u} - \omega/k}\right) \qquad (18.12)$$

where at each point along the ray,

$$m = \pm \frac{N}{f}\left(\frac{\bar{q}_y}{\bar{u} - \omega/k} - \frac{f^2}{4N^2 H^2} - k^2 - l^2\right)^{1/2}$$

Such a generalization of the ray tracing set out in Section 9.7 is cumbersome and does not readily lend itself to simple interpretation. However, there is a graphical way of describing rays in the meridional plane which gives more insight. It is the topic of the next section.

This theory of ray tracing emphasizes the importance of the parameter

$$K_T = \left(\frac{\bar{q}_y}{(\bar{u} - \omega/k)}\right)^{1/2} \qquad (18.13)$$

This parameter has the dimensions of a wave number and is analogous to the refractive index in optical ray tracing. Rays are refracted away from smaller values of K_T and towards larger values of K_T. In places where $\bar{u} < \omega/k$, K_T becomes imaginary and here no propagation is possible. The line defined by $\bar{u} = \omega/k$ marks a critical line. Here, according to linear theory, the wave packets slow down and their scale perpendicular to the critical line collapses. This behaviour is the two-dimensional

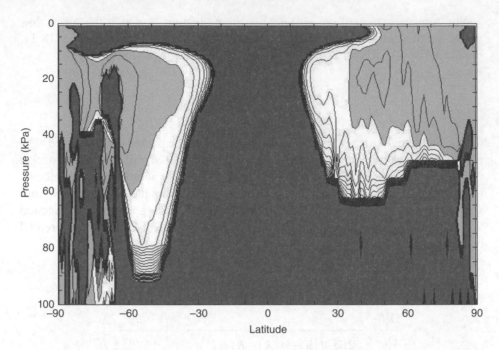

Figure 18.2   Showing the Rossby refractive index for the December–January–February season, assuming a phase speed of $14\,\mathrm{m\,s^{-1}}$. Contour interval 2 units with values less than 6 indicated by light shading; dark shading denotes an imaginary refractive index and therefore no propagation. Note that the vertical coordinate in this diagram is pressure

extension of the one-dimensional behaviour discussed for barotropic Rossby waves in Section 9.7.

Figure 18.2 shows this refractive index calculated for the mean December–January–February season. The large noisy values near the surface at high latitudes should be disregarded. They arise from difficulties in calculating the potential vorticity gradient accurately in these regions, particularly where atmospheric variables have been extrapolated beneath the ice surface of Antarctica. But in the middle and upper troposphere and in the stratosphere, the refractive index is fairly clearly defined. The smallest values are on the poleward side of the jets in middle and high latitudes near the tropopause. A critical line surrounds the jet cores. The exact location of this critical line depends upon the choice made for the phase speed ω/k. Since long Rossby waves are highly dispersive, the proper choice of ω depends upon the wave numbers of the Rossby wave. For the purposes of this discussion, a phase speed typical of the midlatitude transients, such as can be deduced from the Hovmöller plots shown in Figure 1.8, has been used. Therefore vertically propagating waves in middle latitudes will be guided in the tropopause region towards the subtropics.

Figure 18.3 shows an example of ray tracing, applied to steady zonal wave number 1 disturbances, that is, to disturbances with $\omega/k = 0$. The zonal mean flow corresponds to mean northern hemisphere winter conditions. Rays were initiated at $45°\mathrm{N}$

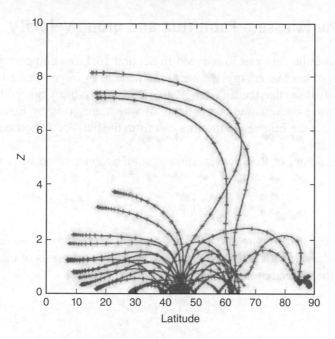

Figure 18.3   An example of a ray tracing calculation from Karoly and Hoskins (1982). Rays for zonal wave number 1, initiated at 45°N in the upper troposphere in mean December–January–February conditions, initially propagating in different directions, are shown. The crosses mark the daily position of the wave packets. From Karoly and Hoskins, 1982, with permission from the Meteorological Society of Japan

and at a height of 0.7 H. The different rays correspond to different values of m/l, that is, to different directions on the meridional plane in which the initial wave packet propagated. Many rays were confined to the troposphere and lower stratosphere, quickly turning to propagate nearly horizontally. However, some rays do pass on the polar side of the tropopause and penetrate into the upper levels of the stratosphere, turning towards the equator at heights of around 50 km. For these rays, the polar night jet acts as a waveguide, with vertical propagation near the core of the jet between 50° and 60°N. This is possible because it is so strong that K_T is a maximum in it, as is just visible near 65N in Figure 18.2.

Rossby ray tracing in two and three dimensions lacks the intuitive simplicity of the barotropic ray tracing discussed in Section 9.7. Furthermore, the rays are very sensitive to the fields of $\bar{q}_y$ and $\bar{u}$. The potential vorticity gradient depends upon the second derivative of the zonal wind, and so even small changes to the zonal wind field, well within the uncertainty of the observations, can lead to major changes in the refractive index and so to the trajectories of the rays. The zonal state used to generate Figure 18.3 was based on a polynomial approximation to the observed zonal flow and was therefore smoothed. Using unsmoothed data introduces considerable uncertainty into the rays. To some extent, these difficulties can be circumvented using the approach outlined in the next section.

18.4   The Eliassen–Palm flux and group velocity

The Eliassen–Palm flux was introduced in Section 16.4 as a compact summary of the transport of internal energy and zonal momentum by large-scale eddies. In this section, we shall see that the Eliassen–Palm flux is also closely related to the group velocity of wave packets, the components of which are given by Equation 18.10. This means that the Eliassen–Palm flux can form the basis of a short cut to Rossby wave ray tracing.

The linear theory of Rossby wave propagation assumes plane wave solutions of the form

$$\psi' = \mathrm{Re}\left(\Psi e^{z/2H} e^{i(kx+ly+mz)}\right)$$

where ψ' is the geostrophic stream function, and the factor $e^{z/2H}$ accounts for the variation of density with height. Then the horizontal components of the wind associated with the wave are related to ψ' by

$$u' = -\frac{\partial \psi'}{\partial y}, \quad v' = \frac{\partial \psi'}{\partial x}$$

The hydrostatic relationship relates ψ' to the buoyancy fluctuation associated with the waves:

$$b' = f_0 \frac{\partial \psi'}{\partial z}$$

Making use of the assumed plane wave form of the Rossby wave solution to the governing equations, these relationships relate the eddy fluxes associated with the propagating Rossby wave to the structure and scale of the wave. The poleward momentum flux is

$$\overline{u'v'} = -\frac{1}{2}|\Psi|^2 \frac{\rho_R}{\rho_{R0}} kl$$

In the same way, the poleward buoyancy flux is

$$\overline{v'b'} = \frac{f_0}{N^2}|\Psi|^2 \frac{\rho_R}{\rho_{R0}} km$$

But compare these expressions with those for the group velocity of Rossby waves, Equation 18.10. The meridional components of the group velocity can be written as

$$c_{gy}\mathbf{j} + c_{gz}\mathbf{k} = \left(\frac{4\overline{q}_y}{\rho_{R0}|\Psi|^2 K_T^4}\right)\left(-\rho_R \overline{u'v'}\,\mathbf{j} + \frac{\rho_R f_0}{N^2}\overline{v'b'}\,\mathbf{k}\right) \qquad (18.14)$$

The second bracketed term on the right-hand side is the Eliassen–Palm flux defined in Section 16.4. The Eliassen–Palm flux is therefore parallel to the local group velocity in the meridional plane. This implies that a simple way of constructing Rossby rays is to draw Eliassen–Palm lines, that is, lines which are always parallel to the local Eliassen–Palm flux vector. In practice, this is easily done by eye.

Figure 18.4 shows Eliassen–Palm sections for the December–January–February season, broken into the contributions from the transient eddies in (a) and the steady eddies in (b). These diagrams represent the accumulated effect of many

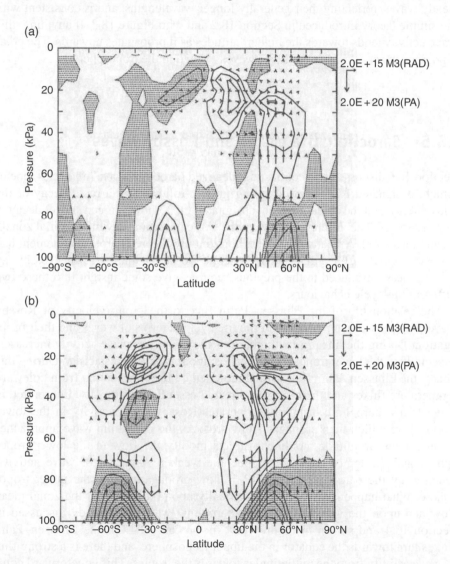

Figure 18.4 Eliassen–Palm sections for the DJF season, showing the Eliassen–Palm flux vectors and contours of its divergence, contour interval $10^{15}\,\mathrm{m}^3$, positive values are shaded. (a) Contribution from the transient eddies and (b) contribution from the steady eddies

individual events over several seasons. The flux vectors reveal upward propagation of Rossby waves in the lower troposphere, with strong divergence near the ground. In the upper troposphere, where the momentum fluxes become large, the vectors turn mainly towards the equator. There is strong convergence around the subtropical tropopause, consistent with a critical line for Rossby waves in that vicinity.

However, some of the Rossby wave activity does not turn towards the tropics but propagates on upwards into the stratosphere. This is particularly so for the steady waves, reflecting their generally longer wavelengths, and is consistent with the simple theory introduced in Section 18.2 and with Figure 18.3. If anything, this wave activity tends towards the higher latitudes as it propagates upwards. The polar night jet can act as a wave guide for waves excited in the troposphere.

18.5   Baroclinic life cycles and Rossby waves

Section 16.6 discussed two contrasting idealized baroclinic wave life cycles, one of which, designated LC1, showed strong growth followed by a rapid decay of the eddy energy, and the other of which, designated LC2, showed much slower decay in its later stages. The difference between the two calculations lay in the initial zonal–mean zonal wind. A relatively modest horizontally sheared barotropic element had been added to the LC1 initial state to generate the LC2 initial state. The Rossby ray tracing ideas introduced in the previous sections give some insight into these two different life-cycle behaviours.

The relationship of the Eliassen–Palm flux to the group velocity of Rossby waves suggests an interpretation of the life cycle in terms of waves and their propagation. During the linear growth phase of the life cycle, wave activity increases, especially in the lower troposphere, as a result of baroclinic instability. During this phase, the Eliassen–Palm fluxes are dominated by the contribution from poleward temperature fluxes and therefore are directed essentially vertically. Interpreted in wave terms, baroclinic instability generates Rossby wave activity in the lower troposphere which then propagates upwards. As the baroclinic wave approaches its maximum amplitude, nonlinear effects modify the flow in the lower troposphere and the temperature fluxes become weaker. A packet of wave activity, released by the baroclinic instability, is now propagating into the upper troposphere. What happens next depends critically upon the details of the zonal mean flow and upon the wavelength of the baroclinic wave. As already discussed in Section 16.4 and shown in Figure 16.9, in the case of LC1, the Eliassen–Palm fluxes turn towards the equator in the upper troposphere, and there is a strong flux of wave activity from the midlatitudes towards the tropics. This equatorward component marks the development of a strong poleward momentum flux, and in terms of the energetics discussed in Section 16.4, to the rapid decay of the eddy energy.

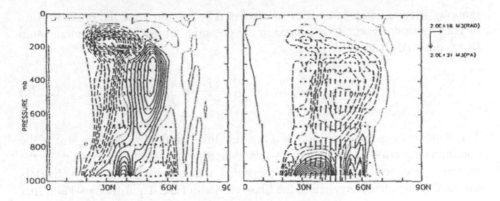

Figure 18.5 Eliassen–Palm sections for the decay phase of (a) the LC1 life cycle and (b) the LC2 life cycle, both at day 8 of the life cycle. From Thorncroft *et al.* (1993), with permission from the Royal Meteorological Society

Figure 18.5 shows the Eliassen–Palm fluxes for the LC1 and LC2 life cycles at day 8, the time of rapid decay of the LC1 wave. During the decay phase of the LC1 life cycle, the Eliassen–Palm fluxes develop a strong horizontal component, and wave activity propagates parallel to the tropopause towards the tropics. Associated with this equatorward propagation is the rapid decay of the baroclinic wave. In contrast, there is little equatorward propagation in the case of the LC2 life cycle. Rather, the wave activity is refracted towards high latitudes and upper levels, and the Eliassen–Palm fluxes develop little horizontal component. In terms of the energetics, there is much less decay of the wave activity; instead, the eddy energy decays essentially on a frictional timescale rather than the much faster dynamical timescale of the LC1 life cycle.

18.6   Variations of amplitude

Much of the thinking in this section has been based on linear ideas. A background flow, typically the zonal mean flow, determines the properties of the medium through which waves propagate. But of course, the propagating waves modify the zonal mean flow itself. The problem is nonlinear, with feedbacks between eddies and zonal flow. These feedbacks are summarized by the equation for the evolution of the zonal mean potential vorticity. As discussed in Section 16.4, at the level of quasi-geostrophic theory, this is

$$\frac{\partial \overline{q}}{\partial t} + \frac{\partial}{\partial y}\left(\overline{v'q'}\right) = 0, \tag{18.15}$$

ignoring friction and heating. But, as discussed in Section 16.4, the poleward flux of eddy potential vorticity is related to the potential temperature and momentum fluxes due to the waves:

$$\overline{v'q'} = \frac{1}{\rho_R}\left\{\frac{\partial}{\partial y}\left(-\rho_R\overline{u'v'}\right) + \frac{\partial}{\partial z}\left(\frac{\rho_R f_0}{N^2}\overline{v'b'}\right)\right\} = \frac{1}{\rho_R}\nabla\cdot\mathbf{F} \qquad (18.16)$$

This follows from the definition of q in a few lines of algebra, making use of the continuity equation in the form $u_x = -v_y$ and the thermal wind relationship. So the feedback between the eddies and the zonal mean potential vorticity field is summed up by the divergence of the Eliassen–Palm flux. The Eliassen–Palm flux takes into account the effects both of momentum and temperature transports by the eddies.

But there is more. Take the linearized eddy potential vorticity equation, Equation 18.3, multiply by $\rho_R q'/\overline{q}_y$, take the zonal mean and find

$$\frac{\partial}{\partial t}\left(\frac{\rho_R\overline{q'^2}}{\overline{q}_y}\right) + \rho_R\nabla\cdot\mathbf{F} = 0 \qquad (18.17)$$

The quantity

$$A = \frac{\rho_R\overline{q'^2}}{\overline{q}_y}$$

is called the wave action density and is arguably the most fundamental measure of the vigour of Rossby waves. Equation 18.17 is a conservation law for A; the Eliassen–Palm flux, which is parallel to the Rossby wave group velocity, is the flux of wave action density. Wave action density propagates with the Rossby group velocity, which has been shown earlier in parallel to $\mathbf{F}$. Since A is conserved, $\overline{q'^2}$ must vary as the wave packet propagates into regions of different density or potential vorticity gradient.

One interpretation of the wave action density A is in terms of the horizontal displacement of fluid elements. For small amplitude waves, the perturbation potential vorticity can be written as

$$q' = \overline{q}_y y' \qquad (18.18)$$

where y' is the meridional displacement of the fluid element from its initial latitude. Using this relationship, the wave action density can be re-written in the following form:

$$A = \rho_R\overline{q}_y\overline{y'^2} \qquad (18.19)$$

The wave action density is therefore proportional to the mean square parcel dispersion for a given density and potential vorticity gradient. Now suppose a wave packet propagates from a low level to a higher level where the density is smaller. In order to conserve A, the parcel dispersion must increase. Similarly, if the packet propagates into a region of small potential vorticity gradient, the parcel dispersion must increase to compensate for the decrease in potential vorticity gradient. There is an analogy with the behaviour of surface water waves as they propagate into regions of shallow water. As the depth of water decreases, the surface displacements associated with the waves increase. Eventually the waves break, and linear descriptions of wave motion break down. In the same way, the horizontal displacements associated with Rossby waves which propagate into regions of low density or small potential vorticity gradient will increase until linear theory breaks down.

The reasoning leading to Equation 18.19 was based on a linear, small amplitude description of Rossby waves. Equation 18.18 assumed that the potential vorticity gradient was unaffected by the passage of the Rossby wave and that parcels of fluid in the Rossby wave felt the same potential vorticity gradient regardless of the magnitude of their displacement. Such assumptions inevitably break down as the displacements increase. Certainly, this picture of the wave is completely inappropriate when the parcel displacements become comparable with the wavelength of the Rossby waves, that is, when

$$\overline{y'^2}^{1/2} \sim k^{-1} \tag{18.20}$$

This is equivalent to saying that the line of fluid elements aligned along a latitude circle in the undisturbed state now makes significantly large angles with the latitude circle. This is the point at which breaking started to occur in the baroclinic wave life cycle LC1 described in Section 16.3. Such large distortions of lines of fluid elements are also a necessary precursor for the development of blocking which was discussed in Section 17.5.

18.7   Rossby waves and potential vorticity steps

The preceding development was based on the slowly varying assumption that the background state varies on a spatial scale which is larger compared with the wavelength of the Rossby waves. This is a questionable assumption. In particular, potential vorticity and its gradient vary sharply on small scales. In this section, we consider an alternative Rossby paradigm, in which waves propagate on a discontinuity in potential vorticity rather than on a smooth gradient of potential vorticity.

Such a configuration is ubiquitous. In Chapter 17, potential vorticity maps in the 'middle world', the set of isentropes which intersect the tropopause, are characterized by a sharp jump in the potential vorticity by a factor of around four where the

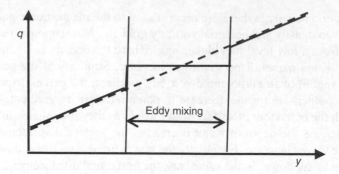

Figure 18.6 Schematic illustration of the effect of eddy mixing on the distribution of potential vorticity. The initial monotonically increasing distribution is shown by the dashed line, and the step-like distribution after eddy mixing by the solid line

θ-surfaces cross the tropopause. At higher levels, in the winter stratosphere, the potential vorticity increases abruptly in the vicinity of the polar night jet, with much higher values of potential vorticity in the polar vortex than in the middle latitudes. In the lower troposphere, potential vorticity is generally small through the subtropics and increases markedly in the region of the subtropical jet.

The key to the formation of such potential vorticity steps is the conservation of potential vorticity on synoptic timescales. Figure 18.6 illustrates the configuration envisaged. Imagine a restricted region filled with vigorous eddies. As they circulate, these eddies transport potential vorticity both polewards and equatorwards. In the zonal mean, the eddies will mix potential vorticity and weaken the potential vorticity gradients. But at the edge of the eddy mixing region, the distribution of potential vorticity will jump more or less abruptly to its unmixed value. In this way, steps of potential vorticity will form on either side of the eddy region. Figure 17.8 shows this process happening in baroclinic wave life cycles. In reality, heating and friction may act to offset the eddy mixing. But the timescales of these non-conservation processes will often be rather long compared with circulation timescale of the eddies, and so they will only have a minor effect on the potential vorticity distribution.

As an example, we will consider barotropic flow in a channel width Δy on a β-plane. Suppose the initial flow is a uniform zonal flow U_0. Mixing concentrates the gradient of potential vorticity to a step in the centre of the channel, the magnitude of which is $\beta \Delta y$. Away from the step, the potential vorticity gradient is zero, so that

$$\bar{q}_y = \beta - \frac{\partial^2 U}{\partial y^2} = 0$$

Integrate this to determine the zonal flow profile in the channel. The constants of integration are determined from the boundary conditions

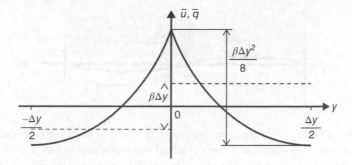

Figure 18.7   A potential vorticity step in a barotropic channel flow (dashed line) together with the corresponding zonal wind profile (solid line)

$$\frac{\partial U}{\partial y}=0 \text{ at } y=\pm\frac{\Delta y}{2}$$

and from conservation of zonal momentum, so that

$$\frac{1}{\Delta y}\int_{-\Delta y/2}^{\Delta y/2}U(y)dy=U_0$$

Then

$$U(y)=U_0+\frac{\beta}{2}y^2+ay+b \quad \text{for } y>0 \tag{18.21}$$

and

$$U(y)=U_0+\frac{\beta}{2}y^2+cy+d \quad \text{for } y<0 \tag{18.22}$$

The first two terms ensure that $\bar{q}_y=0$. To satisfy the condition on $\partial U/\partial y$ at $y=\pm\Delta y/2$ -a=c=$\beta\Delta y/2$, and to make U continuous at y=0 and the average flow U_0, b=d=$\beta\Delta y^2/12$. At y=0, $\partial U/\partial y$ is discontinuous, changing from $\beta\Delta y/2$ for negative y to $-\beta\Delta y/2$ for positive y. It follows that there is a δ-function in $\bar{q}_y$ at y=0, with $\int_{0-}^{0+}\bar{q}_ydy=\beta\Delta y$. The resulting zonal wind profile, shown in Figure 18.7, consists of a sharp jet, speed $U_J=U_0+\beta\Delta y^2/12$ at the potential vorticity step, falling away as a quadratic with y to $U_0-\beta\Delta y^2/24$ at the edges of the channel, a speed difference across the channel of $\beta\Delta y^2/8$.

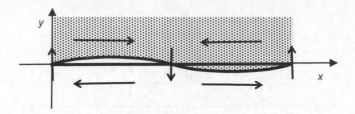

Figure 18.8   A Rossby wave on a potential vorticity step. The boundary between a region of large potential vorticity (shaded) and lower potential vorticity (clear) has a wavy distortion. The arrows indicate the circulation associated with the potential vorticity anomalies

For typical terrestrial midlatitude values, a zonal speed difference between the jet and the edges of the channel would be around 20–40 m s^{-1}. If the pole-to-equator potential vorticity difference took the form of a number of such steps, the corresponding zonal wind profile would be a series of sharp jets separated by broad minima of zonal wind. This is indeed very reminiscent of the patterns of easterly and westerly jets observed by tracking cloud motions in the atmospheres of the giant planets Jupiter and Saturn.

Just as a smooth variation of zonal mean potential vorticity supports Rossby waves, so does a step in potential vorticity. However, the properties of Rossby waves on a step differ somewhat from those of Rossby waves on a β-plane. Qualitatively, the basic mechanism is the same as for a Rossby wave on a uniform gradient of potential vorticity, as discussed in Section 9.5. Figure 18.8 shows an interface between a region of large uniform potential vorticity, shown shaded, and a region of smaller potential vorticity, shown unshaded. If the interface is distorted in a wave-like pattern, then there will be a series of positive and negative anomalies of potential vorticity along the interface. The arrows indicate the circulation associated with these anomalies. The perturbation meridional wind is in quadrature with the distorted interface, with the result that the whole pattern will migrate along the negative x-axis, relative to any mean zonal flow along the interface.

Let us put this analysis on a more quantitative basis. The starting point is the linearized perturbation potential vorticity equation

$$\frac{\partial q'}{\partial t} + U \frac{\partial q'}{\partial x} + \frac{\partial \bar{q}}{\partial y}\frac{\partial \psi'}{\partial x} = 0$$

Away from the line $y=0$, this equation is satisfied trivially, since both q' and $\partial \bar{q} / \partial y$ are zero in these regions. It follows that away from $y=0$, the stream function perturbation satisfies

$$\frac{\partial^2 \psi'}{\partial x^2} + \frac{\partial^2 \psi'}{\partial y^2} = 0$$

Seeking wave-like solutions in x and time of the form

$$\psi' = \Psi(y)e^{i(kx-\omega t)}$$

it follows that

$$\frac{\partial^2 \Psi}{\partial y^2} - k^2 \Psi = 0$$

and so, since $\psi' \to 0$ as $y \to \pm\infty$, the perturbation stream function has the form

$$\psi' = Ae^{-ky}e^{i(kx-\omega t)} \quad \text{for } y > 0$$
$$\psi' = Ae^{ky}e^{i(kx-\omega t)} \quad \text{for } y < 0$$

(18.23)

Notice that both ψ' and $\partial\psi'/\partial x$ are continuous across the step at $y = 0$, but that $\partial\psi'/\partial y$ is discontinuous.

We now turn our attention to the step itself. Here, q' is $\pm \Delta q/2$ within a tiny distance of the $y=0$ line, and this is not small. But the integral of q' across the step is small; it follows from Equation 18.23 that

$$\int_{0-}^{0+} q'dy = \left.\frac{\partial\psi'}{\partial y}\right|_{0-}^{0+} = -2kAe^{i(kx-\omega t)}$$

Here, $y=0-$ or $0+$ denotes an infinitesimal distance on either side of the step. Integrate the potential vorticity equation across the step to give

$$\frac{\partial}{\partial t}\left(\int_{0-}^{0+} q'dy\right) + U_J \frac{\partial}{\partial x}\left(\int_{0-}^{0+} q'dy\right) + \frac{\partial\psi'}{\partial x}\left(\bar{q}\big|_{0-}^{0+}\right) = 0$$

Here, U_J denotes the mean zonal wind along the step, that is, at the jet maximum. In terms of the stream function, this equation is

$$i(U_J k - \omega)(-2k)Ae^{i(kx-\omega t)} + ik(\beta\Delta y)Ae^{i(kx-\omega t)} = 0$$

The dispersion relationship is therefore

$$\omega = U_J k - \frac{\beta\Delta y}{2}$$

(18.24)

or, in terms of the phase speed,

$$c = \frac{\omega}{k} = U_J - \frac{\beta \Delta y}{2k}$$

As in the case of Rossby waves on a β-plane, the waves propagate from east to west relative to the zonal flow along the step. The long waves propagate westwards most rapidly, while the phase speed of short waves differs only slightly from the zonal flow speed. For a given zonal wave number, $c - U_J$ is proportional to the strength of the potential vorticity step. The details of U away from the step play no part in the dispersion relation.

The two components of the group velocity are

$$c_{gx} = \frac{\partial \omega}{\partial k} = U_J; \quad c_{gy} = \frac{\partial \omega}{\partial l} = 0$$

Wave packets are confined to the potential vorticity step, where they move at the same speed as the jet. There is no meridional propagation. Since the group velocity does not depend upon the wave number k, a packet of any arbitrary shape will preserve its identity as it propagates, even though the individual troughs and crests will pass along the packet. Figure 18.9 illustrates this behaviour in a simple example.

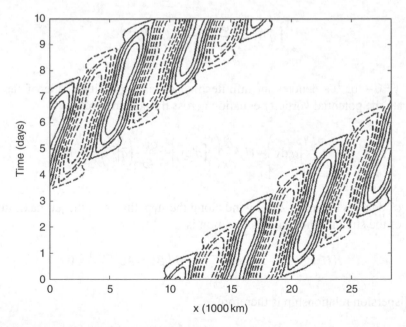

Figure 18.9   A longitude–time plot, showing a packet of Rossby waves on a potential vorticity step. The packet moves from west to east at speed U_J. The waves making up the packet move considerably more slowly.

This analysis reveals that a potential vorticity step supports low-frequency Rossby wave motions, but that the dynamically active part of the flow is confined to the vicinity of the step. Away from the step, where the potential vorticity gradient is zero, the disturbances simply decay with distance from the step. The step itself is elastic or resilient; when disturbed in some way, packets of Rossby waves will propagate along the step at the jet speed. The greater the potential vorticity step Δq, the faster the jet speed and so the packets will move more quickly. Packets of Rossby waves can be excited if, for example, a vortex impinges upon the potential vorticity step. A weak vortex will simply excite Rossby waves which will propagate and eventually dissipate. A sufficiently strong vortex may disrupt the step and so succeed in crossing it.

The problem of Rossby wave propagation on a potential vorticity step has implications for a more general insight into the large-scale dynamics of the atmosphere. Most of the discussion in earlier chapters has assumed a smooth variation of potential vorticity, on a scale larger than that of the waves or vortices within the flow. Yet observations and modelling studies all suggest that the field of potential vorticity is highly structured down to very small scales. Extended streamers of potential vorticity develop as eddies are strained. Sharp gradients of potential vorticity separate regions of well mixed, relatively uniform potential vorticity. The single potential vorticity step introduced in this section is an idealized example of such a configuration. A more realistic model would consider a succession of steps.

If potential vorticity steps are widely separated, with the distance between them large compared with k^{-1}, then waves on each step will be independent of each other. The analysis of this section is relevant to Rossby wave propagation on each step separately. But if the spacing is small compared with k^{-1}, Equation 18.23 shows that a disturbance on one step will still have a significant amplitude at the adjacent step. A disturbance on one step will excite disturbances on neighbouring steps. In the limit of a series of small closely spaced potential vorticity steps, we recover the problem of waves on a smoothly varying distribution of potential vorticity.

18.8   Potential vorticity steps and the Rhines scale

In the previous section, we have introduced a distinction between situations in which the potential vorticity is distributed smoothly and situations in which sharp changes of potential vorticity separate regions of well-mixed, relatively uniform potential vorticity. What factors determine which of these configurations is likely to be observed? When potential vorticity steps do develop, what factors determine their spacing and magnitude? These are the questions which this section sets out to address. For simplicity, our discussion will be in terms of the simple barotropic configuration of the previous section. The results are readily generalized to more realistic configurations.

Two different timescales characterize the development of the flow. First, there is the overturning timescale for individual vorticity anomalies. Denoting the typical eddy velocity as U_E and the horizontal scale of the eddy as L, the eddy overturning timescale is

$$T_E \sim \frac{L}{U_E},$$

which is simply a restatement of the kinematic result of Section 2.3 which relates the relative vorticity to the spin of fluid elements. The second important timescale is the period of Rossby waves. An isolated vorticity anomaly can be regarded as a packet of Rossby waves. It will propagate and disperse on a timescale which is related to the Rossby wave period. The frequency of Rossby waves in a frame of reference moving with the background flow $\bar{u}$ is

$$\omega - \bar{u}k = -\frac{\bar{q}_y k}{K^2} \qquad (18.25)$$

Here, K is the total horizontal wave number; for the purposes of this discussion, the typical horizontal scale of the eddy is $L \sim K^{-1}$. The typical Rossby dispersion timescale for the anomaly is therefore

$$T_\beta \sim \frac{K}{\bar{q}_y} \sim \frac{1}{\bar{q}_y L} \qquad (18.26)$$

If this Rossby timescale is short compared with the overturning timescale, that is, if $T_\beta \ll T_E$, then the anomaly will have dispersed before it has time to overturn and mix potential vorticity. In this case, potential vorticity mixing is suppressed. In the opposite case, when $T_\beta \gg T_E$, potential vorticity will be mixed before Rossby wave propagation can take place. The condition for effective mixing can be written in terms of the intensity of the eddies; mixing is likely if

$$U_E \gg \bar{q}_y L^2 \qquad (18.27)$$

One interpretation of this criterion is in terms of the nonlinearity of the flow: when U_E is large enough to satisfy Equation 18.27, the eddies advect potential vorticity and the linear assumptions made in deriving the Rossby dispersion relation (18.25) break down.

One example of the condition for potential vorticity mixing, Equation 18.27, comes from the vertical propagation of long Rossby waves from the tropopause into the stratosphere, discussed in Section 18.1. Even if the potential vorticity gradients are constant, the amplitude of the wave increases with height as the density decreases, a variation built into the form of the linear solution given by Equation 18.4. At lower levels, $U_E < \bar{q}_y L^2$, so that Rossby wave propagate upwards in accord with the linear theory. But at some level, as the amplitude of the wave increases, $U_E > q_y L^2$. As the eddy overturning time becomes short compared with the Rossby wave period, the waves break and mix potential vorticity. There are of course important nonlinear feedback mechanisms at work here. As the eddies mix potential vorticity and so change the distribution of potential vorticity, the propagation of subsequent wave packets will be altered.

The principle summarized in Equation 18.27 can also be expressed in terms of the length scale of the eddies. Re-write the condition for mixing, Equation 18.27, in the form

$$L^2 \ll L_\beta^2 \quad \text{where } L_\beta = \left(\frac{U_E}{\bar{q}_y} \right)^{1/2} \tag{18.28}$$

The length scale L_β is called the 'Rhines scale', and it plays an important role in determining the character of a geophysical flow. For the Earth's midlatitude troposphere, where a typical value of $\bar{q}_y$ is $1.5 \times 10^{-11}\, \text{m}^{-1}\, \text{s}^{-1}$, and U_E can be taken as around $15\, \text{m s}^{-1}$, the Rhines scale is around $10^6\, \text{m}$, comparable to the synoptic scale. So for a typical terrestrial weather system, both vortex dynamics and Rossby wave propagation will be important.

In Section 11.3, we showed that for two dimensional turbulent flows, energy cascades upscale, that is, from large to small wavenumbers, and from high frequencies to lower frequencies. The arguments in that section were based on homogeneous, isotropic flow on an f-plane. As the scale of eddies increases, the effects of potential vorticity gradients become more important. When Rossby wave dispersion becomes faster than eddy overturning, the cascade of energy to larger scales is blocked. So, the Rhines scale gives an effective maximum scale for the eddies. However, the dispersion relation, Equation 18.25, indicates that the frequency of Rossby waves in a frame of reference moving with the local zonal flow depends upon the angle α between the wave crests and the zonal direction:

$$\omega - \bar{u}k = -\frac{\bar{q}_y \cos(\alpha)}{K}$$

The energy cascade can still proceed to lower frequencies if α tends towards $90°$, that is, if the zonal wave number k tends to 0 while the total wave number $(k^2 + l^2)^{1/2} = k_\beta$, where $k_\beta = L_\beta^{-1}$ is the Rhines wave number. Figure 18.10 is a schematic illustration of this flow of kinetic energy. Eddy energy is generated at some small scale: in a baroclinic atmosphere, this might be comparable to the Rossby deformation radius. It cascades to smaller wave numbers until the total wave number is comparable with k_β, at which point the cascade continues only in the zonal wave number. The end state is one of parallel zonal jets, with a typical width of π/k_β.

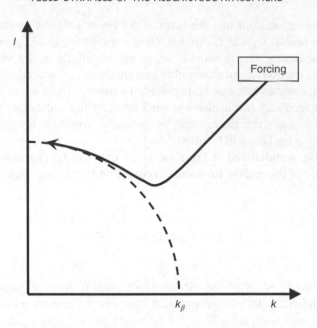

Figure 18.10 Schematic illustration of the energy cascade for two-dimensional turbulence in the presence of a β-effect.

Figure 18.11 shows a calculation from Williams (1978) which illustrates this cascade and its end state of jets, separating regions of more or less uniform potential vorticity. Small-scale forcing is applied to a spherical layer of initially motionless barotropic fluid: this forcing might represent eddies developing by baroclinic instability, or, in a low Richardson number atmosphere, from convection. Figure 18.11a shows the effect of this forcing a short time after the start of the integration. Figure 18.11b shows the eventual development of the flow, with several parallel jets separating mixed regions. The jet separation is comparable to the Rhines scale.

For terrestrial parameters, the Rhines scale implies a jet width somewhat less than the distance from the pole to the subtropics. In such case, the end state of the turbulence could be a single or even a double jet. Two jets are in fact found in the zonal average for the Southern Hemisphere winter. They are also found at some longitudes in the northern hemisphere winter, but the zonal average is a single jet. In contrast, the gas giant planets such as Jupiter and Saturn are much larger so that the pole–equator distance is several times that of the Rhines scale. Tracking cloud features reveals that at the level of the cloud tops, there are indeed a number of alternating easterly and westerly jets. Figure 18.11 illustrates these jets in the atmosphere of Jupiter.

Figure 18.11b certainly bears a strong resemblance to the banded structure observed on Jupiter and other gas giant planets. Winds at the cloud top levels in Jupiter's atmosphere have been mapped by tracking identifiable features, using

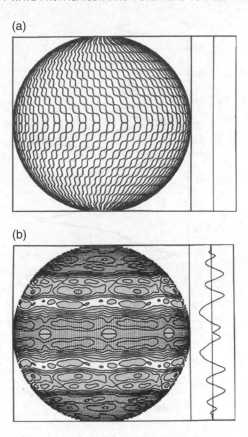

Figure 18.11 Results from a barotropic numerical model of two-dimensional turbulence on a sphere. (a) Day 4.6, showing weak eddies at the forcing scale, and (b) day 294.4, late in the integration, showing the development of multiple jets. From Williams (1978), with permission from the American Meteorological Society

high-resolution images from space probes and the Hubble space telescope. Figure 18.12 shows wind profiles derived in this way. They display multiple sharp westerly jets, separated by broader regions of weaker flow. These too are features of the zonal flow for an idealized potential vorticity step, Equations 18.21 and 18.22. It seems that in Jupiter and other gas giant planets, the flow at the cloud top levels consists of a 'potential vorticity staircase', with a number of sharp changes of potential vorticity separating regions of relatively well-mixed potential vorticity. Embedded within these bands are smaller-scale circulating weather systems which have variously been interpreted as baroclinic eddies, or convectively driven systems analogous to terrestrial hurricanes.

Let us conclude this chapter, and this book, by summarizing a conceptual model of midlatitude atmospheric dynamics which repeatedly arises out of much of the material we have covered. That conceptual model is founded on potential vorticity and its two fundamental properties, conservation and invertibility. First, potential

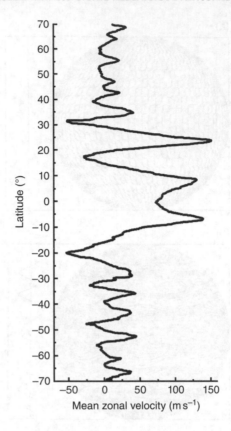

Figure 18.12 Zonal winds on Jupiter, deduced by tracking cloud features observed from the Hubble Space telescope. From Marcus and Shetty (2011), with permission from the Royal Society

vorticity is a conserved variable which is advected by the flow. Patches of potential vorticity or 'eddies' remain identifiable features for synoptic timescales. Invertibility refers to the elliptic relationship between potential vorticity and the stream function of the flow. The most important consequence of such an elliptic relationship is that even isolated eddies modify the flow at distances large compared to the size of the eddy itself. As a result, eddies interact with each other, they are advected and usually deformed by the flow induced by distant eddies. A far reaching consequence of invertibility is that in all except the very simplest situations, lines of constant potential vorticity and lines of constant stream function are not parallel. In other words, the flow is unsteady and continually evolves. Furthermore, its dynamics is inherently nonlinear: the distribution of potential vorticity determines the flow field which in turn advects and deforms the potential vorticity.

On the large scale, a paradigm of low-frequency motion in the midlatitude atmosphere is the Rossby wave, a wave which can exist whenever there are large-scale gradients of potential vorticity. The properties of Rossby waves are intrinsically linked to the conservation and invertibility properties of potential vorticity. Processes such as heating, cooling and friction act to set up gradients of potential vorticity, while, left to itself, the flow dynamics ultimately tend to mix potential vorticity, generating regions of more uniform potential vorticity, where the dynamics of individual eddies are dominant, separated by regions of strong potential vorticity gradient. Here, Rossby wave propagation is dominant. When the gradients of potential vorticity are sufficiently strong, such zones behave as elastic barriers, soaking away the wave activity associated with weaker disturbances by propagating it to remote parts of the fluid. Such potential vorticity barriers are impervious to all but the strongest disturbances impinging upon them. The ideas discussed in this section suggest how potential vorticity mixing and step formation are related and what factors govern their size and strength.

The theoretical basis of potential vorticity has been well known since the pioneering work of Rossby and Ertel. Only much more recently, with the increasing sophistication of synoptic observing and analysis schemes, and the complementary development of high-resolution numerical models of the atmosphere, has it become possible to relate this theory to the development and even forecasting of observed weather systems. Now, any understanding of midlatitude atmospheric dynamics has to be grounded in potential vorticity and its properties.

Appendix A: Notation

This list is not exhaustive, but is intended to summarize notation which is used in several different chapters. Generally, symbols which are only used in a few lines in a particular argument will not be included in this list. Inevitably, a few symbols have been used to represent different quantities in different places; generally, the context will make clear which meaning is intended.

Variable	Meaning and Units	Equation
a	Radius of planet, m	
A	Available potential energy per unit volume, $J\,m^{-3}$	7.41
b	Buoyancy, $m\,s^{-2}$	7.11
B	Bernoulli potential, $m^2\,s^{-2}$	2.45
Be	Dimensionless β, $\beta L^2/U$	9.7
Bu	Burger number	5.39
c	Phase speed, $m\,s^{-1}$	9.19
c_p	Specific heat of air at constant pressure, $J\,kg^{-1}\,K^{-1}$	
c_v	Specific heat of air at constant volume, $J\,kg^{-1}\,K^{-1}$	
$\mathbf{c}_g$	Vector group velocity	9.25
C	Circulation, $m^2\,s^{-1}$	8.2
D	$=\partial u/\partial x+\partial v/\partial y$, horizontal divergence, s^{-1}	
	Typical vertical scale of weather system, m	
e	Vapour pressure of water, Pa	
e_s	Saturated vapour pressure of water, Pa	
f	Coriolis parameter, $2\Omega\sin(\phi)\,s^{-1}$	4.37
$\mathbf{F}$	Eliassen–Palm flux, $kg\,m^{-1}\,s^{-2}$	16.36
Fr	Froude number	5.43
g	Magnitude of the acceleration due to gravity, $m\,s^{-2}$	
g_e	Magnitude of the effective acceleration due to gravity, a function of position, $m\,s^{-2}$	

(*Continued*)

Fluid Dynamics of the Midlatitude Atmosphere, First Edition. Brian J. Hoskins & Ian N. James.
© 2014 John Wiley & Sons, Ltd. Published 2014 by John Wiley & Sons, Ltd.

Variable	Meaning and Units	Equation
$\mathbf{g}$	Vector acceleration due to gravity, $\mathrm{m\,s^{-2}}$	
$\mathbf{g}_\mathrm{e}$	Effective acceleration due to gravity at the Earth's surface	
H or H_p	Pressure scale height, m	7.3
Hi	'Hide number' $h_{max}/\mathrm{Ro}D$	9.7
$\mathbf{k}$	Wavenumber vector $k\mathbf{i}+l\mathbf{j}+m\mathbf{k}$, $\mathrm{m^{-1}}$	
K	Total wavenumber, $|\mathbf{k}|$, $\mathrm{m^{-1}}$	7.36
	Kinetic energy per unit volume, $\mathrm{J\,m^{-3}}$	
L	Latent heat of condensation of water vapour, $\mathrm{J\,kg^{-1}}$	
	Typical horizontal scale of a weather system, m	
L_R	Rossby radius of deformation, NH/f, m	
M	Montgomery potential, $\mathrm{m^2\,s^{-2}}$	6.37
N	Brunt–Väisälä frequency $(g\partial(\ln(\theta))/\partial z)^{1/2}$, $\mathrm{s^{-1}}$	
p	Pressure, Pa	
$p_R(z)$	Pressure profile of reference atmosphere, Pa	
p_0	Standard pressure (often 100 kPa)	
P	Rossby–Ertel potential vorticity, $\mathrm{K\,m^2\,kg^{-1}\,s^{-1}}$	10.4
q	Heat per unit mass entering or leaving a fluid element, $\mathrm{J\,kg^{-1}}$	2.19
	Specific humidity, dimensionless	2.54
	Quasi-geostrophic potential vorticity, $\mathrm{s^{-1}}$	12.19
Q	Arbitrary scalar variable (Chapter 2)	
$\mathbf{Q}$	'Q-vector' determining vertical circulation, $\mathrm{s^{-3}}$	13.27
r	Distance to origin in polar coordinates, m	6.5
	Pseudo-density, various units	
R	Gas constant for dry air, $\mathrm{J\,kg^{-1}\,K^{-1}}$	
	Radius of curvature of parcel trajectory, m	
$\mathbf{r}$	$=x\mathbf{i}+y\mathbf{j}+z\mathbf{k}$; position vector	
Re	Reynolds number	11.1
Ri	Richardson number	11.19
Ro	Rossby number	5.13
s	Specific entropy, $\mathrm{J\,K^{-1}\,kg^{-1}}$	
	Distance along a fluid trajectory, m	
	Perpendicular distance to rotation axis, $r\cos(\phi)$, m	
S	Source term in ω-equation, $\mathrm{m^{-1}\,s^{-1}}$	13.4
t	Time, s	
T	Temperature, K	
$T_R(z)$	Temperature profile of reference atmosphere	
u	Zonal component of velocity, $\mathrm{m\,s^{-1}}$	
U	Typical horizontal velocity fluctuation	
$\mathbf{u}$	$=u\mathbf{i}+v\mathbf{j}+w\mathbf{k}$; velocity vector	
v	Meridional component of velocity, $\mathrm{m\,s^{-1}}$	
$\mathbf{v}$	$=u\mathbf{i}+v\mathbf{j}$; horizontal component of velocity vector, $\mathrm{m\,s^{-1}}$	
$\mathbf{v}_\mathrm{a}$	Ageostrophic wind, $\mathrm{m\,s^{-1}}$	
$\mathbf{v}_\mathrm{g}$	Geostrophic wind, $\mathrm{m\,s^{-1}}$	5.24 and 12.1
$\dot{\mathbf{u}}, \dot{\mathbf{v}}$	Acceleration due to friction, $\mathrm{m\,s^{-2}}$	
w	Vertical component of velocity, $\mathrm{m\,s^{-1}}$	
x	Zonal co-ordinate, m	
(X, Y, Z, T)	Semi-geostrophic coordinates	16.5

Variable	Meaning and Units	Equation
y	Meridional coordinate, m	
z	Vertical coordinate, m	
β	Poleward gradient of Coriolis parameter, $m^{-1}\,s^{-1}$	4.39
Γ	Lapse rate, $K\,m^{-1}$	
Γ_a	Adiabatic lapse rate, $K\,m^{-1}$	
χ	Velocity potential, $m^2\,s^{-1}$	2.7
ϕ	Latitude, rad	
Φ	Gravitational potential, $m^2\,s^{-2}$	
κ	R/c_p, dimensionless	
Λ	Vertical shear $\partial u/\partial z$, s^{-1}	14.29
μ	Dynamical coefficient of viscosity, $kg\,m^{-1}\,s^{-1}$	2.1
ν	Kinematic coefficient of viscosity, $m^2\,s^{-1}$	2.41
θ	Potential temperature, K	1.3
$\theta_R(z)$	Potential temperature profile of reference atmosphere	
ξ	Vertical component of relative vorticity, s^{-1}	
ξ_g	Geostrophic vorticity, s^{-1}	
ζ	Vertical component of absolute vorticity, s^{-1}	
ζ_g	Geostrophic absolute vorticity, s^{-1}	
$\boldsymbol{\zeta}$	Vector absolute vorticity	
π	Ratio of circumference to diameter of a circle	6.36
	Exner function	
ρ	Density, $kg\,m^{-3}$	
ρ_0	Constant reference density, $kg\,m^{-3}$	
$\rho_R(z)$	Density profile of reference atmosphere	
σ	Growth rate of unstable wave, s^{-1}	
τ_D	Drag timescale, s	
τ_E	Radiative equilibrium timescale, s	
ψ	Streamfunction, $m^2\,s^{-1}$	2.10
ψ_g	Geostrophic streamfunction	
Ψ	Amplitude of a stream function wave	14.10
ω	Pressure vertical velocity, $Pa\,s^{-1}$	
	Frequency of wave, s^{-1}	
Ω	Rotation rate of planet, $rad\,s^{-1}$	
$\boldsymbol{\Omega}$	Vector rotation rate of planet	

Appendix B: Revision of vectors and vector calculus

B.1 Vectors and their algebra

Many physical quantities have a direction as well as a magnitude. Examples include position, velocity, force, magnetic field, angular momentum and so on. A vector can be described in terms of its components in three orthogonal directions, that is, in terms of its projection onto three orthogonal axes. We write this as

$$\mathbf{a} = \left(a_1, a_2, a_3\right) = a_1\mathbf{i} + a_2\mathbf{j} + a_3\mathbf{k} \tag{B.1}$$

Here, $\mathbf{i}, \mathbf{j}, \mathbf{k}$ are 'unit vectors' parallel to each coordinate axis. A unit vector is a vector with a direction and a magnitude of one dimensionless unit.

The magnitude (or 'modulus') of a vector is

$$|\mathbf{a}| = \sqrt{a_1^2 + a_2^2 + a_3^2} \tag{B.2}$$

Note that while $\mathbf{a}$ and $|\mathbf{a}|$ are independent of the coordinate system used, the individual components depend upon the arbitrary choice of a coordinate system (Figure B.1). The direction of $\mathbf{a}$ is given by the unit vector $\mathbf{l}$ where

$$\mathbf{l} = \frac{a_1}{|\mathbf{a}|}\mathbf{i} + \frac{a_2}{|\mathbf{a}|}\mathbf{j} + \frac{a_3}{|\mathbf{a}|}\mathbf{k} \tag{B.3}$$

Vectors have an algebra which is similar to that of ordinary numbers. For example, a vector can be multiplied by a scalar to give

$$\mathbf{b} = \alpha\mathbf{a} = \alpha a_1\mathbf{i} + \alpha a_2\mathbf{j} + \alpha a_3\mathbf{k} \tag{B.4}$$

Fluid Dynamics of the Midlatitude Atmosphere, First Edition. Brian J. Hoskins & Ian N. James.
© 2014 John Wiley & Sons, Ltd. Published 2014 by John Wiley & Sons, Ltd.

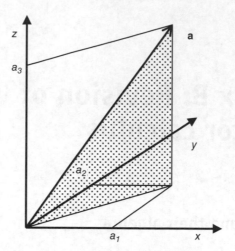

Figure B.1   The vector **a** and its components

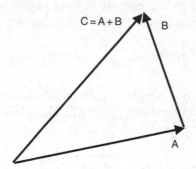

Figure B.2   The 'triangle' construction for the sum of two vectors

This operation corresponds simply to changing the length of a vector by the factor α but leaving its direction unchanged. The sum of two vectors is obtained by adding their components:

$$\mathbf{a} + \mathbf{b} = \left(a_1 + b_1\right)\mathbf{i} + \left(a_2 + b_2\right)\mathbf{j} + \left(a_3 + b_3\right)\mathbf{k} \qquad (B.5)$$

A geometric construction for this operation is shown in Figure B.2.

B.2   Products of vectors

Vectors can be combined in two ways to form a 'product', analogously to the multiplication of scalar numbers.

The 'scalar product' is denoted by

$$s = \mathbf{a} \cdot \mathbf{b} \qquad\qquad (B.6)$$

and is defined as

$$s = a_1 b_1 + a_2 b_2 + a_3 b_3 \qquad\qquad (B.7)$$

If the vectors $\mathbf{a}$ and $\mathbf{b}$ make an angle θ with each other, then it is easy to show that

$$s = |\mathbf{a}||\mathbf{b}|\cos(\theta) \qquad\qquad (B.8)$$

Hence, the scalar product is the product of the magnitude of $\mathbf{a}$ multiplied by the component of $\mathbf{b}$ parallel to $\mathbf{a}$ (Figure B.3). The scalar product obeys the commutative and distributive rules of multiplication, for

$$\mathbf{a} \cdot \mathbf{b} = \mathbf{b} \cdot \mathbf{a}; \quad \mathbf{a} \cdot (\mathbf{b} + \mathbf{c}) = \mathbf{a} \cdot \mathbf{b} + \mathbf{a} \cdot \mathbf{c} \qquad\qquad (B.9)$$

The 'vector product' is denoted by

$$\mathbf{c} = \mathbf{a} \times \mathbf{b} \qquad\qquad (B.10)$$

and is defined as

$$\mathbf{c} = \begin{vmatrix} \mathbf{i} & \mathbf{j} & \mathbf{k} \\ a_1 & a_2 & a_3 \\ b_1 & b_2 & b_3 \end{vmatrix} = (a_2 b_3 - a_3 b_2)\mathbf{i} - (a_1 b_3 - a_3 b_1)\mathbf{j} + (a_1 b_2 - a_2 b_1)\mathbf{k} \qquad (B.11)$$

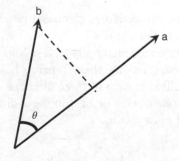

Figure B.3   Illustrating the scalar product

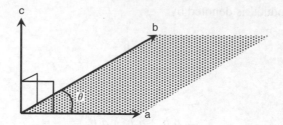

Figure B.4 Illustrating the vector product

Although this looks complicated, the vector product has a simple geometrical interpretation. The magnitude of **c** is

$$|\mathbf{c}| = |\mathbf{a}||\mathbf{b}|\sin(\theta) \tag{B.12}$$

which is simply the area of the parallelogram formed by **a** and **b** (Figure B.4). The direction of **c** is at right angles to both **a** and **b**. Notice that the associative rule of algebra applies, for

$$\mathbf{a} \times (\mathbf{b} + \mathbf{c}) = \mathbf{a} \times \mathbf{b} + \mathbf{a} \times \mathbf{c} \tag{B.13}$$

But the commutative rule does not apply. Instead

$$\mathbf{b} \times \mathbf{a} = -\mathbf{a} \times \mathbf{b} \tag{B.14}$$

B.3   Scalar fields and the grad operator

A physical field which has a value at each point in space is called a 'field'. Fields may be either scalar or vector. A scalar field is denoted by

$$s = s(\mathbf{r}) \tag{B.15}$$

Examples include pressure, temperature, electrical potential and charge density. Vector fields are discussed in Section B.4.

The field $s(\mathbf{r})$ can be represented pictorially by a set of surfaces (or, in two dimensions, a set of curved contours) linking places where s has the same value. The rate of change of s can be specified in each of three directions parallel to the x-, y- and z-axes. Thus, $\partial s/\partial x$ is the rate of change of s in the x-direction with y- and z-held fixed. The total gradient of s is a vector:

$$\frac{\partial s}{\partial x}\mathbf{i} + \frac{\partial s}{\partial y}\mathbf{j} + \frac{\partial s}{\partial z}\mathbf{k}$$

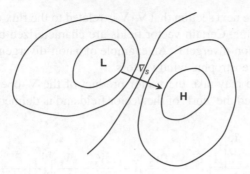

Figure B.5 Contours of s and an example of ∇s

It proves convenient to regard this expression as the result of a vector operator, denoted by ∇, acting upon s:

$$\nabla s = \frac{\partial s}{\partial x}\mathbf{i} + \frac{\partial s}{\partial y}\mathbf{j} + \frac{\partial s}{\partial z}\mathbf{k} = \left(\frac{\partial}{\partial x}\mathbf{i} + \frac{\partial}{\partial y}\mathbf{j} + \frac{\partial}{\partial z}\mathbf{k} \right)s \tag{B.16}$$

∇ is the 'gradient operator', sometimes called 'del' or 'nabla'. In many ways, we can treat ∇ as a vector quantity, using the algebraic rules set out in the last section. There is an important exception to this. $s\nabla$ is an operator, while ∇s is a vector field. They have entirely different meanings.

The vector ∇s is easily visualized. It is a vector, directed perpendicularly to surfaces of $s=$ constant, pointing away from low values and towards large values of s (Figure B.5).

B.4 The divergence and curl operators

A complete specification of the rate of change of a vector with respect to position would require the rate of change of each of the vector components with respect to each of three spatial coordinates. The result would be nine numbers, which may be arranged as a 3×3 tensor. However, a part-way house uses the ∇ operator notation defined in the last section to select certain terms from the full derivative tensor and to combine them in physically important ways.

The first is the 'divergence', which may be thought of as the scalar product of the ∇ operator with a vector field $\mathbf{V}$:

$$\nabla \cdot \mathbf{V} = \frac{\partial V_1}{\partial x} + \frac{\partial V_2}{\partial y} + \frac{\partial V_3}{\partial z} \tag{B.17}$$

We shall show in the next section that $\nabla \cdot \mathbf{V}$ is related to the flux of $\mathbf{v}$ emerging from an element of volume. Certain vector fields are characterized by $\nabla \cdot \mathbf{V} = 0$. Such a field is said to be 'non-divergent'. An example of a non-divergent field in meteorology is the field of geostrophic wind vectors.

Alternatively, we may take the vector product of the ∇–operator with a vector field $\mathbf{V}$. This is called the 'curl' of the vector field and is defined as

$$\nabla \times \mathbf{V} = \begin{vmatrix} \mathbf{i} & \mathbf{j} & \mathbf{k} \\ \dfrac{\partial}{\partial x} & \dfrac{\partial}{\partial y} & \dfrac{\partial}{\partial z} \\ V_1 & V_2 & V_3 \end{vmatrix} = \left(\frac{\partial V_3}{\partial y} - \frac{\partial V_2}{\partial z} \right)\mathbf{i} - \left(\frac{\partial V_3}{\partial x} - \frac{\partial V_1}{\partial z} \right)\mathbf{j} + \left(\frac{\partial V_2}{\partial x} - \frac{\partial V_1}{\partial y} \right)\mathbf{k} \quad (B.18)$$

Notice that between them, the $\nabla \cdot$ and $\nabla \times$ operators include all the elements of the full derivative tensor. The curl of a vector field is related to the degree of circulation which the field exhibits, a result we shall demonstrate in the next section. A field for which $\nabla \times \mathbf{V} = 0$ is described as 'irrotational'. An important example of an irrotational field is the gravitational acceleration.

B.5   Gauss' and Stokes' theorems

These two theorems are useful in manipulating vector fields and also in giving a physical interpretation of the curl and divergence operators.

Consider first a cuboidal region of space, sides δx, δy and δz, illustrated in Figure B.6. Consider the flux of a vector field $\mathbf{V}$ through the faces of this cuboid. Provided the cuboid is sufficiently small, the flux is the normal component of $\mathbf{V}$ out of the surface multiplied by the area of the face. So the flux through face A is $-V_1 \delta y \delta z$. The minus sign indicates that this is an inward flux. The flux out of face B is similarly $(V_1 + \delta V_1)\delta y \delta z$. The net flux leaving the cuboid in the x-direction is therefore $\delta V_1 \delta y \delta z$. Similar results follow for the y- and z-directions so that the total net flux emerging from the cuboid is

$$\oint_s \mathbf{V} \cdot d\mathbf{A} = \delta V_1 \delta y \delta z + \delta V_2 \delta x \delta z + \delta V_3 \delta x \delta y$$
$$= \left(\frac{\delta V_1}{\delta x} + \frac{\delta V_2}{\delta y} + \frac{\delta V_3}{\delta z} \right) \delta x \delta y \delta z \quad (B.19)$$

In the limit, as $\delta x, \delta y, \delta z \to 0$, this equation becomes

$$\oint_s \mathbf{V} \cdot d\mathbf{A} = (\nabla \cdot \mathbf{V})\delta x \delta y \delta z \quad (B.20)$$

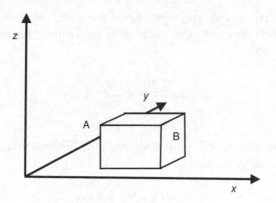

Figure B.6   Illustrating Gauss's theorem

In other words, for such a cuboid, the divergence of **V** may be thought of simply as the outward flux of **V** per unit volume.

Now suppose that two such cuboids are placed in contact. Then, the flux out of one cuboid through the interface between them will be an inward flux to the second cuboid. The total flux will simply be the net outward flux from the combined volume, and this according to Equation B.20 will equal the sum of the divergence in each cuboid multiplied by the volume. The argument can be extended to a large number of such cuboids, forming a finite volume. Then, as their size tends to zero, a generalization of Equation B.20 for an arbitrary volume τ enclosed by a closed surface S is obtained:

$$\oint_S \mathbf{V} \cdot d\mathbf{A} = \int_\tau (\nabla \cdot \mathbf{V}) d\tau \qquad (\text{B.21})$$

This result is called 'Gauss' theorem'.

A parallel result gives a physical interpretation of the curl operator, relating the curl of a vector field to its circulation around a closed loop. The circulation C of any vector field is defined as

$$C = \oint_L \mathbf{V} \cdot d\mathbf{l}$$

where L is an arbitrary closed loop in space, and $d\mathbf{l}$ is an element of that loop. The sign convention is that positive circulation is anticlockwise. To start with, consider a rectangular loop with sides δx and δy lying in the x–y plane. Then, assuming the components of **V** vary smoothly from place to place, the circulation around this loop may be written as follows:

$$\begin{aligned} C_{xy} &= V_1 \delta x + (V_2 + \delta V_2) \delta y - (V_1 + \delta V_1) \delta x - V_2 \delta y \\ &= -\delta V_1 \delta x + \delta V_2 \delta y \end{aligned} \qquad (\text{B.22})$$

Care has to be taken with the signs here: for each segment of the loop, the anticlockwise contribution to the circulation is positive. Equation B.22 can be rewritten as follows:

$$C_{xy} = \left(-\frac{\delta V_1}{\delta y} + \frac{\delta V_2}{\delta x} \right) \delta x \delta y \qquad (B.23)$$

But the right-hand side of this result is simply the z-component of $\nabla \times \mathbf{V}$ multiplied by the area enclosed by the loop.

$$C_{xy} = (\nabla \times \mathbf{V}) \cdot \mathbf{k} \delta x \delta y$$

Similar results hold for the circulation about rectangular loops in the x–z and y–z planes:

$$C_{yz} = (\nabla \times \mathbf{V}) \cdot \mathbf{i} \delta y \delta z; \quad C_{xz} = (\nabla \times \mathbf{V}) \cdot \mathbf{j} \delta x \delta z; \quad C_{xy} = (\nabla \times \mathbf{V}) \cdot \mathbf{k} \delta x \delta y \qquad (B.24)$$

Now the area of a small loop can be regarded as a vector whose direction is perpendicular to the plane of the loop. So for the circulation around a small loop orientated in an arbitrary direction, we may write

$$C = (\nabla \times \mathbf{V}) \cdot \delta \mathbf{A} \qquad (B.25)$$

This expression gives a physical interpretation for $\nabla \times \mathbf{V}$. The curl of a vector is a vector directed perpendicularly to an element of area, whose magnitude is simply the circulation around the loop enclosing that area per unit area.

The argument now proceeds in much the same way as for Gauss' theorem. A closed finite loop of arbitrary shape and orientation can be built up by adding the contributions from any number of infinitesimal loops placed adjacent to one another. The contributions to the circulation along the common segments of adjoining loops cancel out, and so only the segments on the periphery of the loop contribute to the total circulation. Integrating over the entire loop gives

$$\oint_L \mathbf{V} \cdot d\mathbf{l} = \int_A (\nabla \times \mathbf{V}) \cdot d\mathbf{A} \qquad (B.26)$$

This result is called 'Stokes' theorem'.

Equation B.26 reveals what at first sight now seem a curious and counter-intuitive result. Consider any two surfaces A_1 and A_2 which terminate on a loop L. Then,

$$\int_{A_1} (\nabla \times \mathbf{V}) \cdot d\mathbf{A} = \int_{A_2} (\nabla \times \mathbf{V}) \cdot d\mathbf{A}$$

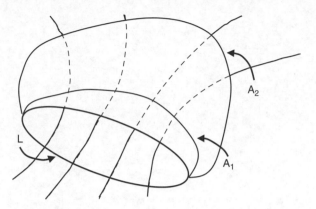

Figure B.7   Two surfaces A_1 and A_2 which terminate on the closed loop L. Lines parallel to $\nabla \times \mathbf{V}$ must pass through both surfaces

This result holds no matter how different the surfaces are, provided they are bounded by the same loop. The reason for this result is simply that $\nabla \cdot (\nabla \times \mathbf{V}) \equiv 0$ for any vector $\mathbf{V}$. In the language of nineteenth-century physics, 'lines of force' (or 'vortex lines') of $\nabla \times \mathbf{V}$ can neither start nor end between the surfaces A_1 and A_2; equal numbers of lines must pass through both surfaces. Figure B.7 illustrates.

B.6   Some useful vector identities

Here, ϕ, ψ, $\mathbf{A}$ and $\mathbf{B}$ are supposed to be continuous, differentiable functions of the position vector $\mathbf{r}$. Some of these identities follow directly from the basic definitions given earlier. Others are a bit more complicated.

Identity 1 $\nabla(\phi + \psi) = \nabla\phi + \nabla\psi$
Identity 2 $\nabla \cdot (\mathbf{A} + \mathbf{B}) = \nabla \cdot \mathbf{A} + \nabla \cdot \mathbf{B}$
Identity 3 $\nabla \times (\mathbf{A} + \mathbf{B}) = \nabla \times \mathbf{A} + \nabla \times \mathbf{B}$
Identity 4 $\nabla \cdot (\phi\, \mathbf{A}) = (\nabla\phi) \cdot \mathbf{A} + \phi(\nabla \cdot \mathbf{A})$
Identity 5 $\nabla \times (\phi\, \mathbf{A}) = (\nabla\phi) \times \mathbf{A} + \phi(\nabla \times \mathbf{A})$
Identity 6 $\nabla \cdot (\mathbf{A} \times \mathbf{B}) = \mathbf{B} \cdot (\nabla \times \mathbf{A}) - \mathbf{A} \cdot (\nabla \times \mathbf{B})$
Identity 7 $\nabla \times (\mathbf{A} \times \mathbf{B}) = (\mathbf{B} \cdot \nabla)\mathbf{A} - \mathbf{B}(\nabla \cdot \mathbf{A}) - (\mathbf{A} \cdot \nabla)\mathbf{B} + \mathbf{A}(\nabla \cdot \mathbf{B})$
Identity 8 $\nabla(\mathbf{A} \cdot \mathbf{B}) = (\mathbf{B} \cdot \nabla)\mathbf{A} + (\mathbf{A} \cdot \nabla)\mathbf{B} + \mathbf{B} \times (\nabla \times \mathbf{A}) + \mathbf{A} \times (\nabla \times \mathbf{B})$
Identity 9 $\nabla \cdot (\nabla\phi) = \nabla^2\phi$
Identity 10 $\nabla \times (\nabla\phi) = 0$
Identity 11 $\nabla \cdot (\nabla \times \mathbf{A}) = 0$
Identity 12 $\nabla \times (\nabla \times \mathbf{A}) = \nabla(\nabla \cdot \mathbf{A}) - \nabla^2\mathbf{A}$

Index

absolute acceleration, 56
absolute velocity, 56
absolute vorticity, 130, 131
action at a distance, 153
adiabatic lapse rate, 2
advection, 37, 38
ageostrophic circulation, 238, 310
ageostrophic wind, 88, 137, 211, 218
air mass, 16
anelastic approximation, 92, 112
angular momentum, 72
anti-geostrophic, 85
aspect ratio, 79, 89
available potential energy, 124, 332

backing of wind, 213
balance, 1, 83, 211
baroclinic initial value problem, 278
baroclinic instability, 175, 245
baroclinicity, 245
baroclinic wave, 246
barotropic instability, 174, 259
barotropic vorticity equation, 150
basic equations
    height coordinate, 73
    pressure coordinate, 101
basic equation sets, 121
Bergen school, 15
Bernoulli potential, 45
Bernoulli's theorem, 45
beta effect, 75
beta plane, 75, 149
Blasius boundary layer, 192
blocking, 350
body force, 27, 41
boundary condition, surface, 100, 248

boundary layer, 3, 139
boundary Rossby wave, 257
boundary temperature, 186
Boussinesq approximation, 101
Brunt–Väisälä frequency, 2, 93, 114, 186
buoyancy, 101
    flux, 326
    oscillation, 301
Burger number, 93, 217, 339, 355
butterfly effect, 201

cascade, energy, 194
cavitating flow, 46
centrifugal force, 56, 57
centripetal acceleration, 56, 57
Charney balance equation, 241
Charney model growth rate, 273
Charney model of baroclinic
    instability, 271
Charney model structure, 271
Charney–Stern condition, 252,
    256, 265, 271, 281
circular cylinder, flow around, 189
circular vortex, 183
circulation, 127, 145
circulation equation, 146
Clausius–Clapeyron equation, 48
clear air turbulence, 202
closure hypothesis, 204
cold front, 318, 321
compressible fluid, 28
condensation, 48
conditional instability, 3
confluence model, 307
continuity equation, 40, 67, 91
continuum hypothesis, 20, 193

Fluid Dynamics of the Midlatitude Atmosphere, First Edition. Brian J. Hoskins & Ian N. James.
© 2014 John Wiley & Sons, Ltd. Published 2014 by John Wiley & Sons, Ltd.